New Armies from Old

New Armies from Old

Merging Competing Militaries after Civil Wars

ROY LICKLIDER, EDITOR

Georgetown University Press
Washington, DC

© 2014 Georgetown University Press. All rights reserved. No part of this book may be reproduced or utilized in any form or by any means, electronic or mechanical, including photocopying and recording, or by any information storage and retrieval system, without permission in writing from the publisher.

Library of Congress Cataloging-in-Publication Data

New armies from old : merging competing militaries after civil wars / Roy Licklider, editor.
    pages cm
 Includes bibliographical references and index.
 ISBN 978-1-62616-043-9 (pbk. : alk. paper)
 1. Armies—Reorganization—Case studies. 2. Civil war—Case studies. 3. Postwar reconstruction—Case studies. 4. Combined operations (Military science)—Case studies. 5. Conflict management—Case studies. 6. Armies—Africa—Reorganization—Case studies. 7. Civil war—Africa—Case studies. 8. Conflict management—Africa—Case studies. Licklider, Roy E., editor of compilation. II. Title: Merging competing militaries after civil wars.
 UA10.N44 2014
  355.3—dc23   2013026134

∞ This book is printed on acid-free paper meeting the requirements of the American National Standard for Permanence in Paper for Printed Library Materials.

15 14   9 8 7 6 5 4 3 2   First printing

Printed in the United States of America

To my teachers:

Mrs. Hamilton, Miss Kaiser, Charlie Orme, Stan Gloss, Buck Hart, Murray Levin, William Yale, William Newman, Andrew Milnor, Brad Westerfield, Karl Deutsch, Bruce Russett, Harold Lasswell, Neil McDonald, Jim Rosenau, Chuck Tilly, and all the others.

Thank you so much.

CONTENTS

List of Tables ix

Foreword, Bruce Russett xi

Acknowledgments xiii

A Note on Abbreviations xv

1 Introduction 1
   Roy Licklider

2 Mixed Motives? Explaining the Decision to Integrate Militaries at Civil War's End 13
   Caroline A. Hartzell

PART I. EARLY ADOPTERS

3 Sudan, 1972–1983 31
   Matthew LeRiche

4 Military Integration from Rhodesia to Zimbabwe 49
   Paul Jackson

5 Merging Militaries: The Lebanese Case 69
   Florence Gaub

PART II. AUTONOMOUS DEVELOPMENT

6 From Failed Power Sharing in Rwanda to Successful Top-Down Military Integration 87
   Stephen Burgess

7 From Rebels to Soldiers: An Analysis of the Philippine Policy of Integrating Former Moro National Liberation Front Combatants into the Armed Forces 103
   Rosalie Arcala Hall

8 South Africa 119
   Roy Licklider

CONTENTS

PART III. INTERNATIONAL INVOLVEMENT

9   Half-Brewed: The Lukewarm Results of Creating an Integrated Military in the Democratic Republic of the Congo   137
    Judith Verweijen

10 Merging Militaries: Mozambique   163
    Andrea Bartoli and Martha Mutisi

11 Bosnia-Herzegovina: From Three Armies to One   179
    Rohan Maxwell

12 Bringing the Good, the Bad, and the Ugly into the Peace Fold: The Republic of Sierra Leone's Armed Forces after the Lomé Peace Agreement   195
    Mimmi Söderberg Kovacs

13 Military Integration in Burundi, 2000–2006   213
    Cyrus Samii

PART IV. ALTERNATIVE PERSPECTIVES

14 The Industrial Organization of Merged Armies   231
    David D. Laitin

15 Military Dis-Integration: Canary in the Coal Mine?   245
    Ronald R. Krebs

16 So What?   259
    Roy Licklider

    References   269
    Contributors   293
    Index   297

TABLES

1.1	Dimensions of Success in Integrating Fighters into a Postwar Military	4
1.2	Potential Military Integration Cases	7
1.3	Initial List of Questions asked about All Cases	8
2.1	Inclusion of Military Integration as a Term of Civil War Settlements, 1945–2006	16
2.2	Logit Analysis of Military Integration as a Component of Civil War Settlements, 1945–2006	23
3.1	Anyanya Integration into the Government Services	39
5.1	Approval Rates of the Lebanese State and Its Institutions in Lebanese Society	81
15.1	Country Military Integration Details	246
15.2	Country Military Integration Outcomes	247
16.1	Causal Links to a Reduced Probability of Civil War, by Case	262

FOREWORD

Bruce Russett

This volume fills a serious gap in our understanding of civil wars and their possible resolution. Especially since the end of the Cold War, study of the causes and consequences of civil wars has become a huge academic and policy-wonk industry. This research delves into how and why civil wars end, why they often soon resume between the original combatants or new ones, the role of institution building after civil wars, civil–military relations, the disarmament and demobilization of combatants, and the integration of combatants into more heterogeneous postwar military structures—the particular subject on which this volume focuses.

The United Nations and other international organizations, as well as national government agencies, frequently carry out retrospective examinations of their operations for "What lessons are to be learned?" studies. Some of those exercises have included aspects of military integration and security reform, more or less defined as here. But all too often such exercises focus on a just a few operations—or, if the sample is expanded, the basis for choosing examples may be opportunistic, unsystematic, and subject to intentional or more likely unintended bias in selecting cases. Similarly, the cases they consider may be very hard to compare if they were compiled by different writers, at different times, with different key assumptions and concepts. Until recently, much of the work on military integration has been mostly case specific, lacking an established comprehensive framework with which to systematically compare the costs and benefits of trying to carry out integration in ever-changing circumstances.

Military integration has become accepted, over time, as a key element in negotiations to end civil wars. For this book the editor and his contributors have developed—through a process requiring a great deal of financial support, time, trial, and error—a careful and explicit framework for selecting and comparing their cases. This could not have been an easy task for area specialists from different intellectual and cultural backgrounds. In effect they were looking at the impact of experiments in peace-building carried out by individuals who were both agents and observers. The authors explore the role of support, by governments and international organizations as well as internal groups, of efforts to create integrated military forces. By doing the individual case studies within a common framework they can probe into micro-political behavior to support casual inferences. Relevant outcomes include whether major violence is resumed and how fast, and if not, whether "peace" was imposed by a faction that emerged as most powerful, or became a peace with democratic or at least somewhat liberal and pluralistic institutions. What are the limits as well as the possibilities of imposing a policy like military integration, and what may be the costs to the international organizations and governments that try to carry them out? Aid for reconstruction and development is scarce, and military integration is not cheap; is it the best use of resources? Most of all, what are the costs incurred from these efforts by the people of these post–civil war states?

I have followed this project from its inception. As usual with any serious academic venture—or for that matter life venture—the answers offered here are incomplete and contestable even between the contributors. The book is nonetheless a landmark in exploring terrain that has been undermapped and poorly understood, yet one that can offer great rewards to the explorers and to those many in the world who could greatly benefit from the riches evident here. Lessons-learned studies too easily wind up in bureaucratic oblivion. Maybe the analysts learned much, but policymakers learned little if anything. This book is different, and it deserves a far better reception.

ACKNOWLEDGMENTS

One of the joys of academic research is how helpful people are, from your friends and family to total strangers. Clearly, my primary debt is to the authors who responded to my critiques and comments, met my deadlines, ignored it when I missed my deadlines, and remained cheerful through it all. It has been a pleasure working with them.

I started this project with very little knowledge of either security-sector reform in general or any of the countries in particular. I was aided with incredible generosity by more people than I can remember, but they certainly include Sandy Africa, Robert Beecroft, Steve Burgess, Gavin Cawthra, Jakkie Cilliers, John Drewienkiewicz, Véronique Dudouet, Tim Edmunds, Abel Esterhuyse, Karen Guttieri, Ernie Harsch, Lindy Heinecken Macartan Humphreys, Paul Jackson, Mary Frances Lebanoff, Mark Malan, Rob Muggah, Lara Nettlefield, Desireé Nilsson, Ingrid Samset, Annette Seegers, Tim Sisk, Judy Smith-Höhn, Susan Woodward, and Marie-Joell Zahar.

For nineteen years I was a member of Charles Tilly's weekly Workshop on Contentious Politics (as it was eventually called), first at the New School for Social Research and then at Columbia University. It changed my life. This project was first articulated there and was shaped by the intense, focused, and supportive discussions that were a specialty of that group, both in the seminar room and then over dinner. Chuck supported and critiqued this project, as he did for so many activities for so many people over the years. I like to think he would have approved of the results, and I miss not having his suggestions for pushing the work further.

Several institutions have made major contributions to this book. The research working group on security-sector reform of the Folke Bernadotte Academy, headed by Birger Heldt, oriented me to the field of study and introduced me to many of the individuals who would make the project a reality. The work was funded by Grant BCS 0904905 from the Social and Behavioral Dimensions of National Security, Conflict, and Cooperation, a joint program of the National Science Foundation and the Department of Defense/Department of the Army/Army Research Office, better known as the Minerva Program, an extraordinary attempt to bring academic rigor to policy issues. The US Institute of Peace also offered to support the project and was probably relieved to be able to use its too scarce resources elsewhere. The Peacekeeping and Stability Operations Institute of the US Army War College graciously provided hospitality and support for the conference, which was central to the effort; special thanks are due to Raymond Millen and Colonel Stephen T. Smith. Obviously, the views expressed here are those of the authors, not these organizations. In this age of paranoia, it is perhaps worth noting that there was never a hint of interference with the research or its conclusions.

ACKNOWLEDGMENTS

Don Jacobs at Georgetown University Press has been enormously supportive and helpful, from our first contact at a convention book display to the final product. I also thank the anonymous reviewers for their guiding feedback. I also acknowledge the work of the staff at Georgetown University Press for all the work given to bring this book to fruition. I give major credit for bringing the mauscript to a coherent volume to my indefatigable copy editor S. B. Kleinman—awesome, Sadie (that can be correct with or without a comma!).

Closer to home, Virginia and Andrew, and more recently Nancy and Malcolm, kept me focused on the future, and Patricia kept me well loved, well fed, and reasonably sane.

A NOTE ON ABBREVIATIONS

In those chapters with a number of abbreviations, the main abbreviations are spelled out in a list following the chapter text.

Chapter 1

INTRODUCTION

Roy Licklider

Until the end of the Cold War, it was conventional wisdom that civil wars necessarily ended in military victories (Iklé 1971, 95; Bell 1972, 218; Modelski 1964, 125–26; Pillar 1983, 24–25). Nonetheless, more than twenty negotiated settlements of civil wars have been reached since 1989 in places as disparate as El Salvador and South Africa. Some of these compromise settlements have ended civil wars and resulted in postwar regimes that are substantially more democratic than their predecessors.

These settlements have usually involved power sharing among the former contestants and other sectors of society. Many of these agreements have, as a central component, provisions to merge competing armed groups in a single national army. How can people who have been killing one another with considerable skill and enthusiasm be merged into a single military force?

Other than a few scattered case studies and some contradictory aggregate data analyses, we have very little information about the process of military integration. Why has it been used? What strategies have been most effective? Does integration help prevent renewed civil war? Following up on earlier pioneering analyses (Simonsen 2007; Glassmyer and Sambanis 2008; Hoddie and Hartzell 2003, 2005; Hartzell and Hoddie 2003, 2007; Hartzell 2009; Burgess 2008; Jarstad and Nilsson 2008; Gaub 2007, 2011; Knight 2009; DeRouen, Lea, and Wallensteen 2009), this project is an attempt to answer these questions.

THEORIES

Why has military integration suddenly become the new normal? Many (although not all) examples of military integration are linked to negotiated settlements of civil wars. Such settlements, in turn, have become more common because military victories are increasingly difficult to achieve for several reasons: (1) The issues in dispute now tend to involve identity rather than ideology. In ideological wars a military victory can be followed by conversion of the vanquished to the victor's position (China is an excellent example), but often identity cannot be changed, so the underlying conflict remains, despite temporary military outcomes. (2) Genocide and ethnic cleansing have become increasingly difficult to implement, making a military stalemate increasingly likely. (3) The end of the Cold War reduced external support for many states in the developing world, making them less able to win quick victories. A stalemate does not necessarily result in a negotiated settlement (the classic statements are by Zartman 1993, 1995a, 1995b), but it makes them more attractive; some studies show that long wars are more likely than short ones to end in negotiated settlements—although, given wide differences in the coding of onset and termination dates in different data sets (Sambanis 2004, 855), these findings must be used with care. (4) The peace industry, the new complex of international and nongovernmental organizations dedicated to encouraging the end of mass violence, has also contributed to this trend (Licklider 2001, 698–99; Hartzell and Hoddie 2007; Hironaka 2005).

The literature suggests that military integration might be a response to three common problems of negotiated settlements. The first is the issue of security. It is generally agreed that ending a civil war involves disarmament, but how do you persuade people to put themselves and their families at the mercy of untried security institutions controlled in part by people who have been their deadly enemies? Barbara Walter (2002, 103–5) notes that security issues consume the vast majority of time and attention in negotiating settlements. Security is, of course, the central problem of any state, but the issue is particularly important after civil wars when the combatants must live side by side indefinitely; very few civil wars have ended in partition. (For arguments about the utility of partition in ending civil war, see Kaufmann 1996, 1998, 1999; and Licklider and Bloom 2006).

Military integration seems a plausible solution to the security problem. Having a substantial number of one's fellows in the new security force can be reassuring, both because they may be less likely to injure you and also because, if the settlement collapses, you will at least have some fighters with whom to start the next round of warfare and perhaps some early warning of the event (Call and Stanley 2003, 212). The hope, of course, is that over time this sense of vulnerability will decline and this political insurance will be less imperative. We know that countries that have experienced civil wars are at risk of new ones; military integration is thus seen as a means of making renewed civil war less likely by reducing fear of former antagonists.

The second common problem of negotiated settlements is that merging armies is one way to reduce the number of former fighters who have to be disarmed and integrated into the society. Most settlements include provisions for disarmament, demobilization, and reintegration (DDR) of fighters into civil life, but at best this is a lengthy and expensive process, usually taking place in countries that cannot easily afford it. Taking some of these people into the military could presumably improve the situation. However, in practice relatively few people are usually involved (although Uganda is an interesting counterexample [Mutengesa 2013]), because a country usually needs to reduce the overall size of its military after a civil war, and indeed the necessity for armed forces often comes into serious question.

The third common, longer-term problem of negotiated settlement is how to create a nation out of competing groups. Creating a working state, a governmental apparatus that can collect taxes and deliver public goods to society, is hard enough after a civil war; creating a nation, a group of people who feel that they are part of a common loyalty group, is more difficult by an order of magnitude. Conversely, it is not impossible; most of the major states in the current international system have had to do this at one point or another, although usually after victorious wars—Britain after its civil war, and then in integrating Scotland, Wales, and Ireland; France after the French Revolution; Germany after the wars of German unification; the Soviet Union and China after their respective revolutions; and so on. The United States has done it twice, after its Revolutionary War and its Civil War.

HISTORY AND BACKGROUND

The idea that the army can bind competing people together has a long history. Alexander the Great tried to integrate Persians into his Macedonian army and had problems that are similar to those in some current cases (Kuranga 2011a). The colonial empires established by the European powers from the seventeenth to the twentieth centuries relied on native troops, often recruited from those groups that had most strongly resisted colonialism. Britain, for example, controlled India for more than a century with a force in which Indian soldiers outnumbered British soldiers by roughly twenty to one (Kuranga 2011b). One of the most profound changes of the twentieth century was the rise of nationalism in the former colonies, undermining the ideological basis of this kind of relationship.

More recently, regimes from czarist Russia and Meiji Japan to the Soviet Union and Israel have sought to use their militaries to create and strengthen nationalism. Ronald Krebs (2004, 2005, 2006) has argued that advocates of this practice assume that a national military will change the values of individuals, either by showing members of different groups cooperating or by bringing people from different backgrounds into contact with each other. Though Krebs concluded that there is in fact very little evidence that these processes actually work much of the time, many people disagree, and so the idea of a national army composed of people from previously competing militaries seems a worthy goal to many. A similar disjunction between fact and belief has been found in developing countries (Dietz, Elkin, and Roumani 1991).

RESEARCH DESIGN

The preceding brief survey suggests at least three questions worth exploring: (1) Why has military integration been used? (2) What particular strategies seem to work better under what circumstances? (3) Has successful integration made the resumption of civil war less likely? Each of these questions is more complicated than it first appears.

We first need to establish what we are talking about. It is certainly possible for the victorious side of a civil war to allow substantial numbers of its former enemies into its military over time, but they are usually not restored to their prior ranks. *Integration means that individuals are brought into the new military in positions similar to the ones they occupied in prior organizations.* It is not impossible for such integration to take place after a military victory, but most of the recent examples come from negotiated settlements to civil wars after the Cold War.

Answering the question "When is military integration more or less likely to succeed?" involves first defining success. In fact, the literature suggests several different dimensions (table 1.1). First, military *efficacy* is simply the extent to which the new military can perform tasks. It may, for example, be able to continue to exist, primarily as a symbol of national unity, but not be reliable in combat. It may be able to wage war against foreign adversaries at a level appropriate to its resources; it may be able to put down domestic disturbances involving members of the groups most strongly represented in the military; it may be able to do both or neither. Moreover, different countries have different needs, so success may mean different things at different times and in different places. South Africa, for example, does not need a strong military capable of facing a serious external military threat; conversely, several neighboring states threaten to exert control over parts of the Democratic Republic of the Congo.

TABLE 1.1 **Dimensions of Success in Integrating Fighters into a Postwar Military**

1. Demonstrate military efficacy
   a. Remain integrated in peacetime and not kill its own members in large numbers
   b. Perform as well as other militaries with comparable resources in combat against foreign enemies
   c. Perform as well as other militaries with comparable resources against domestic opponents from the groups represented in the military
2. Accept civilian control
   a. On budget and personnel issues
   b. When ordered into use against foreign adversaries
   c. When ordered into use against domestic opponents from groups represented in the military
3. Reduce the likelihood of the resumption of large-scale domestic violence
   a. Improve the security situation by making the military less threatening to members of integrated groups within the society
   b. Employ ex-combatants who might otherwise either encourage or participate in renewed hostilities
   c. Provide security for the society, making possible political and economic development

Second, with regard to which strategies are most effective, much of the recent literature on security-sector reform suggests that the key issue is whether or not the new military is subject to *civilian control* (Bermeo 2003, 163; Vankovska and Wiberg 2003; Hendrickson and Karkoszka 2005). Given the tendency for military coups to occur in countries in conflict, the desire for civilian control is understandable, although it may be worthwhile to ask which civilians we are talking about (Luckham 2003).

Third, from the point of view of policymakers and theorists interested in the resolution of civil wars, the critical question is whether military integration makes *the resumption of large-scale violence* more or less likely. This question is not answered for a given case by the fact that such violence did or did not resume, although this is sometimes used as a proxy in statistical studies. Instead, it requires serious study of the counterfactual, what would have happened to the country if this particular tactic had not been used. Cross-national analyses of many different cases can inform such analysis but cannot substitute for it.

The literature suggests at least three ways in which an integrated military might prevent a resumption of civil war in the short term: (1) It may improve the security situation because the presence of individuals (and sometimes entire units) from different factions makes the new military less threatening to each of them. (2) It may also employ some of the many demobilized fighters on all sides of the conflict who might otherwise be available to carry on renewed hostilities (Spear 2006, 67; Williams 2005; Chuter 2006; Salomons 2005). (3) It may also make a return to war more difficult by providing the security for the society as a whole that postwar analysts believe is central to political and economic development (Stedman 2002, 668; Salomons 2005, 19–20). (Clearly, this last possibility is related to military efficacy, but it is not solely a measure of that efficacy. Efficacy is itself a measure of the *capability* of a military, but that is quite separate from the issue of *behavior*, how that capability is or is not used. Similarly, whether or not a military is under civilian control has no necessary connection to whether it will actually provide real security for some or all members of the society.)

Several cross-national studies on the relationship between military integration and renewed civil war reached rather similar conclusions. Barbara Walter found that agreements on quotas in the new military force (one particular subset of military integration agreements) increased the willingness of both sides to participate in negotiations but had no impact on willingness to sign or implement agreements (Walter 2002, 63–86). Monica Toft found that post–civil war countries with security-sector reform were significantly less likely to have new civil wars than countries without such changes. Military integration is just one component of security-sector reform, but she uses Walter's definition for her data analysis so it does not seem a stretch to imply that military integration makes renewed civil war less likely (Toft 2010, 12–13, 58–60).

Matthew Hoddie and Caroline Hartzell (2003) found that the implementation of military integration made the resumption of civil war ended by negotiated settlement less likely. Katherine Glassmyer and Nicholas Sambanis (2008) found that military integration did not make renewed civil war less likely once other factors such as power-sharing agreements were controlled for. However, they noted that many of their cases were so poorly structured and implemented that "the label 'military integration' is often meaningless" and that they could not demonstrate that competent integration did not have some positive effects. DeRouen, Lea, and Wallensteen (2009) focused on cases in formal power-sharing agreements, which presumably were more likely to be well thought through, and found that military power-sharing agreements were positively linked to postwar peace.

Four of the five studies find a positive relationship between military integration and length of postwar peace. The fifth finds no relationship but adds that it cannot exclude the possibility that competent integration might be helpful. The studies use different measures of both dependent and independent variables and cover different time spans; the findings can be fairly called robust. However, they do not specify why military integration has this effect or what qualities of integration (size, balance, military capabilities, visibility, etc.) will affect what sorts of targets (military elites, other elites, mass publics, people in close proximity to militaries, et al.). Indeed, they all focus on agreements to integrate, rather than on integration itself. They also do not show that the causal arrow is not reversed, that countries with good prospects for peace are able to better integrate their militaries than those that are less fortunate. Given these results, the logical next step seemed to be to get inside cases to specify the processes by which military integration was adopted or not and trace the impact of military integration on the likelihood of resumption of civil war in some cases order to develop some theories about this relationship which could then be tested in more precise, large-N analyses. The obvious strategy was to do comparative case studies, by matching similar cases where civil war termination did and did not involve military integration to see what difference it made. Unfortunately, I was unable to do this, for several reasons.

Matching cases assumes that we know what variables are important so we can select the cases (King, Keohane, and Verba 1994). Given the lack of theory in this area, that was a challenge. Moreover, the number of cases was fairly small. But the real problem was finding people who had done research on the militaries in some very obscure places and were interested in participating in the study. Clearly this would have to be a group project; no one (at least not I) could do this sort of detailed analysis on many countries. But it quickly became apparent that I could not simply specify the cases and get appropriate researchers, even if I could find appropriate pairings.

I therefore went to a second strategy of recruiting researchers on the topic of military integration and allowing them to focus on the cases in which they were interested. The object of the project thus became to generate theory rather than to test theory. In order to develop some general ideas about how military integration worked, we needed to get inside some of the cases where it had been attempted to specify the processes by which military integration had actually been adopted and implemented and its consequences. I deliberately tried to get a fairly large number of cases, assuming that there would be some attrition, and indeed researchers interested in Uganda, Angola, Liberia, and South Africa withdrew from the project for various reasons. The obvious drawback to this strategy was that, because cases without military integration were not included, it could not directly address the issue of whether military integration made renewed civil war less likely. The more modest goal was to lay down a baseline and suggest hypotheses for future large-N and comparative case studies.

Recent cases were excluded because we needed some idea of the outcome of a process in order to evaluate it; thus a number of really interesting current cases like Iraq and Nepal were not included. A list of twenty-one possible cases was prepared, based on lists from the research of Caroline Hartzell and Matthew Hoddie (2003) and Stephen Burgess (2008); the list is given in table 1.2.

I circulated the list of cases fairly widely, advertised for authors, and selected cases based on who was willing to do the work. I ended up with highly qualified people for eleven country case studies. When the South African specialist had to drop out of the project, I decided to do that chapter myself.

Following the usual strategy to make the cases structured and focused (Bennett and George 2005), I prepared a list of questions to which all authors were asked to respond, which is given in table 1.3. Authors were given the initial questions and asked to prepare drafts for a conference to be held from August 31 to September 1, 2010, at the Peacekeeping and Stability Operations Institute of the US Army War College in Carlisle, Pennsylvania. In order to facilitate cross-case comparisons, several senior scholars were invited to participate in the conference and to submit comparative chapters for the book.

TABLE 1.2 Potential Military Integration Cases

Case	Source
Sudan, 1971	Added by Licklider
Angola, 1975	Burgess
Zimbabwe, 1980	Hartzell/Hoddie; Burgess
Namibia, 1988	Burgess
Lebanon, 1989	Hartzell/Hoddie
Angola, 1991	Hartzell/Hoddie
Cambodia, 1991	Hartzell/Hoddie
Georgia/South Ossetia, 1992	Hartzell/Hoddie
Mozambique, 1992	Hartzell/Hoddie, Burgess
Rwanda, 1993	Hartzell/Hoddie, Burgess
Angola, 1994	Hartzell/Hoddie
Djibouti, 1994	Hartzell/Hoddie
Mali, 1994	Hartzell/Hoddie
Chad, 1996	Hartzell/Hoddie
Bosnia, 1995	Added by Licklider
Philippines, 1996	Hartzell/Hoddie
Sierra Leone, 1996	Hartzell/Hoddie, Burgess
South Africa, 1997	Hartzell/Hoddie, Burgess
Sierra Leone, 2000	Burgess
Dem. Rep. of Congo, 2003	Burgess
Burundi, 2004	Burgess

HOW MIGHT MILITARY INTEGRATION MAKE RENEWED CIVIL WAR LESS LIKELY?

At the conference the papers were presented and debated. Ronald Krebs argued that the initial questions did not really get at our central issue, whether military integration could make civil war resumption less likely, which was, after all, the major point of the project, and that we needed to explore possible causal pathways connecting these two variables. After the conference I prepared a set of four such pathways; a fifth was suggested by Andrea Bartoli based on his work on Mozambique.

Costly Commitment

The first causal pathway is derived from the notion that military integration makes renewed civil war less likely because it is a costly commitment by both sides, each showing to the other that it is seriously committed to peace. Mixing personnel makes it less likely that the new force can be readily used against either, which raises the cost of reversing the policy for both you and your opponent. It is costly because you run the risk of weakening yourself vis-à-vis your opponent

TABLE 1.3 **Initial List of Questions Asked about All Cases**

Origins
What was the historic and cultural role of the military in this country?
When did the issue of merging competing militaries arise in negotiations?
Which individuals or groups supported it, which were opposed, which were uninterested?
What compromises were involved in the final outcome?

Creation
Who determined what the new military would look like?
How were people selected for entrance?
    Quotas
    Evaluating military experience
    Screening for human rights violations
How were officers and noncommissioned officers selected?
Who did the training?
What strategies were used during training?
What problems were encountered during training?

Outcome
Who really controls the force?
What can the new force actually do?
    Remain in existence without its members killing one another in large numbers?
    Deploy to different parts of the country for symbolic reasons?
    Defend the state against foreign attack as well as can militaries in comparable states?
    Use force against all groups in society when asked to do so by legitimate political authority?
Is the force the "right size" for the state?
Can the state maintain this force indefinitely, economically and politically?
What set of circumstances could have produced a drastically different outcome?
Has the force made the resumption of large-scale violence less likely?

and also risk being outbid by opponents within your own group. By showing that the leaders are willing to take risky steps for peace, integration makes other elites more likely to be willing to make compromises for a settlement (Hoddie and Hartzell 2003).

Security

The second causal pathway focuses on the ability of the new military to provide security to the postwar society. It assumes that the new military has sufficient coercive capacity to provide security, at least for elites, and to punish potential spoilers (Toft 2010, 36–38). Military integration means that all sides have reassurance that this power is unlikely to be used against them. As a result, elites and masses feel able to take risky steps toward setting up effective government institutions and resolving underlying disputes, making renewed civil war less likely. This increased security also makes economic development more likely (Stedman 2002, 668; Salomons 2005, 19–20), which in turn makes renewed civil war less likely.

Employment

The third causal pathway involves the ability of the new military to employ individuals who might otherwise be available for renewed civil war. Even if the new military does not have sufficient coercive capacity to provide security, it employs the most dangerous former fighters on both sides (Glassmyer and Sambanis 2008). This increases security enough that elites feel able to take risky steps toward setting up effective government institutions and resolving underlying disputes, which makes renewed civil war less likely and makes economic development more likely, which in turn makes renewed civil war less likely.

Symbol

The fourth causal pathway suggests that the major impact of the new military is to symbolize integration for the entire society, regardless of its coercive capacity. The ability of the military to exist as an integrated institution in a divided country becomes a symbol of national unity. The existence of the military is important; its actual capabilities are irrelevant. The military may well be the most integrated sector of society because its culture makes it less likely to have internal conflicts than other elements of government (Gaub 2011, 137–40). The fact that the individuals who had been killing one another are now working together encourages elites to move on to other issues with some confidence that agreement can be reached, making renewed civil war less likely (Higgs 2000, 48).

The continued existence of the integrated military creates and strengthens trust among military personnel. Military personnel transmit this feeling of confidence and trust to civilians in their own reference groups. These civilians in turn are more willing to support elites who advocate cooperation with their former enemies.

Negotiation Process

The fifth causal pathway is the negotiation process. The process of negotiating military integration increases trust among elites of different sides so that they can move on to other issues with some confidence that agreement can be reached, making renewed civil war less likely. The actual existence of the military is irrelevant. Indeed, the agreement may be to have no military at all (the Costa Rican solution). All the authors were asked to see which of these pathways, if any, applied to their cases.

CHAPTER 1

FINDINGS AND PLAN OF THE BOOK

The chapters are arranged in roughly chronological order; later ones are sorted by degree of international involvement. The categories are:

- Early adopters—Sudan, Zimbabwe, and Lebanon;
- Autonomous development—Rwanda, Philippines, and South Africa;
- International involvement—Mozambique, Bosnia-Herzegovina, Sierra Leone, Democratic Republic of the Congo, and Burundi.

After the present introductory chapter 1, Caroline Hartzell steps back from the eleven cases in the book in chapter 2, which precedes part I of the book, and analyzes all cases of civil wars from 1945 to 2006 to ask why some end with the integration of rival militaries while others do not. The primary explanation of mergers seems to be international involvement rather than any obvious qualities of the civil war or the country itself. This raises a very important ethical issue for the international community: If we are going to push people in other countries to undertake this very expensive and potentially risky procedure, we had better have good evidence that the strategy actually makes civil war renewal less likely. We are, after all, taking chances with other peoples' lives and fortunes. We return to this issue in chapter 16.

Part I of the book focuses on the early adopters, which had little experience with which to work. In chapter 3 Matthew LeRiche shows that the 1971 Sudanese peace agreement was extremely vague but that, despite many problems, it held together for eleven years before Nimieri deliberately violated the terms of the agreement, causing renewed civil war. (LeRiche sees this as a failure; I am inclined to think that eleven years of peace is a success.) In chapter 4 Paul Jackson argues that the military integration of two competing rebel groups and the remnants of the Rhodesian army in 1980 went fairly smoothly but was subsequently undermined by Mugabe's decision to eliminate his rivals; the ensuing corruption created the militaristic kleptocracy that controls Zimbabwe to this day. And in chapter 5 Florence Gaub tells how the Lebanese army reinvented itself in 1989 by integrating Muslims to become an important symbol of a united Lebanon, despite not having the military capability to defend the country against its neighbors Syria and Israel or to defeat the forces of Hezbollah in southern Lebanon.

Autonomous developers, the focus of part II, devised their own strategies with little or no international support; in the current jargon, *local ownership* was strong in these cases. Perhaps the biggest surprise in these cases comes from Stephen Burgess's description in chapter 6 of the integration by the Tutsi government, during and after the civil war, of thousands of Hutus, including many who had participated in the genocide. Indeed, a majority of the current military, one of the strongest in Africa, is now Hutu. In chapter 7 Rosalie Arcala Hall argues that the integration of several thousand members of the Moro National Liberation Front in 1996 had little effect other than to marginally strengthen the Philippine army in its continuing conflict with other Moro groups. Chapter 8, which examines South Africa, focuses on the generally successful integration by 1996 of no fewer than eight separate military groups, ranging from a modern army to guerilla forces, whose members did not even have a common language.

Part III, which comprises the largest group of chapters, includes cases with deep international involvement. In chapter 9 Judith Verweijen argues that military integration in the Democratic Republic of the Congo produced only meager results due to the difficulties of bringing the relatively autonomous wartime power networks into the central state. Ironically, these weak results made it possible for the elites to agree on a negotiated peace, since they opposed a strong central government. In chapter 10 Andrea Bartoli and Martha Mutisi show how the principals of the two sides in Mozambique first reached an agreement in 1992 with extensive outside assistance, and then had to improvise to face a series of unforeseen obstacles, from limited resources from the United Nations to a general reluctance to enter the new army. In chapter 11 Rohan Maxwell shows how the international community pushed and cajoled the three major factions in Bosnia-Herzegovina into a military integration that initially had little local support and still faces implementation challenges. It seems to be working, but until the outsiders withdraw it will be hard to judge its success. In chapter 12 Mimmi Söderberg Kovacs notes that the military integration in Sierra Leone in 2000 was one in a series of similar projects and explains that this one worked because of strong international support, particularly from Britain. Burundi initially looks like a hard case, with a history of massacres on both sides. Unlike Rwanda, neither side could win a military victory. In chapter 13 Cyrus Samii shows how the combination of a highly detailed peace agreement between the government and some smaller rebel groups in 2003 and considerable international pressure led the major rebel group to join on the same terms a few years later; the new combined army was promptly used successfully against the remaining rebel holdouts and seems to be doing fairly well.

In part IV, our cross-case chapters provide wider, alternative perspectives. In chapter 14 David Laitin's comparison of the military mergers presented in this volume to corporate mergers and acquisitions suggests that the risks of failure are perhaps overblown. Laitin observes that, whereas the literature on business mergers predicts unfortunate outcomes, military integration actually has some advantages that may explain what he calls the "moderate successes of the cases in this volume."

The two final chapters offer somewhat different conclusions. In chapter 15 Ronald Krebs argues that the evidence in this volume does not support the contention that military integration reduces the likelihood of civil war. Partially, this can be explained by the research design, which deliberately did not include comparable cases in which military intervention was not used. But he also notes that much of the evidence cited is anecdotal and cannot be validated and suggests that military integration seems to be the result rather than the cause of peace. He points out that military integration is a local political problem; it is not a technocratic solution to political problems that outsiders can import and impose. However, he suggests it may be useful as an indicator of political trust—if military integration fails, trouble is likely to ensue.

In chapter 16 I do not dispute many of Krebs's claims, but I do suggest that the glass is half full rather than half empty. I give greater credence to judgment calls by reasonably impartial outsiders on the symbolic impact of military integration (at least until someone else gets close enough to the critical political operators to demonstrate that it does not matter). I suggest some generalizations about the process and their implications for policy. I then suggest some broader issues, asking whether it is ethical for outsiders to push for military integration when it is clear that local opinion and institutions by themselves will not produce this outcome and raising two troubling issues about the overall wisdom of military integration in postwar countries: its sustainability and its impact on civil–military relations and democracy. The chapter concludes with suggestions for further research.

Chapter 2

MIXED MOTIVES? EXPLAINING THE DECISION TO INTEGRATE MILITARIES AT CIVIL WAR'S END

Caroline A. Hartzell

Sixty-four countries fought and ended 128 civil wars between 1945 and 2006.[1] Forty-six of the settlements ending those conflicts, or roughly 40 percent of the total, called for some form of integration of the government and nonstate actors' military forces. Laos's first civil war (1959–73), which ended in a negotiated agreement, did not include any terms for the integration of the armed adversaries' military forces. However, the settlement following the country's second civil war (1975), which concluded with the Pathēt Lao's military victory, did include such provisions. Burundi, which has experienced a civil war in each of the decades since the country declared its independence in 1962, has called for rival militaries to be integrated in half of the six war-ending settlements arrived at in that country. The Republic of the Congo followed much the same pattern, with arrangements for military integration outlined in the settlements following two of its four civil wars. Neither Croatia's nor Cyprus's civil war settlements, conversely, stipulated the use of this mechanism as a means of ending either of the two intrastate conflicts fought in each country.

These facts raise some interesting questions about the reasons some conflicts, but not others, end with settlements that stipulate the use of military integration. Although a small (but growing) number of works has sought to assess the impact that the integration of military forces has on the postconflict environment, few studies have analyzed the determinants of the choice by armed opponents to merge their militaries. Identifying the factors that shape such a decision is important for a number of reasons. First, knowing what variables have an impact on the choice to integrate militaries can be helpful in ascertaining whether or not selection effects are at work.[2] If we find that military integration tends to be employed in the most difficult conflict management cases, for example, that should alert policymakers to the possibility of underestimating the positive effects this measure has on the postconflict peace.[3] Second, if studies indicate that the integration of militaries does help to secure constructive postconflict outcomes, it would be useful to know whether there are factors susceptible to influence that can be used to encourage warring parties to agree to merge their fighting forces. Third and finally, a focus on the merging of militaries could help to account for trends in the use of this measure as a means of ending civil wars.

Drawing on the case studies in this book, as well as a number of theoretical and empirical works that focus on civil war termination, I identify several potential factors affecting the decision to call for the merging of adversaries' militaries as part of a civil war settlement. These include the balance of military power between the government and rebel groups, the state's capacity for accommodation, economic motivations, and the role of external mediation. Employing a logit model, I test a number of propositions regarding the effects these factors have on agreements to integrate militaries. I find no support for arguments regarding the positive relationship between relative rebel strength and military integration, nor for economic incentives as drivers for this type of agreement. Somewhat unexpectedly, I find that states with a higher accommodative capacity are more likely to accede to a military merger. In addition, I find that the international community appears to play a particularly significant role in enhancing the likelihood that military integration will be agreed to as part of a civil war settlement.

In the first section of this chapter I provide a brief overview of the concept of military integration and discuss trends in its use as a component of civil war settlements. Next I examine the theoretical literature relevant to the terms of civil war settlements, focusing on what it does and does not tell us about the likelihood that warring groups will agree to merge their forces. In the third section I conduct an empirical analysis of the hypotheses derived from the theoretical overview. I conclude with a discussion of the results and their implications for the use of military integration as a war-ending strategy.

MILITARY INTEGRATION AS A COMPONENT OF CIVIL WAR SETTLEMENTS

I adopt the perspective advanced by the bargaining model of war: that all wars end in an agreement, whether explicit or implicit, between the parties to the conflict. According to the bargaining model, armed conflict is part of a process of bargaining that takes place between actors seeking to achieve the most favorable allocation possible of some value or good. A war ends when the belligerents negotiate settlement terms that they prefer to continuing the war. The terms of the settlement, according to the bargaining model, specify what each of the parties gets from the conflict. As such, settlement terms "specify who gets what and thereby determine the benefits of peace and the incentives to return to war" (Werner and Yuen 2005, 262).

Although scholars who employ the bargaining model of war note that the terms of war-ending settlements vary in nature, bargaining theorists generally have not developed different means of categorizing or differentiating among types of settlement terms. This may be because the bargaining and conflict literature has focused primarily on the terms of settlements of *interstate* wars. Because the issues, values, or goods contested by sovereign states vary so widely, it can be difficult to meaningfully distinguish among settlement terms on the basis of the "gets what" part of the "who gets what" equation.[4]

This task is easier in the case of civil war settlements because the central issue at stake in civil wars is the control of state power. Whenever a nonstate armed actor militarily challenges a government, it does so with the goal of claiming power at the political center or in some region of the country. The fundamental question that *civil war* settlements resolve is who is to have access to the levers of state power. It is precisely because there are several possible answers to this question, with some settlements allocating all elements of state power to one belligerent while others distribute state power among the various adversaries in a conflict, that it is possible to distinguish among the terms of civil war settlements.

Hartzell and Hoddie (2003, 2007) have identified four components of state power as being of central interest to adversaries negotiating the terms of civil war agreements.[5] State power can be conceptualized as having political, military, territorial, and economic dimensions. Control of any one of these dimensions of state power can add to a conflict actor's overall level of power, thereby serving to enhance that actor's potential to defeat its adversaries in any future hypothetical conflict. Civil war rivals thus have an overriding interest in exercising sole control of each of the various dimensions of state power or, at a minimum, in denying their opponents unfettered control of any one of them.

Military power, particularly the extent of the state's control of the legitimate means of violence, is, as noted above, a central dimension of state power. The question of who controls the military and other state security forces in the aftermath of armed conflict is of particular concern to the parties to a civil war. Leaders and followers are likely to fear that, once they disarm, those who control the state's coercive forces may use those forces to eliminate rival groups or otherwise secure a victory that they did not achieve on the battlefield.

Adversaries in civil wars should thus be motivated to gain control of the security sector apparatus of the state as part of a civil war–ending agreement. Any party to the conflict that believes it cannot secure that outcome as part of a settlement should strive to attain its second-best option, which is to impede its rival's gaining control of the military. The most feasible means of doing that is to engage in some form of military power sharing.[6]

Military power sharing focuses on the distribution of authority within the coercive apparatus of the state. The most straightforward means of sharing military power is to integrate government and nonstate actors' forces into a unified state security force. This can be accomplished either on the basis of some formula (e.g., a proportional formula that reflects the relative sizes of the armed factions) or through strictly equal division of troops among the contending parties. A second avenue for distributing military power involves focusing not on merging troops but on integrating the leadership of nonstate actors' forces into the state's security forces; these actors are appointed to key leadership positions in the security sector, such as those of general, commander, director, and defense minister. Third and last, in rare instances, striking a balance among the militaries of rival forces may involve allowing opposing sides to remain armed or retain their own security forces (Hartzell and Hoddie 2007).

In light of this book's focus on military integration, I concentrate on the first two forms of military power sharing described above in analyzing the terms of civil war settlements. As table 2.1 makes clear, the use of military integration in civil war settlements has grown in the decades since the end of World War II. Whereas none of the civil war–ending agreements in the first decade and a half following the end of that conflict called for integrating rival forces' militaries, for the next five decades a

TABLE 2.1 Inclusion of Military Integration as a Term of Civil War Settlements, 1945–2006

1945–49	1950–59	1960–69	1970–79	1980–89	1990–99	2000–2006
7 civil war settlements	11 civil war settlements	10 civil war settlements	21 civil war settlements	14 civil war settlements	47 civil war settlements	18 civil war settlements
0 provisions for military integration	0 provisions for military integration	1 provision for military integration	5 provisions for military integration	5 provisions for military integration	47 civil war settlements	18 civil war settlements
0% of settlements include military integration as a term of settlement	0% of settlements include military integration as a term of settlement	10% of settlements include military integration as a term of settlement	24% of settlements include military integration as a term of settlement	36% of settlements include military integration as a term of settlement	49% of settlements include military integration as a term of settlement	56% of settlements include military integration as a term of settlement

steadily increasing percentage of all settlements included a provision for some form of military integration.

This upward trend in the use of military integration as part of the terms of civil war settlements suggests that factors that vary with time may well be exercising some influence on warring actors' decisions regarding military integration. Three possibilities come quickly to mind. One is the change in the means by which civil wars end. Whereas 82 percent of civil wars in the 1950s ended in military victory, the percentage of wars ending in that manner has declined during every decade since then, reaching a low of 11 percent in the period from 2000 to 2006. Second is the role of outside actors. Although all the decades analyzed in this chapter saw mediators involved in efforts to end civil wars, mediation activity reached unprecedented levels in the 1990s, with mediators involved in 81 percent of the civil wars in that decade, and again in the years from 2000 to 2006, when 89 percent of the conflicts saw mediation activity. Finally, the end of the Cold War also appears to have played a role in growing calls for military integration, perhaps in part through affecting mediation activity and the means by which civil wars are now being terminated.[7]

Although the foregoing trends associated with civil war settlements are suggestive, they do not encompass the range of potential explanations for intrastate adversaries' decisions to merge their militaries. A more complete analysis of this choice requires that we consider what various theories have to say about war termination and the terms of settlements. I now turn to these.

EXPLAINING MILITARY INTEGRATION

In light of the growing reliance on military integration as part of civil war settlements, a clearer understanding of the drivers of this choice would be useful for both analytical and policy purposes. As noted in the introduction to this chapter, knowledge of the conditions that lead civil war rivals to decide to merge their militaries can be useful in helping to determine whether selection effects are at work. It is highly unlikely that the conflict actors in all civil wars will be equally likely to opt to merge their militaries, and military integration might be a more likely choice either in easier-to-resolve civil war environments or in particularly challenging civil war circumstances. If unaccounted-for factors that influence the selection of military integration as a settlement term also have an impact on outcomes of interest such as the duration of the peace, there is a risk of either overestimating or underestimating the effects that military integration has on the dependent variable in question.

As a first step toward identifying the factors that influence conflict actors to agree to integrate their militaries, I draw on a number of theoretical and empirical studies regarding the terms of civil war settlements. I use these to identify four testable propositions.

The Bargaining Model of War and the Balance of Military Power

The reigning theory of conflict analysis, the bargaining model of war, conceives of bargaining as taking place during all stages of war. As noted by Reiter (2003, 29): "Fighting breaks out when two sides cannot reach a bargain that both prefer to war. Each side fights to improve its chances of getting a desirable settlement of the disputed issue. The war ends when the two sides strike a bargain that both prefer to continuing the war, and the outcome is literally the bargain struck. Finally, the duration of peace following the war reflects the willingness of both sides not to break the war-ending bargain."

Opponents reach an agreement to end a war when their beliefs regarding the consequences of continued fighting converge sufficiently so as to make it possible for them to agree "which side needs to make concessions . . . [and] how large the concessions need to be" (Werner and Yuen 2005, 264). Beliefs converge because of the information that is revealed through the process of fighting. According to this perspective, combat reduces adversaries' uncertainty regarding their abilities to mete out and/or absorb costs by providing information about the balance of power, military effectiveness, and resolve (Reiter 2003). With this information in hand, antagonists should be able to arrive at a distribution of benefits—a settlement—that corresponds to their expectations about the consequences of continued fighting (Werner 1999).

As discussed in the previous section of the chapter, the terms of civil war settlements can be expected to center on the issue of who is to exercise control over various components of state power. Several hypotheses regarding the distribution of these "goods" can be derived from the bargaining model of war. First, as Cunningham (2011) posits in a recent study of the content of negotiated civil war settlements, "combatants with a higher likelihood of victory should get more of what they want in an agreement than those with a lower probability of victory." Both governments and rebel groups, it can be assumed, would like to exercise full control over the state. If governments have a higher likelihood of victory, they should favor the status quo and continue to exclude rebel forces from the national military. Military integration is thus likely to take place only when rebel groups have a high enough probability of victory that they can pressure the government to agree to such terms. Nonstate actors are most likely to be able to pressure governments to agree to share power, argues Gent (2011), when they are militarily as strong as or stronger than government forces. This logic provides the basis for the first hypothesis to be tested:

H_1: *Military integration is most likely to be included as part of a civil war agreement when rebel groups are at military parity with, or are stronger than, the government.*

State Capacity for Accommodation

State capacity plays a central role in political opportunity models of rebellion. Scholars such as Fearon and Laitin (2003) have emphasized the state's repressive capacity as fundamental to the onset of civil conflict. As noted above, the military capacities of belligerents play a central role in the bargaining model's predictions regarding the terms of war-ending agreements. But coercive capacity is arguably not the only type of capacity that can play a role in shaping settlement terms. In addition to a capacity for repression, states also have a capacity for accommodation. This capacity, based on a state's ability to redistribute resources and power, can also have an impact on the likelihood of renewed rebellion. As Hendrix (2010, 273) observes,

"If the state is capable of accommodating grievances via institutionalized channels, such as redistribution . . . or the incorporation of dissident movements, . . . then the motivation for violent rebellion will be lessened and conflict will be less likely."

There are reasons to expect that states with a higher capacity to redistribute resources and power will be more likely to agree to merge militaries as part of a civil war settlement. First, these states are the ones most likely to have the ability and the wherewithal to accommodate such a demand on the rebels' part. Integrating militaries can be a technically demanding and potentially costly proposition—that calls for, among other tasks, equilibrating military and rebel forces' ranks, providing literacy and skills training, and, in some cases, equipping large numbers of former rebel troops. South Africa, for example, found itself faced with the "Herculean task of integrating all members of the eight separate military forces" in the country at the time of its peace settlement—a task that state, which has a higher capacity than many other countries in which civil wars have been fought, was able to accomplish, albeit with the assistance of external actors (see chapter 8 in this volume).[8] The belief that they are capable of handling the challenges involved in integrating rebel troops into the military may help to make higher-capacity states more inclined to agree to such measures.

Second, more capable states may be more willing to agree to merge military forces based on the reasoning that such a concession will have only a limited effect on overall levels of state power. Generally speaking, agreements calling for military integration rarely involve the complete reconstruction (or "transformation," as it is described in chapter 11) of the military; rather, rebel troops are usually integrated into existing military structures (the strategy of "continuity," per chapter 10), which is likely to mean some degree of continued government dominance of the military. In addition, relatively few agreements are like Burundi's and Mozambique's, both of which stipulated a 50/50 allocation of posts in the military to the respective warring parties (Samii 2013; chapter 10 in this volume). Settlements calling for military integration instead more often resemble that of the Philippines, according to which a limited number of Moro National Liberation Front fighters and their proxies (5,750) were integrated into the state's armed forces (see chapter 7). Finally, high-capacity states are also likely to calculate that their control of other resources and forms of state power provides them with the ability to balance or check whatever power rebels have gained from an agreement to integrate militaries. Our second hypothesis for analysis is as follows:

H_2: *Military integration has a higher likelihood of being agreed to as part of a civil war settlement in high-capacity states.*

Economic Motivation

In one of the few studies to focus specifically on the question of what induces parties to agree to military integration (rather than treating power-sharing measures more generally), Glassmyer and Sambanis (2008) hypothesize that economic incentives are an important determinant of the choice to merge militaries at the conclusion of a civil war. Military integration, they posit, is used as a means of providing jobs for rebels in poor countries who lack other employment options for demobilized actors. Governments in these countries embrace this option as a means of preventing rebel remobilization; rebels support it as a means of ensuring that they have a regular source of income. Glassmyer and Sambanis's reasoning serves as the basis for the following hypothesis:

> H_3: *The poorer countries are, the more likely they are to agree to the merging of military forces as part of a civil war settlement.*

Mediation

During the past few years, scholars have paid increasing attention to the effects of mediation on conflict resolution in civil wars. Generally speaking, studies have focused on the effect of mediation on whether or not a negotiated settlement is reached, as well as its effect on the durability of such agreements (e.g., see Beber 2009; DeRouen, Bercovitch, and Pospieszna 2011; Gartner 2011). Much less attention, however, has been focused on the influence, if any, that mediators may have on the terms of civil war settlements. The only work I am aware of that has addressed this issue is a study by Svensson (2009), in which he examines the effects that both biased and neutral mediation have on the content of peace agreements. Noting that military power sharing provides a form of insurance to minority groups that fear future exploitation, he hypothesizes that mediators biased in favor of rebel groups will be more likely to try to convince the conflict parties to agree to some form of military power sharing.

Although I do not categorize mediators on the basis of their neutrality or bias in this study, I follow the lead of both Svensson (2009) and Cunningham (2011) in arguing that mediation can influence the content of civil war settlements. Although he does not have any theoretical reason to expect that mediation will have different effects on different types of settlement terms, Cunningham (2011, 26) does find that when mediators are involved in a conflict, "rebels are more likely to be integrated into the military, but not necessarily more likely to receive political or territorial concessions." He concludes that this result is driven by mediator preferences rather than the adversaries' attributes. Noting that much of the work on conflict resolution emphasizes the risks of excluding rebels from military power, he surmises that this might induce mediators to influence combatants to agree to a merger of militaries as one of the terms of war-ending settlements. This logic serves as the basis for the final hypothesis to be tested:

> H_4: *Civil wars in which mediators become involved have a higher likelihood of ending in settlements that call for military integration.*

It should be noted that of the hypotheses presented above regarding military integration, only the economic motivation and mediation hypotheses actually develop theoretical propositions regarding the factors that motivate warring parties to agree to merge their fighting forces. Although the hypotheses about the balance of military power and about state capacity for accommodation can help us to identify conditions under which warring parties should be more likely to agree to some form of power sharing—including, potentially, the merging of militaries—as part of a war-ending settlement, they do not provide a specific rationale for expecting that the parties will agree to military integration.

Control Variables

Also included in the analysis of the factors explaining military integration are several control variables that scholars have suggested may have an impact on the terms of settlements. These include conflict duration, the history of conflict between the parties, whether or not the conflict was an identity conflict, and whether or not the conflict was settled during the post–Cold War period. Parties to a war that is long or that is the latest in a series are thought to be more likely to agree to measures for the sharing of power, because these factors indicate the low likelihood each has of prevailing militarily. Settlements agreed to in the post–Cold War period are also posited to have a higher likelihood of producing a power-sharing agreement, the warring actors being less beholden to the competing interests of the major powers. The relative rigidity of identities, conversely, has been identified as a factor that makes power-sharing arrangements less likely to be agreed to as a means of ending civil wars (Hartzell and Hoddie 2007).

Although the control variables discussed above are hypothesized to have an impact on the likelihood that warring parties will agree to some form of power sharing, none of them speaks directly to the question of whether—and why—rival groups will agree to integrate their military forces. However, one control variable that might be expected to have an effect on the likelihood that adversaries will agree to merge their militaries is territorial conflict. Groups engaged in a conflict over territory are likely to seek territorial concessions as part of a settlement. If they receive such concessions, these may act as a substitute for a military integration measure, because territorial autonomy arrangements provide nonstate actors with an area they can patrol or a physical buffer zone of sorts in which they can protect themselves. Able to provide for their security in this fashion, nonstate actors are likely to see less need for military integration, and governments may be less inclined to provide them with additional means of guaranteeing their physical safety.

Data and Measurement

I employ a data set of my own design that covers civil wars fought and ended from 1945 through 2006. I classify intrastate conflicts as civil wars if they meet the criteria employed by Small and Singer (1982) in the Correlates of War project: The conflict produces at least 1,000 battle deaths per year; the central government is a party to the conflict; both the national government and its opponents put up an effective resistance during the course of the conflict; and the conflict occurred within a defined political unit. This classification yields 128 civil wars, which were fought in 64 countries.

The dependent variable in this study, *military integration*, is a dichotomous variable. Analyzing the settlements reached in each case of civil war, I determined whether or not the parties either agreed to some proposal for merging state and nonstate actors' forces, or opted to integrate the leadership of nonstate actors' forces into the state's security forces. Instances in which parties agreed to either of these options were scored 1 for military integration.

I rely on Cunningham, Gleditsch, and Salehyan's (2009) summary ordinal measure of the military strength of rebels relative to the government for the variable *relative rebel strength*. I employ their coding rules to update the data through 2006. I also alter the data from the dyadic form they employ to conflict-level data by scoring the variable 1 if any rebel group in the conflict in question was at parity with, or stronger than, the government. I use the Penn World Tables' government consumption share of purchasing power parity converted gross domestic product (GDP) per capita for the year when the war ended as a measure of *state capacity* (Heston, Summers, and Aten 2011).[9] The Penn World Tables also serve as the data source for purchasing power parity converted *GDP per capita* for the final year of the war, the measure used to test Glassmyer and Sambanis's hypothesis regarding the relationship between national income and the commitment to military integration. *Mediation* is a dichotomous variable. I utilize Regan, Frank, and Aydin's (2009) data set on diplomatic interventions, as well as DeRouen, Bercovitch, and Pospieszna's Civil Wars Mediation data set, to determine whether or not mediators were involved in negotiations to end a civil war.[10]

The coding for the *history of conflict, conflict duration, identity conflict,* and *post–Cold War settlement* variables is my own. I employ Toft's (2010) coding of the variable *territorial conflict*.

DATA ANALYSIS

Because the dependent variable is dichotomous, I employ a logit model. Model 1, shown in table 2.2, reports the results of a stripped-down model that consists solely of the control variables. Here we find support for the hypothesized impact of all the control variables except *identity conflict*. In this instance, not only is the variable not statistically significant, but the coefficient is positively rather than negatively signed.

I test hypotheses 1 through 4 by adding the explanatory variable relevant to each hypothesis to Model 1. Model 2 employs *relative rebel strength* in order to test the insight of the bargaining model of war regarding the effect that the balance of military power at the time of a settlement has on the likelihood that the warring parties will agree to integrate their militaries. Although the positively signed coefficient lends some support to the notion that adversaries are more likely to agree to merge their militaries when nonstate actors are at parity with or stronger than the government's forces, the variable does not prove to be statistically significant. Model 3 tests whether the *state's capacity for accommodation* has any impact on the likelihood of an agreement to merge militaries. The positively signed and statistically significant (at the $p < 0.05$ level) coefficient provides support for this hypothesis. The lack of statistical significance for the *GDP per capita* variable in Model 4 indicates a lack of evidence for economic incentives as a motivating factor for military integration. Finally, Model 5 provides statistical confirmation for the hypothesis that external *mediators* play a role in encouraging warring groups to agree to merge their militaries.

I bring each of these explanations, along with the control variables, together in

TABLE 2.2 Logit Analysis of Military Integration as a Component of Civil War Settlements, 1945–2006

	Model 1: Control Variables	Model 2: Bargaining Model of War	Model 3: State Capacity	Model 4: Economic Incentive	Model 5: Mediation	Model 6: Integrated Model
Relative rebel strength		0.523 (0.509)				-0.211 (0.536)
State capacity			0.040** (0.018)			0.061*** (0.020)
GDP/pc, final year of war (logged)				-0.000 (0.000)		-0.000 (0.000)
Mediation					1.430** (0.648)	2.187*** (0.798)
History of conflict	1.109** (0.458)	1.242*** (0.474)	0.946** (0.474)	1.113** (0.497)	0.930* (0.523)	0.465 (0.598)
Conflict duration (logged)	0.353** (0.149)	0.375** (0.147)	0.270 (0.164)	0.281* (0.166)	0.335** (0.158)	0.165 (0.167)
Identity conflict	0.158 (0.402)	0.116 (0.407)	-0.117 (0.447)	0.106 (0.412)	-0.021 (0.410)	-0.498 (0.477)
Territorial conflict	-1.128** (0.574)	-1.206** (0.564)	-1.039 (0.653)	-1.146* (0.651)	-1.283** (0.585)	-0.844 (0.705)
Post–Cold War settlement	1.614*** (0.481)	1.569*** (0.477)	1.813*** (0.598)	1.647*** (0.536)	1.242** (0.514)	1.433** (0.652)
Constant	-2.841*** (0.642)	-3.119*** (0.679)	-3.100*** (0.722)	-2.350*** (0.741)	-3.365*** (0.717)	-3.594*** (0.868)
N	128	126	111	112	128	111
Pseudo R^2	0.261	0.267	0.256	0.250	0.302	0.340
Log pseudo-likelihood	-62.214	-61.653	-55.149	-55.933	-58.705	-48.882
Prob > chi^2	0.000	0.000	0.000	0.002	0.000	0.000

Note: Robust standard errors, reported in parentheses, are adjusted for clustering by country. *$p < 0.1$; **$p < 0.05$; ***$p < 0.01$, two-tailed tests.

Model 6. This integrated model provides further evidence of the positive effects that both state capacity and mediation have on securing warring parties' agreement to integrate their militaries. Each of these variables now proves to be statistically significant at the $p < 0.01$ level. Interestingly, only one of the control variables, post–Cold War settlement, remains statistically significant in this model.[11]

SUBSTANTIVE EFFECTS

As a final step, I calculate the substantive effects of the three variables in Model 6 that proved statistically significant. I begin by generating the baseline predicted value for military integration with all continuous variables in the model set at their mean values and the dichotomous variables set at their modal values. The predicted value in this instance is .6356. Next, I shift the continuous variable *state capacity*, first raising it by 1 standard deviation. This yields a larger predicted value for military integration of .7653; decreasing *state capacity* by 1 standard deviation produces a lower value of .4827 for military integration. Altering the value of the *mediation* variable from its modal value of 1 to 0 lowers the predicted value of military integration to .1638, while shifting the *post–Cold War settlement* variable from its modal value of 1 to 0 yields a predicted value of .2940. These results suggest that the efforts of mediators have a particularly marked effect on agreements by warring parties to integrate their militaries.

DISCUSSION, IMPLICATIONS, AND CONCLUSION

Why do governments and rebel groups agree to integrate their militaries in the wake of civil wars? Realist theory suggests that both sets of actors should be loath to give up any degree of control over their respective coercive forces, an outcome that is presumably inherent in the merging of militaries. And yet, as the analysis in table 2.1 makes clear, a growing percentage of civil wars settled during the post–Cold War period have been characterized by agreements to integrate government and nonstate actors' militaries.

This chapter has attempted to account for this rather surprising trend by examining four potential explanations for warring parties' decisions to integrate their militaries as part of a civil war settlement. The empirical analysis indicates that two of these explanations—those involving state capacity and mediation—provide some insight regarding adversaries' motives for agreeing to merge militaries, whereas two others—those involving the balance of military power and the role of poverty as a motivating factor—do not. These results suggest that mixed motives regarding the decision to integrate militaries do exist—but, somewhat surprisingly, that it is not rebel groups and governments that play the most important roles in making this decision, but rather governments and external mediators.

The finding that rebel military strength does not translate into nonstate actors' ability to secure an agreement to merge militaries presents something of a challenge to the bargaining model of war. The bargaining model suggests that when they are the militarily stronger actors in a conflict, rebel groups should be able to secure a concession of this nature from the government. Although the failure of this variable to emerge as significant in Model 2 and Model 6 is unexpected, it is not without precedent. Cunningham (2011) and Gent (2011), both of whom included measures of rebel military strength in their respective studies of peace agreement concessions and power sharing in civil wars, both found the variable to be statistically insignificant as an explanation for the outcomes on which they focused.

The fact that the "balance of military power" explanation for military mergers did not pan out should not be taken to imply that what rebels want with respect to settlements does not matter or does not at least exercise an indirect effect on the peace process. It may well be, for example, that nonstate actors rely on mediators to represent their interests in this regard. The finding of significance for the mediation variable suggests the importance of learning more about the interests of the mediators who play a role in civil war settlements. If mediators see military integration as an important component of peace-building, as Cunningham (2011) suggests, they are likely to push for it even in those instances in which relatively weak rebel groups might not normally expect to wrest such a concession from the government. Whatever mediators' motivations may be, the growing use of mediation in civil war settlements appears to be an important factor in trends in the use of military integration since World War II.

Perhaps the most interesting finding to emerge from the analysis is that states with a higher capacity to engage in accommodation are the ones most likely to agree to military integration. The decision by governments of high-capacity states to agree to such a measure is somewhat puzzling. Why do these states not simply rely on the resources and power they command to push through agreements that do not include such concessions? One possibility is that the governments of high-capacity states see military integration as a low-cost strategy in comparison with continued fighting. Butler and Gates (2009, 334) make this point well:

> As the weaker group [in a civil war] becomes comparatively weaker, it will devote more and more resources to warfare to continue the fight. This . . . has significant implications for negotiation. As the weaker power has a higher marginal benefit from fighting, it expects to get a lot from war and, thereby, expects to get even more out of negotiations. Indeed, the government must overcompensate the rebel group. Given that the government has more to lose from fighting than the rebels, a negotiated deal that favors the rebel group is better for the government than continued conflict.

Another (and possibly more cynical) explanation is that because rebel groups are generally integrated into national military structures, a government may believe that on balance it will continue to exercise control over the state's coercive forces. That belief, in combination with the other resources and power that high-capacity states control, may lead such states to calculate that an agreement to merge militaries does little to alter the balance of power. If this is indeed the logic that motivates governments to agree to military integration, the implications are disturbing. Nonstate actors, I have suggested, are motivated by their security concerns to integrate their forces with those of the government. If military integration does little to alter the government's control of the military, rebel groups' ability to protect their interests may well be compromised.

Knowledge of the conditions under which governments and nonstate actors are most likely to agree to military integration can play a role in helping the international community determine how well this strategy works in helping to keep the peace. In addition, such knowledge can help in identifying factors such as mediation that play a role in fostering agreement on military integration. The next important step is to analyze the circumstances that make it more likely that civil war adversaries will *implement* agreements to merge their militaries. Although agreements to integrate militaries have the potential to exercise an influence on the peace through the costly signaling they involve, the extent to which groups follow through on their agreements is likely to matter as well.

NOTES

1. Author's data.

2. Selection bias arises when analysts fail to take into account factors that may influence both the decision regarding whether or not to agree to military integration as part of a civil war settlement, and outcomes such as the duration of the peace. See Vreeland (2003) for a discussion of issues associated with selection.

3. Glassmyer and Sambanis (2008) make this point in their analysis of the impact that military integration has on the duration of the post–civil war peace.

4. Werner (1998), who categorizes interstate war settlements along a spectrum ranging from the "most benign terms" to the "most punitive terms," does so on the basis of many, although certainly not all, of the potential goods or values whose allocation states might contest, including reparations, territorial and political concessions made by one belligerent to the other, and imposed regime change.

5. It is important to emphasize that *all* civil wars involve bargaining regarding the terms on which the war will be settled. Bargaining is not limited to those instances of civil war that have been categorized as ending via negotiated settlement. See Pillar (1983) for an exposition of this argument.

6. Another potential alternative is the Costa Rican model. Costa Rica dismantled its military following the country's civil war in 1948. The country has a public security force that is responsible for internal security. No other country has demilitarized following a civil war, although Grenada and Panama shed their standing armies following US invasions of the two countries.

7. Cross-tab analyses of the relationship between the call for military integration as part of a civil war agreement and each of these three variables—the means by which wars are ended, mediation activity, and the end of the Cold War—are significant at the $p = 0.000$ level.

8. As Licklider notes, South Africa was assisted in its efforts by the British Military Advisory Training Team.

9. My logic in employing this measure is that states that score higher on it arguably have more resources and power that they can use to accommodate other actors than do lower-capacity states.

10. I thank Karl DeRouen for providing me with access to an early version of the Civil War Mediation data set.

11. Although I do not report the results in table 2.2, I also ran a variant of Model 6 that included the variables *negotiated settlement* and *military victory* as proxies for the means by which the wars ended (the excluded category was *negotiated truces*). Because the majority of agreements to share power, including agreements to merge militaries, emerge from negotiated settlements, it is not surprising that the *negotiated settlement* variable, which was positively signed, was statistically significant (at the $p < .05$ level). The negatively signed *military victory* variable, conversely, did not prove to be statistically significant. The *state capacity* and *post–Cold War settlement* variables, which were significant in Model 6, remained significant in this model as well (at the $p < .05$ level in this instance), as did the *mediation* variable (although only at the $p < .1$ level).

Part I

EARLY ADOPTERS

Chapter 3

SUDAN, 1972–1983

Matthew LeRiche

> The most serious crisis in the implementation of the [Addis Ababa] agreement related to the integration of the Anya Nya into the army.
> —*Mansour Khalid (2003)*

The failure of the attempt to integrate the South Sudanese Anyanya ("snake poison") rebel fighters and the Sudanese military, a major requirement of the 1972 Addis Ababa Peace Agreement, made a return to war in Sudan almost inevitable.[1] When the Addis Ababa Peace period deteriorated into confrontation among Southerners and between Southerners and Northerners, former fighters who had integrated into the new Sudanese Armed Forces (SAF) returned to the bush to fight for their original goal: Southern independence. Midway through the peace period, Jaafar Nimeiri—though heralded as a peacemaker—abrogated the agreement, which provided the impetus for the return to civil war. The case of the Addis Ababa Peace Agreement is important; it was one of the first major agreements to resolve a civil war in postcolonial Africa and included stipulations for military integration.

Although it has been argued that the military integration process was conducted reasonably well, it is clear that many lower-level officers and noncommissioned officers (NCOs) quickly became frustrated (Allen 1991). According to Mansour Khalid (2003), one of the Sudan's most prominent intellectuals and the first major Northern political figure to join the Sudan People's Liberation Movement/Army (SPLM/A): "The implementation of the military protocol, particularly as far as the integration of the forces was concerned, proved to be the most bothersome of all aspects of the agreement. . . . The army command was anxious to neutralize the Anyanya through absorption. Naturally, the rank and file of the Anyanya was suspicious and afraid."

Seeing the disillusionment of their commanders, particularly former Anyanya leader Joseph Lagu, Southerners who had been integrated into the SAF abandoned the peace agreement. By the early 1980s the agreement was collapsing, largely due to intertribal politics in the South and alienation in the integrated military. Jaafar Nimeiri, then president of Sudan, went on to abrogate the agreement by revoking Southern autonomy and moving toward support of Khartoum-based Islamists and their desire to impose sharia law, rescinding the secular-state principle and the religious freedoms enshrined in the agreement. The dismantling through Nimeiri's redivision policy of Southern Sudan's regional autonomy, already a compromise on the Southern rebels' goal of independence, ended the peace for many (Fegley 2011, chaps. 2 and 5). Although some in the South, particularly in Greater Equatoria, supported redivision—the division of Southern Sudan into three states under the government in Khartoum—there was now little holding most Southern soldiers and officers to the Sudanese Army. The trickle of defections increased to a steady flow, and with the major defection of groups in Bor and Pibor in 1983 it was clear the Addis Ababa peace project was finished.

The failure of military integration was central to the return to war. This failure, in turn, was due to Nimeiri's decision to move the Anyanya rebels integrated into the SAF north, away from Southern command and their home areas. These Absorbed Forces (the term given to Anayana who had been integrated into the new army) had expected to maintain their local character and deployments, but Nimeiri chose to disperse the Southern forces to assert his control. This critical issue was never settled because the integration of forces outlined in the Addis Ababa Peace had not been designed with input or influence from a wide range of stakeholders. Moreover, it served Nimeiri's purposes to have Southerners believe something that would keep them loyal in the short term, regardless of his longer-term designs for their roles in the army, bureaucracy, and government; only one Southerner was in Nimeiri's cabinet, no more than three were ever admitted into either the national police or military academies, and few were given civil service posts (Fegley 2011, 158–59).

Former Anyanya rebels apparently expected to remain together as originally organized under the direction of Joseph Lagu, the Anyanya leader who later became the commander of SAF forces in South Sudan. Lagu's initial demand during peace negotiations in Addis Ababa, linked to the goal of Southern independence, was the maintenance of a separate Southern Sudanese Army. His negotiating team withdrew this demand when Ethiopian emperor Haile Selassie, the mediator and facilitator of the talks, suggested integration as a compromise offering qualified autonomy. Along with moving Southern soldiers northward, the process of integration, particularly decisions concerning rank and role, frustrated many of the Absorbed Forces, and compelled some to mutiny; the full abrogation of the agreement in 1983 followed the Bor Mutiny of Battalion 105 under the leadership of Kerubino Kuanyan Bol and John Garang (both founding members of the SPLM/A).

The return to war was not simply a failure of those in the North and South to reconcile; it was also the result of Southern groups' inability to reconcile, unite, and coalesce. Although most Southerners were initially willing to accept the agreement struck by their patrons and self-proclaimed political leaders, their situation did not improve significantly, resulting in dissatisfaction, suspicion, and a general rejection of the peace. The decision to move the Absorbed Forces north was interpreted as an attempt to preclude any Southern influence in Sudan's future, and, like the lack of inclusion in national civil service positions and national academies, it was seen as an intentional slight that proved the lack of interest in affording the rights and respect supposedly provided by the Addis Ababa Agreement. For Southerners, such inclusion meant enfranchisement in the state and government; appointing certain men to certain posts, or moving others from a particular region, was deeply political and signaled the failure of the democratic inclusion that the agreement had promised.

Most important, however, was the fact that the movement of forces was seen as undermining Southern communities' self-defense capacity; armed groups were a kind of insurance these communities were unwilling to lose. Communities have long been the most important feature of Sudanese society. Moving Absorbed Forces north both increased the former rebels' dissatisfaction with the integration process and raised concerns that Khartoum was removing community protection.[2] This communalism was clearly influential in the thinking of those members of the Absorbed Forces who returned to the bush during the peace period and in 1983 to renew their rebellion, and of those youth who went to the bush to fight for the first time.

ORIGINS

The army of the Sudanese government was shaped by the colonial authorities and dominated by Northerners, while the Anayana reflected a local militia tradition. The decision to merge them after the war was driven by outsiders and was highly controversial, setting the stage for future problems.

The Historic Role of the Military

Long at the heart of political power and business, the military in Sudan is used as a means to gain profit and exert the center's control upon the periphery. Key families in the Khartoum area and to the north, particularly current president Omar Bashir's Jaaliyin tribe (Ryle et al. 2011), have dominated the army. Coups have typically come from within the army and, until Sadiq al-Mahdi's use (and Omar Bashir's later refinement) of Islamist paramilitaries, the army was the key force relied upon by Khartoum to maintain control. However, there have been periods during which the army has been more widely inclusive than such institutions as the internal security services and banks. This dates back to the Sudan Defense Forces of the British, who implemented an effective indirect rule with Egyptian assistance; this arrangement was known as the Condominium (Ryle et al. 2011, 10–12). In reality, most Sudanese armed forces are of a local, tribal nature—paramilitary forces, or what might be better described as militias, which are the basis of any armed forces manpower or doctrine. This is especially true in the South, now the Republic of South Sudan, and the Southern areas of Kordofan, Blue Nile, and Darfur, where even the military arranged in formal conventional structures has a strong basis in localized armed formations. Out of this comes the desire to remain in the home area and to engage in raiding or warring primarily in neighboring areas rather than distant ones. Such tendencies are difficult to change, especially in such a short period as was afforded for military integration by the Addis Ababa Peace Agreement.

By the early 1950s it had become clear that the British were going to withdraw from Sudan and leave the country to independence. The Southern and Northern Sudanese disagreed about how independence should be reached. Aware of the imbalance between the central peoples and Southerners in the capacity for administration and governance, Southern leaders hoped to establish a process that would ease Sudan into independence, thus granting the South time to build the capacity needed for inclusion in the new state's administration. This did not occur, however, and despite clear provisions in the terms of independence for referenda and for collaborative efforts to include Southern interests, the British relinquished power to a small core of the elite in Khartoum who controlled the new state's army and economy.[3]

Even before formal independence was gained from British Condominium rule, armed insurrection was building throughout Southern Sudan against the fledgling central authorities in Khartoum. Compelled by a sense of injustice and a history of oppression, and recognizing the opportunity of this historic moment, the Southern Sudanese began to organize various armed forces, most based around tribal and associated political networks. Most of these forces, which became known as Anyanya (a word used in various Equatorian tribes to mean "snake poison"), sprang up in areas along international boundaries. Drawn from Sudan Defence Force troops who mutinied in 1955 and various warriors and young men of Southern tribes, "Anyanya was like a disorganized game. It had no modern material, save for explosives. Nevertheless the British-trained Southern Sudanese Army began to give the Arab government a hard time" (Ga'le 2002, 262).

In the insurrection's final three years, however, the Anyanya forces grew substantially in size and strength and began placing real pressure on the army (Khalid 2003, 137). It is difficult to determine an approximate number of Anyanya at the time of the Addis Ababa Agreement; estimates range from six thousand to twenty thousand. Not long after Sudanese independence in 1956, the SAF probably numbered close to eleven thousand.

The Emergence of Military Integration as an Issue

From the very beginning of the negotiations, the role of the rebel forces in the state that would be created after the peace agreement was a primary issue. Northern negotiators, advised by Abel Alier on this matter, were receptive to the idea presented by Haile Selassie's mediators that a kind of integrated armed force was the appropriate complement to a qualified autonomy. Ethiopia was not interested in any dissolution of its neighbors and had only just rebuilt relations with the Sudanese government in Khartoum after Nimeiri's taking power.

The negotiating parties appear to have conceived of force integration in very different ways. The Sudanese government saw integration as a way to reindoctrinate key Southerners into the establishment and to gain control over the forces causing so much trouble for them. Lagu and his negotiating team, however, saw force integration as a compromise that could guarantee Southern involvement in the governance of the country and, potentially, give Southerners prominence in its most powerful institution. Lagu seems to have engaged in the negotiations with considerable goodwill, while Nimeiri's negotiation team operated in a much more Machiavellian fashion.

Supporters, Opponents, and Neutrals

Lagu's move to agree to peace with Nimeiri was useful in affirming and consolidating Lagu's leadership. By 1971, only a little more than two years after taking power in a coup himself, Nimeiri was facing pressure both from Mahdist revolutionaries led by Imam al-Hadi and on the left from the Sudan Communist Party backed by Iraq's Baath party (Khalid 2003, 137). Although the Anyanya were finally achieving significant success on the battlefield, at the time of the peace talks discord was growing among the various Southern political and military groups. With regional politics swirling, the two leaders found it particularly expedient to reach a settlement as quickly as possible. Pressure was mounting from many directions. Because of Nimeiri's backing of Egypt, the Israelis were supporting the Anyanya, which was a major reason for the group's increasing battlefield success. Uganda was the key rear base of operation for Lagu, but Idi Amin decided that he wanted the Anyanya and Israelis out of his territory. Amin also began manipulating South Sudanese political leaders, playing them off each other as they ran their various meetings in the Ugandan capital. As Khalid (2003, 137–38) has noted, "Whereas there was a genuine desire for peace, there were also dictates of political necessity which could not be ignored." For some, agreeing to the terms of the Addis Ababa peace was a way to help maintain the cohesion of the Southern groups and stave off fighting among the various Southern political and tribal factions.

Joseph Lagu has acknowledged that Selassie and the logic his representatives presented were the crucial factors in his acceptance of force integration. He also believed that integration into the SAF would give many Southerners better access to training and education. Because the field commanders of the Southern rebellion had all but taken over control of the wider Southern political struggle by the late 1960s, Lagu's decision had substantial weight, much to the consternation of many politicians involved in the Southern rebellion. Lagu did, however, face major challenges from both his commanders and Southern politicians. One young officer opposed to the Addis Ababa deal was John Garang; reportedly, upon returning to Sudan from courses in the United States, he opposed the Anyanya, asserting on his first visit to an Anyanya camp that "I will not have anything to do with this racist movement" (Ga'le 2002, 361). Not long after, however, Garang joined and was made Lagu's information officer, and he accompanied Lagu to the talks. According to Masour Khalid, Garang professed his lack of confidence in the negotiations and specific reservations about the terms of the deal itself, and he said to Lagu that no agreement would last unless it established a "fundamental change in Sudan's body-politic" (Khalid 2003, 144–45).

In a symbolic act acknowledging concern about internal Anyanya opposition to the deal, Lagu delayed the ratification ceremony with Nimeiri to convene a meeting with key Southern leaders and political groups in Kampala, purportedly to amend the agreement. This meeting was really an exercise in expressing disgruntled opinion and airing grievances rather than a forum for new ideas and positions—nothing changed in the final text after the meeting between Lagu and the deal's potential detractors. Lagu accorded people time to give their input and then returned to Addis Ababa, ignored most of the issues that had been raised in Kampala, ratified the original agreement, and was immediately made a major general of the Sudanese Army (Allen 1991, 12). Amid the confusion and the fanfare being stirred up by the government, Southern politicians and Anyanya leaders did not oppose the deal; however, as the peace agreement collapsed the issues raised at the meeting in Kampala reemerged. As Allen (1991, 12) points out, "The Southern leaders had to accept what was in fact a defection as a fait accompli."

Compromises in the Process

The deal to which Lagu agreed was essentially the one offered by Alier's negotiating group and was conceived in close contact with the Ethiopian mediators and in consultation with Lagu's people and those close to Nimeiri. It is also worth noting the important role played by the World Council of Churches as facilitators. They also provided some commentary and input. It proposed a semiautonomous Southern region under the control of a High Executive Council headed by a Southerner. Though Lagu expected to be offered this position, it was given to Abel Alier, who was not a member of Anyanya; instead, Lagu was moved into the SAF as a major general and eventually given command of all Absorbed Forces and full command of the SAF in the South.

As it turned out, the decision to leave Lagu in charge of his former fighters was essential for holding the integrated army together. When he left the army to engage in politics, competing to be head of the Southern Regional Government—the position initially denied him by Nimeiri—Lagu became embroiled in combative politics, which meant that he could not act as a leader for the Absorbed Forces. The deal was negotiated without consideration for any dissenting or critical views from either inside or outside Anyanya, and consequently the single greatest compromise with respect to military issues was on numbers. Anyanya received more positions in the army than the government in Khartoum would have liked; six thousand men were to be integrated into the army or nonmilitary government services. Meanwhile, the South compromised politically, allowing the majority of political control to continue to pass through Khartoum, though operating through the Southern Regional Assembly.[4]

Lagu was clearly reassured by the fact that the government's negotiating team included some of his former SAF colleagues. He has indicated that, despite suspicions, he was encouraged by the involvement of Abel Alier, one of the most prominent Southerners working in Khartoum; Alier was a highly educated member of the Southern Sudanese Bor Dinka community. The security protocols included two provisions: one to implement the integration, and the second to monitor the cease-fire. The cease-fire monitoring was to be conducted by a commission that included international observers and monitors. Khalid indicates that Lagu and Nimeiri agreed that no international observers would be required, due to the "spirit of camaraderie" that had developed between the parties during talks.[5] Because he trusted those involved, Lagu accepted the informal guarantee that the Anyanya would be appropriately accommodated and trained without clear specifics as to what that meant. When Nimeiri's plans changed, so too did any commitment to support the Absorbed Forces, even though those very forces had protected him and his position as president in 1976 during a coup attempt by the Islamic political parties.

CREATION

There were deep divisions between the forces and among the Southerners. The agreement was vague on critical issues, and many resented both the process and the outcome of integration.

Shaping the New Military

The language of the Addis Ababa Agreement was largely developed by government negotiators, with significant input and suggestion from Emperor Selassie's aides. Once both negotiating parties and their respective leaders accepted the principle of integrating the Anyanya into the SAF, the details of implementation modalities, force command, and force structure had to be resolved. Article 26 is the agreement's key provision dealing with the construction of the future army and the integration of forces: "Citizens of the Southern Region shall constitute a sizeable proportion of the People's Armed Forces in such reasonable numbers as will correspond to the population of the region. The use of the People's Armed Forces within the Region and outside the framework of national defense shall be controlled by the President on the advice of the President of the High Executive Council. Temporary arrangements for the composition of units of the People's Armed Forces in the Southern Region are provided for in the Protocol on Interim Arrangements."[6]

Despite requiring the inclusion of a significant number of Southerners in the armed forces in Southern Sudan, Article 26 clearly precludes any control of the forces by the authorities in Juba in the Southern High Executive Council, other than in the phrase "on the advice of the President of the High Executive Council."[7] Moreover, the specifics of the integration were "temporary." With the president of the High Executive Council requiring Nimeiri's support, his participation as described in Article 26 provided little in the way of checks and balances to the SAF leadership regarding Southern forces. The agreement also stipulates that the further details specified in the "Interim Arrangements" would only last five years.

However, Lagu and other Southerners involved in the process understood the situation differently. They believed that the integrated forces would maintain their coherence and remain in the South after the interim period, until they agreed to be moved. Moreover, the former Anyanya believed any changes to the arrangements required the assent of the Southern political leaders in the High Executive Council and the Southern Assembly. The security provisions of Addis Ababa did not require any such agreement. Similarly, Southerners believed they would retain some influence over the Southern forces themselves within the army chain of command. During his tenure as commanding officer of Division One, Lagu addressed these issues personally, but eventually the fact that Southern influence was not maintained became problematic. Moreover, the agreement stipulated that in the longer term the demographics of the army should reflect the respective populations of North and South. Rather than change the army to reflect the national character, as specified in Article 26, Southerners were removed, retired, or given postings and orders that all knew would cause them to either resign or desert (Khalid 2003, 148). Article 1 of the interim arrangements reads as follows: "These arrangements shall remain in force for a period of five years subject to revision by the President on the request of the President of the High Executive Council acting with the consent of the People's Regional Assembly."

Lagu and the Anyanya negotiators perceived the details of the integration process as technical issues that the army itself could deal with after the peace negotiations. They considered decisions about deployments, force composition, and the integration of individuals to be the responsibility of the army's leadership and thus did not see a need to stipulate some special Southern authority over this process. Moreover, the expectation that Lagu would continue as a top military officer led to the belief that the process would be conducted in a way they found appropriate.[8] The strong reliance, even faith, in personalities over institutions was dangerous in the long term, although it initially proved useful in holding the force together. They were thus effectively reassured by Presidential Order 41 in April 1972, which enacted the Joint Ceasefire Commission.

Faith in Lagu and other individuals began to erode rapidly, however, and some even quickly characterized the Addis Ababa deal as a "surrender"—a term evoking the peace agreement made by a leader of the Torit Mutiny group in 1955, often referred to as "the Surrender of Lt. Renaldo Oleya" (Ga'le 2002, 377). One of the main criticisms levied at Lagu and the Anyanya deal by key Southern politicians and other leaders of Anyanya and wider Southern resistance groups was that it lacked detail. Joseph Oduho and several other leaders wrote at the time as part of a commentary for Joseph Lagu that the main reasons for fighting in the first place had been abandoned in Addis Ababa: regional autonomy, economic development/control over the economy, and social assistance for those who had suffered from the war. Furthermore, "there was no specification as to what the position of the six thousand Anya-Nya officers and men earmarked for integration into the national army would be," and no communication of the exact terms of the deals nor inclusion of the wider Southern political community (Ga'le 2002, 373). This early frustration motivated many to return to resist, and the largest such group became known as Anyanya 2.

Personnel Selection and Rank Allocation

There was little deliberation over the structure of the new armed forces. Lagu and his immediate officers decided where the Anyanya fighters would prepare for integration; the rest was effectively left to the general staff in Omdurman. Moreover, there was little change in the armed forces. The new elements integrated under Division One were those operating in the South already. They would remain in the South as their own division. The deliberations that had resulted in more interesting engagements were about the positions and ranks key players and their immediate officers were to receive. These deliberations were again undertaken by officers close to Lagu, as well as key military officers in the Sudanese Armed Forces, most of whom Lagu knew well. In an interview Lagu described the technical side of the integration process as having been conducted in a reasonable and effective manner. Abel Alier, too, in his memoirs (1992), suggests that the integration process initially appeared to have been conducted effectively. Tim Allen's assessment supports the claims of the key actors: "With the exception of a few isolated incidents, the ex-guerrillas were smoothly absorbed into government services" (Allen 1991, 17). This initial success was linked to Lagu's management of the integration process. His ability to prevent a mutiny in Juba and elsewhere, when Abel Alier could not, suggests that his personality was a key factor in the initial smooth integration of forces (table 3.1).

While the Anyanya were organizing at their respective collection points, soon to be bases for Division One of SAF, Lagu and some of his top officers were commissioned in the army. Lagu immediately appointed one of these officers, Colonel F. Maggot, to the Joint Military Commission that was to arbitrate and implement the technical terms of Anyanya's integration into the SAF.[9] The colonel's Northern counterparts were two brigadier generals, Migrani Suliman and Abdallah Ateif Dahab. This initial imbalance within the commission is representative of the imbalance that characterized much of the integration process. There was a similar imbalance in the Ceasefire Commission, the other mechanism mandated to implement the negotiated security provisions. Interviews suggest that while the Absorbed Force colonel was involved in decisions about the ranks and positions of the Absorbed officers, the

TABLE 3.1 Anyanya Integration into the Government Services

Mode of Integration	No. of People
Into the army as officers	200
Into the army as noncommissioned officers	865
Into the army as privates	5,012
Into the prison and police services	1,860
Into the civil service (largely as unskilled laborers)	5,489
Into the civil service due to medical rejection from the armed service	2,414
Total	15,840

Source: Data from Alexis Mbale Yango Bakumba, a member of the Ceasefire Commission (Ga'le, p. 393).

Northern brigadier generals had more influence in decision making—which is not surprising,[10] given the imbalance of rank and status deeply entrenched in the army, and perhaps more so in local communities and tribes.[11] This imbalance also shows the influence of military traditions and conduct during British rule. Without a larger Southern membership in the commissions and a balance of rank, the Northern representatives were bound to assume primacy.

Also problematic was the manner in which administration was performed and specific integration decisions were made. No detailed criteria were established to govern the process. Any given soldier's selection as one of the six thousand Anyanya to be integrated was initially based on suggestions from Lagu and his officers. The list of soldiers was vetted by the Military Commission, which consisted of officers Lagu had personally assured of rank. On the advice of Lagu and his staff, the commission then determined ranks and deployments. Although decisions were made at the discretion of the commission, they were supposed to involve input from the Anyanya commander of the individual candidate being considered for integration; apparently, this rarely occurred. Consequently, the process was guided by the intuition and personal opinion of a few leaders surrounding Lagu. He, however, described a process that he believed was reasonable. Interviews with leaders of the insurgency that became known as Anyanya 2 who were also involved in Anyanya and in peace negotiations and implementation indicate that not only the Northern officers but also Lagu and those officers he favored operated with a clear bias. Interviewed men often pointed to the case of Gai Tut, a top Nuer commander in Anyanya, who was not even integrated, let alone given an officer's rank. Other examples include Salva Kiir, the current president of South Sudan, a prominent Dinka leader in Anyanya, who was made only an NCO. Tut would later be brought into the army and Kiir raised to officer rank after their respective tribal groups pressured Lagu. These cases reveal how deeply personal—or, rather, tribal and communal—the selection processes were.

OUTCOME

The integration process resulted in the defection of many former Anyanya in the early 1980s, which was part of the wider collapse of the peace agreement and the return to war. Many of the integrated forces were at the vanguard of the renewed South Sudan rebellion, which eventually resulted, in 2011, in the South Sudanese independence that the Anyanya had always desired. In summary, the prevailing emotion of former Anyanya and many Southerners clearly was fear. They were deeply frustrated with Khartoum due to a long history of repressive rule. Compounding problems was the Army General Headquarters' "reluctance to grant sufficient numbers of former Anyanya higher rank or even to let Southerners into the army—the most coveted place in the Sudanese establishment for those who had been in rebellion."[12]

Political Control

These factors created a situation of deep mistrust; what might otherwise have been isolated problems sparked major incidents requiring the consistent intervention of the top leadership in Southern Sudan. Other issues, such as the introduction of sharia law, reinforced the Southerners' political frustrations and the region's continued lack of economic and social development, making former Anyanya difficult to control. Coinciding with Nimeiri's move to impose more control over the Absorbed Forces by placing more Northern officers in command, Joseph Lagu's entry into politics in 1981 resulted in an environment in which the Absorbed Forces were restless and unwilling to continue as part of the SAF. Little political agitation was required at this stage to convince fighters to step away from the army and return to active insurrection.

Another key factor in the removal of Southern officers and the imposition of "trusted" Northern officers was the beginning of the discovery and development of oil in the 1970s. Under mounting debt, Nimeiri saw oil as his key to survival and thus worked to make certain it was squarely under his control. The areas of discovery were primarily in the South, and the Absorbed Forces were, in effect, there to provide security. In Bentiu, the key site for Chevron's oil project, Salva Kiir, then a young officer, was in command but was replaced with an officer of a "more appropriate loyalty" (Khalid 2003, 147). When some in the Southern leadership, particularly the Southern Regional Government's minister of energy, requested a portion of the prospective oil wealth to help meet development goals, Nimeiri and leaders in Khartoum flatly refused, offering no opportunity for further discussion. Southerners protested, and clashes ensued in which Northern forces were moved in to control security for the Western oil companies beginning major operations.

In the South the officers the fighters trusted, that is to say their former officers, largely controlled the new Absorbed forces in Division 1. The former Anyanya preferred to remain in their previous units, and consequently the force was controlled by its former command structure. In the North the armed forces remained as during the war, although Absorbed soldiers or officers were moved there on a few occasions. Thus the Northern force remained under the same command as during the fighting. Serving under Lagu and several other top officers, the commanders of the former Anyanya were instrumental in mobilizing defection from the army and a return to rebellion in the early 1980s. These defections and rebellions were an expression of revolt driven by the lower- and mid-level leadership of the Absorbed Forces, rather than the result of poor discipline.

Military Capabilities

The rapid expansion of the armed forces is one of the main reasons Sudan has one of the largest debts of any African state. Although they were growing at great cost to Nimeiri's government, the Sudanese Armed Forces were smaller than either the Egyptian or the Ethiopian military. They were, however, large enough and well equipped enough to pose a deterrent to neighbors, including Uganda, Kenya, Chad, and Libya. The integrated military was commanded by a dictatorial leader, so it was as likely to be used to repress as to support legitimate political authority. In the context of East Africa and the Horn, such issues are interwoven into a complex pattern or "way of war." The legitimacy of Sudan's political leadership at the time was frequently challenged, and thus one of the military's primary roles was to reinforce the leadership's claim to power. The Southern component was even crucial in protecting the president and his supporters from coup attempts and other regional political machinations.

Until returning to rebellion, the Southern Division maintained control of various smaller groups that continued to oppose the government and resist the Addis Ababa Agreement. It was instrumental in bolstering Nimeiri in the North—and on one occasion actually essential in putting down a coup attempt in Khartoum by Islamist political factions led by Sadiq al-Mahdi. As long as the Southern forces remained loyal, there was little fighting between the Southern element of the armed forces and Northern troops: Division One, with most of the former Anyanya fighters, remained in the South, and few Northern officers were stationed there until later in the peace period. What was effectively segregation limited opportunity for confrontation or violence between Northern soldiers and the Absorbed Forces. Southern troops could not be deployed to other parts of the country, with two exceptions: a group of Absorbed Forces moved to Khartoum along with Lagu when he took his command there; and another formed part of the Presidential Guard. It was precisely the movement of these forces that triggered the rebellion that resulted in war, because of the profound anxiety and suspicion Southern troops felt about any situation that involved moving forces from their home areas.

In one of his early speeches announcing the rebellion and the Sudan People's Liberation Movement/Army, John Garang cynically claimed that "the honor of precipitating outright armed rebellion" rested with Nimeiri himself. One of the four reasons he gave for launching rebellion was "dishonesty attempting to transfer Southern forces to the North." The others were the dissolution of the Southern Regional Assembly, the redivision of the Southern Region, and what Garang frequently referred to as gerrymandering of the borders where oil was discovered (Khalid 2003, 155). Nimeiri's decision to move the various Southern integrated forces north and to dilute the largely Southern character of the forces stationed throughout Southern Command was thus a major cause of the return to open insurrection by Southern groups and the corresponding creation of the Sudanese People's Liberation Army (SPLA).

The process leading to the final collapse of the Addis Ababa Agreement, and more particularly of the force integration project, was not as direct as this summary might suggest, of course. The agreement had been deteriorating for some time as various mutinies and defections from the army and police occurred from the late 1970s onward. Thus the ongoing collapse of the integration project can be seen as foreshadowing the collapse of the peace agreement as a whole.

Although some officers fared well as Absorbed Forces in the SAF, the ambitions and expectations of many NCOs and enlisted men were frustrated by a lack of training and promotions. Many Anyanya officers, entering the SAF as NCOs, felt from the beginning that they had been demoted or personally slighted and often that they were victims of tribal or communal marginalization. Once integrated, many Southerners also felt they were not receiving the same opportunities as their SAF counterparts, especially in education and training.

Although Southerners were frustrated generally, "the very able former Anyanya military officers . . . eventually caused [the] problems."[13] These officers could harness frustration, confusion, and desperation. The agreement held only until enough of them were willing to move against their commanders and the government in Juba and Khartoum.

The signs of the renewed rebellion were widespread, but it started building serious momentum in Bussere, Bahr el-Ghazal—a central collection point for Anyanya integration into the SAF. Although most mark the Bor Mutiny of Battalion 105, in 1983, as the moment the Addis Ababa Agreement finally collapsed, the Bor Mutiny was "part and parcel of an overall plot to return to the conflict which had started in Bussere in the early years of the agreement" (Madut-Arop 2006, 27). As Madut-Arop explains,

> The Bor incident which ignited the return to the conflict in the Sudan in 1983 was not, in reality, caused and staged by dissatisfied elements within the rank and file of Battalion 105. Nor did the two-month salary delay of March and April of that year prompt the insurrection, as many in the official northern circles presently assert. Rather, the discontent with the peace agreement expressed throughout the ten years of peace and the persistent reluctance of the former Anya Nya officers to fully integrate with their troops into the national army as required by the Agreement were the main causes. (Madut-Arop 2006, 27)

John Garang's role in the agreement's collapse is important; he embodied much of the sentiment of the former Anyanya during the peace period and enhanced that sentiment with a new vigor by expressing grievances in terms of a national liberation struggle. Many in the rank and file date the beginning of their questioning of the Addis Ababa Peace to when they heard, or heard about, Garang's rousing political speeches. A cunning and intelligent young man, he argued that the Addis Ababa

Peace and the situation it created were untenable. According to former Anyanya and SAF officers, Garang's ideas contributed in some ways to many incidents of mutiny and resistance; even when Garang was not in the country, his ideas spread throughout the army's Southern components.

ALTERNATIVE OUTCOMES

In South Sudan, politics and the military are deeply interwoven; therefore, the structure of a military integration is tied to the political and even social features of the peace deal. Contrary to the belief of many observers, it is the military that is politicized, rather than the politics that is militarized. It may be that the armed forces and their actions caused the war or its continuation, and certainly the military leaders engineered peace; however, their actions need to be seen as political. Even anger about rank and pay was linked in Sudan to ethnic and other political constructs. Lagu and most mid-level and even low-level officers were never just fighters; they were political leaders within their respective communities. Therefore, a successful military integration in Sudan is a political process; if that process fails, the failure is one of political resolution. Most prominent among all factors contributing to the outbreak of war, however, was Nimeiri's full abrogation of the agreement's specifications for Southern autonomy and his move away from governing with the support of the Southern factions toward governing using the authority and power of the Islamist political cadres in Northern Sudan.

Though conducting the integration process differently would probably not have prevented an eventual return to war, doing so might have delayed it and allowed space for political engagement. The political nature of the military means that changes in even-macro political issues were rapidly felt throughout the entire army, especially in the South, where the Absorbed Forces were worried that they were under threat from their own president and high command in Omdurman and Khartoum.

DID MILITARY INTEGRATION MAKE LARGE-SCALE VIOLENCE LESS LIKELY?

The integration project was a cornerstone of the Addis Ababa Agreement and thus an important factor in precluding large-scale violence. If the military integration had worked, the peace might have lasted longer. That said, the war began anew for political reasons—because of decisions Nimeiri made and the reactions (both principled and opportunistic) of Southerners to those decisions, rather than solely because of the collapse of the integrated forces.

Some of the causal pathways proposed by the wider project are clearly present in the case of military integration in Sudan during the Addis Ababa Peace period, while others conspicuously are not. That the return to war coincided with the defection of many Absorbed Forces, and with major political events, suggests that the collapse of the military integration project was an important factor. This discussion of causal pathways considers the connection between the military integration project and the return to war.

First, military integration is a costly commitment by both sides to the settlement. Having General Joseph Lagu and several other leaders in influential positions made it more difficult for the security apparatus to be turned against Southerners. With Lagu's retirement from the army and his entry into politics, this commitment eroded—no similarly influential Southerner was given a top post in the Sudanese armed forces. The management of the integration was largely about personalities rather than institutional reform. It thus lacked sustainability and long-term credibility among Southerners.

For Nimeiri, entering into an agreement with the Southerners meant risking further conflict with the Islamic political parties in the North; such groups as Sadiq al-Mahdi's Umma party were long established and had been plotting against Nimeiri since he took power. However, forging an agreement with the South presented an image of leaders willing to take risky steps for peace; consequently, other elites were willing to make compromises for a settlement.

Second, as a result of military integration, elites and masses feel able to take risky steps toward setting up effective government institutions and resolving underlying disputes, making renewed civil war less likely. Because of the initial sense of security that came with the agreement in both South and North, elites and masses felt more able to take risky steps to engage in political and economic life. However, this did not result in the establishment of effective government institutions or the resolution of underlying disputes that would have made renewed civil war less likely. Rather, the riskier political engagement resulted in a return to problematic competition among politicians, often framed in tribal terms. The proclivity of the authorities in Khartoum and Juba toward divide and rule further exacerbated the problem.

This increased security makes economic development more likely, which in turn makes renewed civil war less likely. In South Sudan there was not a significant level of development. The peace allowed some infrastructure developments and major projects, such as the Jonglei Canal, to begin, but they were cut short by the war. It is not clear if greater development in the South might have prevented a return to war, but Sudan's politics definitely precluded such a development, as the central authorities continued either to ignore or to exploit the South. Moreover, this lack of economic development underlay the grievances that justified a return to war.

Third, the new military does not have sufficient coercive capacity to provide security; but because of military integration, the new military gives employment to the most dangerous former fighters on both sides. In Southern Sudan the integrated force was a strong political tool for maintaining support of key individuals because it provided them with positions. Precisely because Nimeiri began to manipulate this approach to co-opting the fighters and military leaders of South Sudan, the return to war eventually occurred.

The integrated armed forces were, for the period of the peace, an effective means of assimilating those who might otherwise pose a threat or instigate rebellion in local areas. Because war began anew shortly after many of the Absorbed Forces defected from the integrated SAF, it is clear that it had employed many risky individuals and groups who would have otherwise caused insecurity. Had the armed forces maintained this assimilation and been able to entrench and expand it, the outcome might have been drastically different.

This increases security enough that elites feel able to take risky steps toward setting up effective government institutions and resolving underlying disputes, making renewed civil war less likely. Integration seems to have been a useful short-term measure for brokering temporary peace rather than a step toward real governance reform. That said, it is difficult to determine whether, if Nimeiri had not abrogated the agreement, the scenario hypothesized by the causal pathway might have come to fruition. The failure of the integration, and of the peace more generally, confirmed the pessimism of individuals like John Garang, and justified their escalated warfare and the more authoritarian, military-focused leaderships of Southern opposition movements. The uncoordinated conduct of Southern politicians, coupled with the complete insincerity of Northern leaders during the peace talks period, engendered a level of suspicion, anger, and determination that led to the SPLM/A and its alternative approach, leading ultimately to South Sudan's independence in 2011.

This increased security makes economic development more likely, which in turn makes renewed civil war less likely. As with the preceding hypothetical causal pathway, the use of the integrated armed forces to co-opt most of the dangerous rebel fighters and leaders did not lead to development that made renewed war less likely. In fact, it reinforced a system of patronage and negotiation strategies that fueled war and that continues to be a marked feature of the Sudanese political landscape. The result was a negotiation process between elites that offered incentives to make war and then reconcile, over and over, making a return to war more, rather than less, likely.

Fourth, the ability of the military to exist as an integrated institution in a divided country becomes a symbol of national unity. The existence of the military is important; its actual capabilities are irrelevant. This is particularly true because the military is less likely to have internal conflicts than other elements of government due to military culture (Gaub 2011, 137–40). The Addis Ababa Peace period and the Anyanya's absorption into the Sudan Armed Forces suggest an alternative to the outcome proposed by this causal pathway. Rather than being a symbol of unity, the army actually entrenched existing divisions. In fact the Absorbed Forces and most Southerners desired those divisions. Anyanya fighters violently resisted any move from their previous areas of operation; integrating the army was of little symbolic effect as long as the deep divisions among most sections of the population continued. There were some glimpses of the military as a symbol of unity, such as when the former Anyanya component based in Khartoum defended the national government from a coup shortly after the peace agreement; however, such examples were few and far between. Moreover, the Anyanya were probably more interested in opposing the Islamists than protecting Nimeiri.

The second aspect of this proposed causal pathway also did not apply; the integrated military had many internal conflicts, both among the Absorbed Forces and between the Absorbed Forces and Northerners. As early as 1974 problems arose with the former Anyanya forces. A string of increasingly serious incidents leading up to the Mutiny of Battalion 105 in Bor foreshadowed the return to war of most of those integrated into the SAF. Joseph Lagu and local leaders were essential in resolving most of the incidents; without the direct influence of their former leaders, the Anyanya were unwilling to work with Northern officers, and often unwilling to work with other Anyanya from competing Southern communities or tribes. Such incidents reinforced this paper's key observation—the movement of the Absorbed Forces away from their home areas caused vehement resistance and was the catalyst for the renewed war; such movement (or threat of movement) created situations in which men from different groups were likely to confront each other.

This increases trust among different elites that they can move on to other issues with some confidence that agreement can be reached, making renewed civil war less likely. Rather than increasing trust among elites, the integration process fostered further competition and suspicion between Northern and Southern leaders (with the distinct exception of some very top leaders whose working relations dated back to before the First Civil War). Overall, the Sudan case shows the opposite of this hypothesized outcome: Integration created opportunities for conflict and entrenched existing divisions rather than abating risks and animosities.

Continued existence creates and strengthens trust among military personnel. As discussed above, there was very little trust among military personnel, either between Southerners and Northerners or among those from various Southern tribes. Though some Southern soldiers from different groups did build better relations during the integration process, this actually translated into an increased likelihood of war; as politics in Khartoum and Juba deteriorated, unified Southern soldiers conspired more easily against those in the North, who were still considered foes.

Military personnel transmit this feeling of confidence and trust to civilians in their own reference group. These civilians, in turn, are more willing to support elites who advocate cooperation with their former enemies. It is unclear whether the civilian population's confidence in government was reinforced by the creation of an integrated army. This may have been so, but given the distance in time from the case, and with a source based in the elites, it is impossible to discern if this causal pathway was a factor in the Sudan.

Fifth, the process of negotiating military integration increases trust among elites of different sides so that they can move on to other issues with some confidence that agreement can be reached, making renewed civil war less likely. The actual existence of the military is irrelevant. Indeed the agreement may be not to have a military at all (the Costa Rican solution). Rather than building trust, the process of negotiation actually created opportunities for anger and mistrust as positions, salary, training, and other prized opportunities were contested. Groups perceived the positions and salaries received by individuals from their respective ethnic or regional groups as the true measure of enfranchisement. Elections for representative assemblies held little relevance for most, and were not seen as connected to material support or the kind of prestige associated with government posts. These competitions for opportunities were transferred to the wider community and increased both general tensions and the likelihood of renewed war. Thus this causal pathway was also not relevant in Sudan.

CONCLUSION

Integrating Anyanya into the SAF as a part of the Addis Ababa Peace Agreement was an ambitious project. As one of the earliest local attempts at reconciliation between previously warring forces in postcolonial Africa, it deserves recognition as an important experiment in peaceful resolutions to civil war. Unfortunately, as the cornerstone of the Addis Ababa Peace Agreement, the integration was flawed from its original conception because those on the government side and the former Anyanya had drastically different understandings of what integration would mean. For Nimeiri and the government in Khartoum, integration was dismantling the Anyanya and localized Southern forces and asserting control through the reindoctrination and dispersal of former Anyanya throughout the SAF. The expectations of the former Anyanya were simple—they would stay as they were, based in their home areas, become the official army, and receive training and regular pay.

The failure of force integration during the Addis Ababa Peace suggests that the Absorbed Forces probably would not have rebelled as they did if they had been left in their home areas. There would have been little support in manpower or resources for a second insurrection launched by Southern politicians opposed to the administrative and political reorganization of Southern Sudan and the institution of sharia law. As it was, however, the former rebels, discontented and confused, were willing participants in the return to war, supporting Southern politicians in their reactions to Nimeiri's decisions and to competition within the Southern Assembly. Thus, the integration of forces must be considered in terms of how the various parties involved perceive it, rather than as some technical concept dealing mainly with the implementation and monitoring of a set of military issues. Force integration is deeply political, at both the strategic and the personal and local levels. Localized perceptions need to be monitored so that expectations can be managed. Certainly this monitoring and management must be paired with the willingness and intent of all parties to build a force meant to prevent a return to war rather than structuring the next conflict and war.

ABBREVIATIONS

Anyanya	"snake poison"—South Sudanese rebels
NCO	noncommissioned officer
SAF	Sudanese Armed Forces—government military
SPLM/A	Sudan People's Liberation Movement/Army—South Sudanese rebels

NOTES

1. This chapter includes interviews with Joseph Lagu, London, May 12 and 13, 2010; Michael Wal Duany, former Anyanya, Juba, South Sudan, May 3, 2010; Emmannuel Waga, former Anyanya, Juba, South Sudan, February 12, 2010; Ayub Phillip, former Anyanya, Juba, South Sudan, February 6, 2010; Gabriel Tang, former Anyanya and later Anyanya 2, by telephone, Khartoum, February, 15, 2010; Gordon Kong, by telephone, February, 15, 2010; SAF general 1, former Anyanya, confidential, Khartoum; SAF general 2, former Anyanya, confidential, Khartoum; SPLA major general 1, confidential interview, Juba, South Sudan, November 2010; SPLA major general 2, confidential Interview, Juba, South Sudan, November 2010; SPLA lieutenant general 1, confidential interview, Juba, South Sudan, November 2010; Salva Mathok, Juba, Southern Sudan, November 2010; Lam Akol, Juba, South Sudan, January 9, 2011; and Bol Gatkouth Kol, Juba, South Sudan, various, March 2010–February 2011.

2. Interview with Ayub Phillip, former Anyanya.

3. The degree to which people from the center of Sudan controlled the army and most other institutions, along with business and trade, is set out in the now-infamous *Black Book: Imbalance of Power and Wealth in Sudan*. It was originally published under the name *The Seekers of Truth and Justice*. It has become clear that it was written by a small group of academics and activists connected with the Darfuri and other opposition groups. It was translated by Abdullahi Osman El-Tom (2011).

4. Interview with Joseph Lagu.

5. Khalid (2003, 148). This was confirmed by some of my interviews.

6. Addis Ababa Agreement, Article 26.

7. Ibid.

8. Interview with Joseph Lagu.

9. Interview with Joseph Lagu. A broader discussion of Lagu's actions is given by Lagu (2006).

10. Interview with Wal Duany.

11. Ibid.

12. Ibid.

13. Interview with Michael Wal Duany, a former Anyanya involved with the integration committees.

Chapter 4

MILITARY INTEGRATION FROM RHODESIA TO ZIMBABWE

Paul Jackson

At the end of the Rhodesian War in 1980, each major faction had a significant military or paramilitary force at its disposal and each leader felt that he would do well in any subsequent election and therefore play a part in developing the new state (Lectuer 1995). However, the history of independent Zimbabwe started with an election victory for Robert Mugabe and his faction, and this faction has systematically destroyed its rivals until, thirty years later, the military and political elite act with virtual impunity. The roots of this power are very much within the integration process and the politicization of the new military, and this case study is one in which an initial integration was systematically undone for political reasons. The case offers a longer-term perspective than many others, and despite problems with some of the available literature, it provides a rich narrative over time, including the development of political power from a base within the security services.[1]

The chapter begins with a brief discussion of the origins of the military integration process and in particular the characteristics of the different forces involved in the conflict itself. The Rhodesian state was a very specific type of ethnic state, and the conflict protagonists were heavily influenced by ethnic rivalry. However, ethnic tensions among blacks were greatly increased by Mugabe's party after the war and led to protracted conflict in Matabeleland. The technical military integration itself was relatively straightforward, but the political coloring of the process and the manipulation of security forces had a strong impact on the outcome, undermining security in the eyes of many Zimbabwean citizens.

The story is of President Mugabe first isolating and then destroying political rivals in the security services, notably his fellow nationalist Joshua Nkomo, and then proceeding to politicize senior military figures. This was initially done mostly outside the formal military, as opposed to the wartime paramilitary groups, but following the departure of the British Military Advisory Training Team (BMATT) in 2000, and with the increased threat of political opposition from the Movement for Democratic Change (MDC) after the presidential elections of 2000, the militarization of Zimbabwean politics proceeded apace. The history of the integration process shows clearly that Mugabe's political party (the Zimbabwe African National Union–Patriotic Front, ZANU-PF) and the military have allied to isolate political opposition since the early years of Zimbabwe.

ORIGINS

The war was essentially between the white-dominated government and three competing resistance movements. Both sides drew on outside assistance, and over time the government's position weakened.

The Historic Role of the Military

Decolonization led in 1964 to the independence of Northern Rhodesia (Zambia) and Nyasaland (Malawi), which had previously been part of the Federation of Rhodesia(s) and Nyasaland. However, the white settler regime in Southern Rhodesia under the Rhodesian Front Party issued a Unilateral Declaration of Independence (UDI) in 1965 in order to maintain white rule. This led to the creation by the white minority Rhodesian state of a formidable military regime aimed at destroying African liberation movements that were attempting to bring about black majority rule and full independence from Britain. Allied to the colonial forces of Portuguese East Africa and the apartheid regime of South Africa, the Rhodesian Security Forces (RSF) constructed a military that has become surrounded by something of a mythology regarding successful counterinsurgency (Bairstow 2012). As Bairstow points out, the Rhodesian military have been held up as an example of how to conduct successful counterinsurgency actions, and militarily they were very successful. In addition, the Rhodesian military have been frequently used as a model for contemporary counterinsurgencies (Hoffman, Taw, and Arnold 1991). At the same time, the weight of international opinion was against the white minority, which numbered less than 300,000 (about 5 percent of the population) at its peak, and in favor of the black majority. UDI, despite Rhodesia's continued desire to be loyal to the British Crown, was viewed as illegal by Britain, the Commonwealth, and the United Nations.[2]

Against this state machinery were ranged three liberation movements. The two armed nationalist parties favored armed struggle following UDI: Robert Mugabe's Zimbabwe African National Union (ZANU) and Joshua Nkomo's Zimbabwe African People's Union (ZAPU).[3] The armed wings of these two movements became known, respectively, as the Zimbabwe African National Army (ZANLA) and the Zimbabwe People's Revolutionary Army (ZIPRA). Both groups initially hoped to insert insurgents into Rhodesia from bases in the frontline states bordering the country and to foment enough violence either to force the Rhodesian government to capitulate or, more likely, to force the British government to intervene and reassert its authority. The third group formed a political party with a wide constituency, the African National Council (ANC), that eventually was led by Bishop Abel Muzorewa. The ANC followed a policy of engagement with the regime and eventually formed a considerable corps of auxiliaries to fight against the other nationalist movements. Favoring engagement and gradual change rather than violent, communist insurrection, Muzorewa and the ANC enjoyed greater popular support than either of the other movements into the late 1970s (Stedman 1993).

The war itself began in 1966 but in reality had little impact until the early 1970s, when ZANLA began a campaign against soft targets and the Portuguese regime in Mozambique collapsed. Before this, ZANLA cadres were denied access to safe havens in Mozambique; the end of the Portuguese war allowed ZANLA cadres to locate within Mozambique and destabilize the Rhodesia/Mozambique border. ZANLA favored a political cadre approach, involving the development of local networks of fighters and political cadres located within communities, while ZIPRA favored the preparation of a conventional army for an eventual invasion of Rhodesia from Zambia.[4] This difference in emphasis would lead to significant rivalries between the two, with ZIPRA cadres being generally better trained and more "professional," while ZANLA cadres emphasized political activities and agitation over military training. These differences are reflected in the opinions of external actors about the integration process and in the attitude of ZIPRA personnel, who regarded ZANLA personnel as poor soldiers (Alexander, McGregor, and Ranger 2000). In fact, rivalry between the two led to significant fighting between ZANLA and ZIPRA in 1976, and the two met each other in the field as enemies rather than allies (Alexander, McGregor, and Ranger 2000).

Although Rhodesia's only real ally was South Africa, both ZAPU and ZANU received significant foreign assistance, and ideologically they followed the lead of their primary sponsors—Russia for ZAPU/ZIPRA, and China for ZANU/ZANLA. As the war progressed, Zimbabwean guerrilla forces also operated on a low level in Botswana and Namibia and particularly Mozambique. External operations by Rhodesian forces to counter this threat became common, and cross-border raids increased significantly between 1972 and 1979 (Cilliers 1985).

Although the Rhodesian Security Forces (RSF) developed very effective counterinsurgency tactics and achieved much military success, by 1977 the Rhodesian Central Intelligence Office believed that it was losing the war (Flower 1987).[5] Although the figures for guerrilla casualties were increasing rapidly, so were those for Rhodesian military and civilians. At the same time, the head of the RSF, Lieutenant General Peter Walls, estimated that guerrilla numbers had increased from about 2,350 in April 1977 to more than 8,000 ZANLA inside Rhodesia by January 1978, and a further 8,000 ZIPRA based in Zambia (Stedman 1993). Against this the white population was falling by more than 10,000 a year. White flight, decreasing South African support, and the increasing inability of the RSF to protect the civilian population meant that though the war had intensified it was unwinnable. Many civilians had lost faith in the RSF's ability to protect them and had become hostile to local paramilitaries that they regarded as little more than bullies; many were at least passively supporting the guerrillas (Alexander, McGregor, and Ranger 2000).[6]

The last years of the war, from 1977 to 1980, were characterized by intermittent peace talks, the installation of Muzorewa as Rhodesia's first black prime minister in 1979, an increased guerrilla presence inside the country, and an acceleration of white flight to about 2,000 a month (Wood 1995). By now ZIPRA was far better trained, organized, and equipped than ZANLA, and it was capable of fighting a conventional war (Wood 1995; Alexander, McGregor, and Ranger 2000). ZIPRA posed a significant threat to Rhodesia and to ZANLA, a fact that was not lost on Mugabe and that helped to shape the subsequent military integration process.

When Muzorewa became the country's first black prime minister in 1979, Nkomo and Mugabe both denounced the new government and increased guerrilla activity. At the same time, peace negotiations continued that had begun with an Anglo-American initiative of 1978. In March 1979, the UN Security Council adopted Resolution 445 condemning the Rhodesian elections. The Conservative victory in the British general election of May 1979 seemed to offer the Rhodesians some hope for continued recognition of their regime, given that there were notable Rhodesian supporters within the UK Conservative Party at that time. But a plan was agreed to within the international community that gave Britain a mandate to mediate between the Rhodesian government and the nationalist groups. This led directly to the Lancaster House Constitutional Conference, which was held in London from September to December 1979. All the nationalist leaders were invited, and some other interested parties also attended. Eventually this conference led to a comprehensive set of agreements; this was the origin of the integration process. UK Ministry of Defence officials became increasingly involved in the conference when it became obvious that the United Kingdom would support military development. In this way the United Kingdom led the process of designing the new military.

Shaping the New Military

The Lancaster House Constitutional Conference succeeded only because, as was later evident, each of the three major signatories (ZANU, ZAPU, and Zimbabwe-Rhodesia) was convinced that it would win sufficient seats in the election to significantly influence the political direction of the new state (Lectuer 1995). When the cease-fire began on December 21, 1979, the three parties moved toward the forthcoming election with the perceived prospect of political victory but also with their respective armed forces intact, in case things did not work out as expected.

At the same time, security was guaranteed by the appointment of a British governor, Sir Charles Soames, and the Commonwealth Observer Group was set up to assist him. There was also the Commonwealth Monitoring Force (CMF), numbering more than 1,300 troops, the majority from the British Army. Guerrillas were to report with their weapons to designated rendezvous (RV) points around the country and would then be escorted to assembly points (APs). The governor's military adviser and commander of the CMF, General Sir John Acland, was responsible for managing the cease-fire and dealing with any breaches, aided by representatives of all the involved military forces.

The CMF and the British understood the demand for a proper demobilization process early on. They would assist in the demobilization of any guerrillas wishing to take that option. At the same time, in the interest of fairness, guerrillas in APs received rates of pay equivalent to those of Africans in the RSF. The CMF found itself looking after a field army of around 22,000 men and women, many with tropical diseases including malaria and cholera (Stedman 1993). For the most part relationships were positive and cooperative, particularly with ZIPRA personnel, who were found to be professional to the point of being able to combine patrols with the RSF (Kriger 2003). Just before the elections Acland received agreement from all three parties to undertake combined training, and both ZANLA and ZIPRA agreed to undergo instruction under the authority of Commonwealth *and* Rhodesian forces. Although military integration was not a component of the settlement, it was seen as a means of facilitating cooperation among all involved. As a consequence 600 ZIPRA and the same number of ZANLA began training at two separate locations.

Following elections on February 14, the withdrawal of Commonwealth peacekeeping forces began. The chief threat at this time was the RSF, which had developed a plan—Operation Quartz—to overthrow a black majority government and had the ability to do so.[7] On March 4 ZANU officially won the elections. Prime minister–elect Mugabe asked General Walls, the commander of the RSF, to take over as army commander, and ZANLA guerrillas started training with the regular army the following day.

The situation in 1980 was complex. RSF numbers at that time are open to some debate. The official view is that total force levels amounted to around 23,000 regulars, territorials, air force, and guards (International Institute for Strategic Studies 1975). However, at least one estimate is as high as 97,800, even with the core of the regular army amounting to around 20,000 (Lectuer 1995).[8] The RSF was supported by a French Foreign Legion company, a group from the Mozambique Resistance Nationale, and several South African units acting as independent units or integrated as volunteers (Lectuer 1995). Muzorewa's party also had a force of around 20,000 nominally under RSF control.

Mugabe and ZANLA had around 16,000 troops at the APs, but an estimated one-third of their total strength remained outside (Goodwin and Hancock 1993). These were supported by 500 Mozambican FRELIMO (Liberation Front of Mozambique) regulars and an unspecified Tanzanian contingent (Johnson and Martin 1986). At the same time ZIPRA had 5,500 cadres in APs, along with some South African ANC cadres, with an additional force of 6,000 to 8,000 remaining in Zambia (International Institute for Strategic Studies 1975).

Immediately following the election, Mugabe disbanded the South African ANC units attached to ZIPRA and returned them to Zambia in order to prevent South Africa from attacking the new government. The continuing presence of the BMATT also provided reassurance to the South Africans, and Zimbabwe remained largely free of direct South African military activity, although South African–sponsored clandestine operations continued for some time (Alexander, McGregor, and Ranger 2000; Catholic Commission for Justice and Peace in Zimbabwe 1997; Shwere 2010). At the same time, the government made provisions for integration among ZAPU, ZIPRA, and "acceptable" elements of the RSF (Lectuer 1995). The government saw the integration exercise as a means of providing internal security but also as a means of providing employment prospects for former combatants (Lectuer 1995).

CREATION

Creating a new military from these diverse sources proved challenging. However, with the assistance of British advisers, it was eventually successful.

Personnel Selection

Almost as soon as the election result was announced, various units of the RSF began to melt away. The Rhodesian Light Infantry, a largely mercenary organization, along with South African and other foreign units, left for South Africa. The disbanding of "unacceptable" white units—including the Special Air Service (SAS), the Selous Scouts, the Rhodesia Defence Regiment, intelligence organizations, the officers of the Rhodesian African Rifles, much of the air force, and the Guard Force—led to a further exodus of professional soldiers; many left for South African, taking their equipment, regimental cash, and arsenals.[9] This exodus was accelerated by the South Africans making known their preferential treatment for former RSF within the South African Defense Force (SADF).[10]

In addition, many of the conscripts simply returned home, had their commitments reduced, and were effectively demobilized. Tribal militia and Security Force Auxiliaries (SFA) were also demobilized, and the Combined Operations HQ of the RSF disbanded. ZANLA and ZIPRA cadres seeking civilian careers were also released. The new government sought to assure the guerrilla fighters that their personal security was guaranteed and that employment would eventually be provided (Lectuer 1995). It dealt with the first issue in part by declaring an amnesty for all combatants, guerrillas, and RSF alike; the second was to prove more difficult, and failure to provide economic security for the former combatants has been a persistent problem throughout the Mugabe regime.

A Joint High Command was established to supervise the command and control of former personnel in the assembly points and barracks. Made up of senior officers from all three groups and a civilian chair, the Joint High Command found itself with a host of administrative and disciplinary problems. An inadequate pay and records system was in place, and there were hundreds if not thousands of "pseudo-guerrillas," mainly family members, friends, and civilian associates of genuine guerrillas, seeking free food, accommodation, and an allowance.

For those who intended to remain within the new army, command and control were gradually implemented, proper camps and barracks were built, rations were provided, and payment was organized. Initially, the BMATT's mission was not to provide a presence throughout all the training establishments where the integration was being conducted but to help select and train leaders and instructors who would then carry out training themselves (Dennis 1992). The original plan suffered from a number of problems, partly poorly disciplined and sometimes mutinous PF cadres, but also the slow pace of the process.

The British were the prime movers, together with the Rhodesians initially, at least in the selection process for officers (see the next section). In terms of organization and structure, Mugabe wanted to adopt a pseudo-Maoist Chinese militia concept, with soldiers also employed in agriculture. The initial plan was four infantry brigades, each comprising three to five battalions.

Rank Allocation

Initially, officers were required to pass the RSF entrance examination, but too many failed due to a lack of education. When this did not work out, the British decided to take those who were nominated to them and train them as officers and NCOs. Initial training was therefore as much about assessment as training, and the BMATT's presence increased confidence in this process.

These men were selected from within their own organizations and therefore had some internal credibility. There were, however, political considerations, and after a time it was noted that the minority ZIPRA was being underrepresented, even before ZANLA launched a purge of the security services and effectively took control.

Training

The initial 450 men from each guerrilla group were taken off to train separately with the ex-RSF soldiers, who were acting as something of a balance, partly because of their relatively apolitical backgrounds but also because they were far better trained than the former ZANLA and ZIPRA cadres. The former guerrillas had only a smattering of basic insurgency tactics but were found to learn quickly, and the BMATT regarded the initial results as very positive. The BMATT also managed to devise a system for identifying and promoting good officers and NCOs while it ran the training.

There was no screening for human rights violations. Although these had clearly been committed during the conflict against both military and civilians, both sides had felt that they were fighting each other rather than targeting civilians per se. This ignored the facts that Mugabe certainly felt that the RSF should have been identified as war criminals after the cross-border raid on the ZANLA camp at Nyadzonya that killed more than a thousand guerrillas and wounded more than two thousand and that RSF nomenclature identified the nationalists as "terrorists" who had perpetrated a number of well-documented and publicized attacks against white families, including children.

Quick yet comprehensive courses in leadership, for those selected as commanders at all levels from battalion to section, were established and taught by BMATT personnel, and specialist courses such as tactics, administration, and drill were introduced, the latter two areas being outside the normal scope of guerrilla forces. As early as July 1980, a group of twelve officers, four men from each of the three forces to be integrated, spent two weeks in the United Kingdom as part of a group cooperative training, assessment, and (hopefully) bonding exercise before visiting the British military training facility at Camberley for a familiarization course on conventional forces functions (Chitiyo and Rupiya 2005).

Initially, the commander of the RSF, Walls, had overseen the implementation of the new defense policy. His appointment was made in part to assuage the fears of whites that the situation might suddenly destabilize and they would become victims of all-out massacres conducted by the former "terrorists" from the Bush War. Walls, however, having been on the receiving end of an increasingly vehement "Balls to Walls!" campaign by a segment of white Rhodesians who considered him a sellout to the new regime, resigned his post in July 1980. His departure was rapidly followed by that of a number of other senior commanders, which led to resignations throughout the former RSF ranks. By the end of August 1980 the white exodus prompted Ian Smith to urge the whites to remain, while he himself continued to remain positive, in public at least, about the majority-rule government. Walls told the BBC in August 1980 that he estimated some 60 percent of the white officers and senior NCOs had resigned, and in September Mugabe temporarily froze white pay and promotions, while acceding to the demands of ZANLA and ZIPRA for pay parity.

The way was now set for rapid promotion among the BMATT-trained African officers and NCOs. The initial plan to include a significant proportion of former RSF personnel together with the ZANLA/ZIPRA mix suffered a setback with the mass resignation of whites, and BMATT officer training had to be stepped up.[11]

By the end of December 1980 Mugabe had announced the plan for a national integrated army numbering 35,000 and made up of two specialist commando and parachute battalions and four infantry brigades as well as the signals, sapper, administration and pay, medical, and logistic units required by a modern military force.[12] There would continue to be some conflict between, on one side, supporters of the new military model envisioned during training and agreed upon with Mugabe, whose ZANLA forces had been trained by the Chinese and who wished to form a people's militia, and, on the other side, the British and the Soviet-trained ZIPRA, who wanted a more conventional, professional army. Eventually, a compromise solution was reached.

The balance of 30,000 former fighters was to be demobilized. An attempt was made to develop a program known as Operation SEED (Soldiers Employed in Economic Development), but this program failed due to a combination of inadequate resourcing, poor investment, and lack of proper planning (Nyambuya 1996).[13] Because it was agriculturally based, those employed in this manner believed they were being marked as unsuitable for military service; this was resented and morale suffered.

A second attempt at demobilization was launched in August 1981. A Demobilisation Directorate was established beneath a ministry to oversee the process, with a budget of about Z$116 million (Chitiyo and Rupiya 2005). Three alternatives were offered: an involuntary nonrenewal of lapsing RSF contracts; a voluntary severance package of four months' salary plus a monthly stipend of Z$185 for two years; and, for special cases, admission to a disabled rehabilitation center. The Zimbabwe Reconstruction and Development Conference (ZIMCORD) was established to encourage combatants to reenter education or to retrain. However, the demobilization directorate was largely theoretical, and failures to come to terms with these issues led in 1991 to the formation of the War Veterans Welfare Organisation, which remains politically active (Chitiyo and Rupiya 2005).

In August 1981 General McLean took over as head of the Zimbabwe Defence Forces. ZANLA's former commander, Lieutenant General Rex Nhongo, was appointed overall commander of the army, and the ex-ZIPRA military chief Lookout Masuku became his deputy commander. By the end of the integration exercise in October 1981, 1,400 leader cadres had been trained, largely by the BMATT, and more than forty battalions were under instruction. Emphasis had been placed on battalion formation and leader training, and staff training courses were well under way. Officers were divided into command and/or unit elements, with the command elements undergoing four-week training courses at the Zimbabwe Military Academy, which was producing platoon, company, and battalion commanders. Nyambuya (1996) notes that command recommendations at all levels were referred to the high command for ratification and arbitration.

The junior commanders, once qualified, were teamed up with their troops, who had been trained by both the BMATT and the first generation of indigenous instructors at the Infantry Training Depots to form battalions. A further four-month training cycle followed, including exercises at platoon, company, and battalion levels, before a battalion was considered operational. Additionally, a start had been made at creating appropriate logistic systems to back up the fighting formations, and these were functioning (Dennis 1992).

However, the failure of demobilization meant that the resulting army of 65,000 was far larger than envisaged and also far larger than any external threat warranted, and although the process of integration and training had initially been positive, discipline and training left a lot to be desired. The exodus of senior and middle-ranking white officers, along with many professional soldiers, weakened the ZNA (Zimbawean National Army), but by 1982 its structures were in place (Dennis 1992).

Initially, the BMATT did most of the training, and there was a BMATT presence in Zimbabwe until 2000. However, the BMATT concentrated on officer training and then "training trainers," initially for other ranks and then for officers. The air force (such as remained) was first officered and instructed by Rhodesians staying on, but after an attack and the subsequent jailing of the air force chiefs for colluding with South African saboteurs, the Zimbabweans used personnel from the Pakistani Air Force.

The nature of the training itself can be summarized as "Start them off basic and then build from there." Officers were trained separately from the others initially and then brought back to take on more complex activities such as field training and exercises. Given the initial lack of training in the guerrilla armies, this training was given great emphasis and included much joint working among former ZIPRA, ZANLA, and RSF troops.

This was a highly risky system that produced a high failure rate. Kriger (2003) suggests that many officers did not really care about the welfare of their men. In interviews with former combatants, new officers were frequently portrayed as being interested mainly in their own activities. The British trainers did feel that there were some very good officers but that the rapid exodus of so many experienced former RSF officers meant that political appointments were made far more easily as the new high command interfered with the appointment process (Dennis 1992).

OUTCOME

The new military had been created fairly successfully in a short period, although obvious problems remained. However, the Mugabe government soon took control of the institution, pushing out former ZIPRA personnel and bringing senior military officers into its political alliance in return for economic benefits.

Political Control

Party politics and the campaigning for the 1980 elections had ethnicized Zimbabwean politics and created a clear division between the two main nationalist parties. Given the hostile history of both ZIPRA and ZANLA in the field, perhaps this is unsurprising, but it began to take on increasingly sinister overtones from 1980 to 1983. In Augusta and September 1980, as political campaigning started for local government elections, the old rivalries reopened. In November 1980, following chance remarks at a rally in Bulawayo that warned ZAPU that ZANU-PF "would deliver a few blows against them," the first Entumbane uprising started a two-day pitched battle between ZIPRA and ZANLA, which was ended by two senior officers, one from each faction.

This was followed in February 1981 by a second uprising, which spread to Glenville in Ntabayezinduna, and to Connemara in the Midlands. ZIPRA troops in other parts of Matabeleland flocked to Bulawayo to reinforce their comrades; former units, including aircraft and armored cars, had to be deployed to bring the fighting to a stop. More than three hundred people were killed. In military terms it was a disaster; three of the nine integrated battalions disintegrated in faction fighting. Many former ZIPRA guerrillas, both within the military and without, felt threatened and insecure, and many deserted and returned to their tribal homelands, where they were then seen as a political (ZAPU) threat to the government. And in a departure from the initial aims of integrating the factions, but in keeping with his Marxist principles, Mugabe established military units outside the integration structure.

By 1983, Mugabe had arrested virtually all the senior military leadership of ZIPRA, and in March 1983 all the senior leadership of ZAPU, including Nkomo, went into exile. The unrelenting harassment of ZIPRA cadres led many to leave the APs, which were still functioning (Alexander, McGregor, and Ranger 2000). This led to widespread violence against former ZIPRA cadres within the ZNA, coupled with segregation, disarmament, disappearances, and an overall downplaying of ZIPRA's role in the liberation struggle that continues to date (Alexander, McGregor, and Ranger 2000). These moves meant that of the initial triumvirate designated to share power in the 1980 agreement, only ZANLA senior officers remained. This effectively cleared the way for the creation of a ZANU-led, politicized security policy that, as in the Chinese model, emphasized the political role of the military. A number of new units then emerged, undermining much of the integration that had taken place.

Painting former ZIPRA cadres as dissidents, effectively making them outlaws, led to the government's claiming that these dissidents were trying to overthrow it. At the same time ZAPU believed (rightly) that Mugabe was trying to eliminate it, and the situation within the ZNA had become so harsh that many felt they had no option other than to leave. At the same time South Africa decided to meddle in Zimbabwean politics and provided support to "Super ZIPRA," thus giving Mugabe political ammunition (Catholic Commission for Justice and Peace in Zimbabwe 1997). This led to significant problems in Matabeleland, where the government was responsible for killing some 20,000 civilians, but also in Mashonaland, where government troops attacked civilians despite two high-profile court cases failing to prove any involvement of ZAPU or ZIPRA (Catholic Commission for Justice and Peace in Zimbabwe 1997; Alexander, McGregor, and Ranger 2000; Shwere 2010).

The creeping politicization coincided with the creation of two sets of security units outside the integration structure: the Fifth Brigade (5B) and the Zimbabwe People's Militia (ZPM).[14] The origins of 5B can be found in an attempt on Mugabe's life in February 1981 and a following meeting in which it was suggested that he develop a presidential guard unit. In October 1981, North Korea offered weapons, equipment, and training worth around £12.5 million and the provision of around 1,065 North Korean instructors led by a brigadier. The Fifth Brigade had been trained and was ready for deployment by 1983, and, disastrously, it was deployed in an internal security role. The ZPM was also trained by the Koreans with officers temporarily reassigned from the ZNA; it was regarded as a paramilitary organization dedicated to rooting out opposition to the ruling ZANU-PF party rather than as a national security institution. During the conflict in Matabeleland, the ZPM became a formidable force, numbering some 20,000 by 1985 (Chitiyo and Rupiya 2005).

Both units were extensively deployed within Matabeleland against former ZIPRA cadres and civilians loyal to ZAPU. However, following the Unity Accord reached between ZANU and ZAPU in December 1987, and then the demise of the apartheid regime in South Africa, the ZPM became marginalized and was then disbanded. The Fifth Brigade, however, continued to operate in an increasingly ZANLA-dominated ZNA, and while the Unity Accord brought an end to explicit violence in Matabeleland, it did not end violence against Mugabe's political opponents.

Chitiyo (2009) eloquently describes the current situation in Zimbabwe as an attempt to ally the political elite with senior military officials in order to prevent the political opposition, the Movement for Democratic Change (MDC), from gaining access to real power. Zimbabwe scores very highly on the Failed States Index and the Index of State Weakness, despite starting from a high base in terms of food security and service provision. It has been progressively degraded from within by a rapacious state increasingly allied to the military. These indicators show very clearly a falling level of legitimacy within the state, whose machinery is used for the benefit of a specific group. How did Zimbabwe get to the current point from the initial integration process?

The answer lies in the political ruthlessness of Mugabe and the waves of politicization that affected the ZDF following the formation of 5B and the destruction of Nkomo and the ZAPU in Matabeleland, along with the external operations against South Africa and Mozambique. Operation Gukarahundi against the suspected dissidents ended with the eventual massacre of an estimated 20,000 to 25,000 civilians in Matabeleland and effectively destroyed internal opposition to ZANLA from ZIPRA.

At the same time, the deployment into Mozambique to shore up FRELIMO forces against RENAMO troops between 1985 and 1993 involved the ZDF in an ideological struggle for African nationalism as well as in securing the vital road and rail links to Beira and the Mozambican ports. The ZDF and other elements of the security forces, notably the Central Intelligence Organisation (CIO), were engaged in counterterrorism and countersubversion against South African military and intelligence incursions aimed at Mozambique and Angola.

All these activities led to a fundamental adherence to a series of political causes, notably an ideology of black African liberation that resonated with the revolutionary Marxism of the former Bush War combatants, but that led to a tightening of the relationship between military and party. A sinister development was the conflation of national security with regime security, and the identity of the state with the party and Mugabe himself. This ideology led to the progressive politicization of the military and the ZDF throughout the 1980s.

Successive crises within Zimbabwe, and the growth of political opposition in the shape of the MDC, led to a close and unhealthy relationship among the party, the military, and those designated as war veterans. Nyathi (2004) points out that the failure to either integrate the ex-combatants or to rehabilitate them has come back to haunt ZANU-PF. This was illustrated when they forced Mugabe to offer them Z$50,000 (about US$1,310) each as compensation for their involvement in the liberation war in 2000. However, the ruling party now appears to be benefiting from the situation, using the ex-combatants to fight its political battles; the ex-combatants who were not integrated are being used to destroy the democracy they strove for (Nyathi 2004).

In successive elections in 2000, 2002, 2005, and 2008, the MDC was a growing threat despite widespread political violence against dissidents. As a result the relationship between the military and the political hierarchy became explicit through the Joint Operational Command (JOC) and through the appointment of several high-ranking military officials to lucrative commercial and government positions. Several of the key ministries of state—including Energy, Transport, Trade, Construction, Information, Foreign Affairs, Prisons, and Railways, as well as the Commercial Bank of Zimbabwe—are headed by senior military figures. Though there have always been close links among the Zimbabwe Defence Forces and the government, these have become stronger since 2000 with an alliance between the military and ZANU-PF designed to prevent the MDC from gaining power (Ndlovu-Gateshi 2009).

The spread of politicization has been accompanied not only by the increasing control of state institutions by military chiefs but also by direct interference in politics by the security forces. This has been seen in forcible appropriation of land from commercial white farmers, Operation Tsuro (Rabbit); attempted control of the 2002 elections including coordination of youth groups, war veterans, and intelligence actors in intimidating voters;[15] and Operation Murmbatsvina (Drive Out Rubbish), which led to forced demolition of high-density urban housing and shacks, partly as collective retribution for urban support of the MDC.[16] Last and most telling was Operation Makovotera Papi? (Who Did You Vote For?), which involved military-style operations by the ZDF, Green Bombers, and the CIO in former ZANU-PF regions of Masvingo, Mashonaland, and Manicaland provinces that had voted for the MDC. Camps were established where villagers were forced to attend indoctrination sessions and where security forces would humiliate and torture those who had not shown loyalty to ZANU-PF. The JOC of the military had effectively become a political organization.

At the same time the economic activities of the JOC were becoming more public. Apart from controlling state institutions, the military has been engaged in diamond mining and control over the diamond trade for some time. Starting in Eastern Zimbabwe, the military has been involved in widespread abuses of diamond, gold, and other mineral mining operations, most of which are small-scale, but many of which, particularly around Marange, are extremely lucrative (Human Rights Watch 2009a). The new Government of National Unity (GNU) is likely to contest the military's control of the diamond fields, probably through contracting security and mining to a private company, but it seems likely that the military will not easily relinquish its control of such a lucrative source of revenue. In fact, the military has extended its economic interests in the diamond trade through its involvement in the DRC.

The ZDF initially deployed to the DRC to support President Kabila in 1997, and along with several other African militaries it intervened to defend the president against the forces of rival countries, notably Rwanda and Uganda. However, a later 2002 agreement with the DRC government effectively changed the ZDF into a conflict entrepreneur (Chitiyo 2009). Deployment into the mineral-rich Mbuji Mayi region of the DRC has catapulted the military-political class of Zimbabwe into the global diamond trade, involving it in both the mining of diamonds and the development of Harare as a major source of and processing center for blood diamonds from the DRC. This huge revenue boost has led to the accumulation of enormous personal fortunes for military and political leaders in Zimbabwe but also to the need for them to cling to power in order to milk the system even more. The military is inextricably entwined with both ZANU-PF and also this illicit trade in mineral wealth, and thus the nature of state fragility in Zimbabwe is not one of lack of control; rather, it is one of a kleptocratic oligarchy run by a military-political elite to the detriment of most of the population.

The nature of the autocratic control effectively means that the main organs of state security—the CIO, the Zimbabwe Republic Police, the ZDF, and the Zimbabwe Prison Service—are all paramilitary or military organs of state repression. There have been numerous and well-documented instances of human rights abuses and an increasingly militarized methodology of repression involving effectively turning Zimbabwe into a military operational zone. These formal state institutions are supplemented by state-sanctioned militias, notably the Zimbabwe National Liberation War Veterans Association and the Youth Brigades, both of which are effectively politicized militias that work in concert with the formal security services in perpetrating political violence. In fact, as Chitiyo (2009) points out, the youth militias have frequently been accused both of employing child soldiers and of acting as recruitment vehicles for the ZDF.

The merger of the guerrilla armies and the RSF was originally suggested by the British mediators during the Lancaster House negotiations and was firmed up in detail only with the creation of the document that formed the basis for the cease-fire. In many ways it was precipitated by the failure of the Zimbabwe-Rhodesia government of Muzorewa to get international recognition, a lifting of sanctions, and the resources necessary to continue the war.

ZANLA and ZIPRA did not actually start to merge until after the cease-fire agreement was in place in 1980, when Commonwealth forces began the process of integration by establishing assembly points within Zimbabwe. When ZANU-PF and ZAPU instructed their respective guerrilla forces to comply (Kriger 2003, 48–59), these groups, political parties, and individuals all supported the merging of competing militaries, at least officially.

This case is one in which an initial integration was deliberately undermined for political reasons. The initial integration produced a superficially effective military, but real control lay with Mugabe. According to the International Regional Information Network (United Nations 2008), the government has systematically used military-style campaigns within Zimbabwe to impose measures ranging from control of hyperinflation to "acts of alleged genocide."

Military Capabilities

Despite the uses to which the force has been put, the ZDF has shown that it can fight an effective counterinsurgency against dissidents in Eastern Zimbabwe. It has also performed well in other African states, including Angola, the DRC, and Mozambique. Its efficacy means that the army is capable of deployment to different parts of the country and overseas. However, the early years of integration were marred by internal violence between former ZIPRA and ZANLA cadres; and since

ZANLA dominance began, the ZDF has increasingly deployed to take control of significant economic resources, particularly diamonds in Eastern Zimbabwe and the DRC. At the same time, the ZDF does not threaten the Mugabe regime because the regime provides it with mineral wealth and significant business interests. These have helped to finance a military that is almost double the size of most other Sub-Saharan countries' militaries and is the nearest rival to South Africa in the region in terms of military capability.

HAS MILITARY INTEGRATION MADE LARGE-SCALE VIOLENCE LESS LIKELY?

The ability to use force against any group in society when asked to do so by a legitimate authority is enshrined as one of the four core principles of the ZDF. In practice this means loyalty to Mugabe, and while this does prevent a return to full-scale conflict, the ZDF and associated security organizations are the main internal perpetrators of violence against the population. The military maintains itself through the sale of diamonds. This means that it no longer relies on the state or the population itself for support, and suffering for the population is a likely outcome. At the same time, external military observers (interviews with UK Ministry of Defence officials 2010) also claim that elements of the formal military are underfunded as a result of additional resources' being allocated to paramilitaries or 5B, Mugabe's personal force. The result of this is that the ZDF has developed a significant capability to deploy as a conflict entrepreneur.

The Zimbabwean regime has clearly been committed to developing security institutions within Zimbabwe. This development involved significant costs early on, particularly on the part of the military forces. Initially, Mugabe's ZANLA was the weakest of the three military forces, and, given that he was certainly not guaranteed to win the 1980 election, ZANLA's participation involved considerable risk. However, in the medium term, integration proved costly to both the former RSF and ZIPRA in the early 1980s as Mugabe sought to develop a system designed to protect the regime. This outcome is somewhat in line with the second hypothesis put forward by this book, in that the ZDF had significant coercive capacity to protect the elite and make civil war less likely. However, this hypothesis does not entirely apply. A civil war took place in the early 1980s as ZANLA and elements outside the integration process (5B) were deployed to destroy rival nationalists in ZIPRA. This is a function of one elite consolidating power and of the agency of the ZANLA leadership in creating a set of institutions that benefited it as a group.

Initially, the consolidation of ZANU-PF power did produce sufficient confidence to develop economic institutions. In the 1990s Zimbabwe was a major producer of grain and tobacco and managed significant natural resources. However, while this could have led to widespread economic development, the dominance of ZANU-PF in the military led to the elite's taking possession of many economic assets and then effectively privatizing itself by accessing natural resources in Zimbabwe and overseas.

Consequently, Zimbabwe's new military now has significant coercive capacity to effectively destroy rivals and consolidate power. The authoritarian nature of the regime means that civil war renewal is unlikely, but it also presents the inevitable issues of alienation and social conflict that build up in authoritarian states. In Zimbabwe these have manifested in the MDC, which is kept out of power only by the integrated military created after the war.

The Rhodesia–Zimbabwe integration process does raise interesting issues regarding the five hypothetical causal paths identified within this volume. First, the history of the Zimbabwe military since 1980 shows the different degrees of commitment to the military by all sides at different points in the process. The initial integration process had considerable commitment from all three security forces with the overall stated aim of each being to develop a national army for the new state. However, the political commitment of ZANU-PF to the process and the associated takeover of the security services, first in defeating ZIPRA and then in making it untenable for large numbers of RSF to stay, meant that ZANU-PF cadres have essentially remained in charge. As such, the associated commitment to the military has been largely geared toward the maintenance of a politicized, but effective military.

At the same time, the risks of such commitments to eventual losers are painfully obvious. The ZANU-PF takeover and the politicization of the military has meant that the RSF has lost out as well as the ZIPRA fighters, with many of the RSF exercising their ability to leave Zimbabwe altogether, leaving ZIPRA to be gradually airbrushed out of the nationalist narrative.

This one-sided result of the integration process has had notable implications in terms of effectiveness. Though the new Zimbabwean security services have been effective in terms of operational ability, their coercive capacity has been geared—and increasingly so since the UK withdrew in 2000—toward regime protection rather than security more broadly. The early experience did not provide any security for opposition groups within the process, with the RSF and ZIPRA being either forced to leave or violently crushed. Since the mid-1980s the Zimbabwean military has been primarily concerned with political protection and personal enrichment. The involvement in the diamond trade that accelerated through the 1990s and into the 2000s has meant that the military has effectively undermined broader economic development as it has sought to maintain its own elite position.

The third hypothetical path is an interesting one in the case of Zimbabwe. Clearly the security services do provide employment opportunities for former fighters. However, these are made up primarily of former ZANA fighters rather than from the other forces. At the same time, the poor results of the rehabilitation programs of the 1980s and creeping political opportunism on the part of the regime have created a highly politicized and prominent war veterans movement. This movement is motivated largely by demands for compensation, employment, and opportunities that have not been realized since the war, despite the maintenance of significant numbers of security service personnel.

The history of integration also mitigates against the fourth potential hypothetical path. The military has certainly not become the symbol of national unity. Rather, the opposite has happened, and the army has become a symbol of an elite regime that enriches itself at the expense of the general population. The military has increasingly become distant from the population as repression has been used as a means of enforcing support and punishing opposition. The military is therefore seen as the very instrument of Robert Mugabe's control over the political landscape of Zimbabwe.

Finally, there are sanguine, even depressing, lessons arising from the process of negotiation as a means to increase trust among military personnel. Within the integration process, the units that were engaged in integration did produce considerable evidence of increasing trust between former combatants. Many British trainers working with the Zimbabwean forces were overwhelmingly positive about the ability of the different groups to work together, and they frequently reported positively particularly on ZIPRA and RSF forces. It is notable that much of the Matabeleland violence of the mid-1980s was perpetrated by the Fifth Brigade, which was deliberately trained outside the integration process. This raises the intriguing possibility that greater inclusion of all parties within the process could have reduced the ability to perpetrate violence by one party against others and so gain political and military power.

CONCLUSION

The story of military integration in Zimbabwe is one of initial success in avoiding both civil war and invasion by South Africa, followed by the consolidation of power by one group of political actors allied to senior military figures. First, ZANLA was able to neutralize elements of the former RSF, many of whom left for South Africa; second, it was able to successfully purge its former political rivals in the African liberation movement, ZIPRA, and its political wing, ZAPU. The initial "white flight" of 1980–83 created conditions enabling ZANLA to colonize the newly formed ZNA, even though the BMATT remained in place until 2000. At the same time, internal harassment of ZIPRA cadres within the ZNA, and then full-scale civil war until 1987 against ZIPRA dissidents, meant that by 1987 Mugabe was in complete control of the security forces, and only ZANLA remained from the initial tripartite power-sharing structure.

In terms of political control, ZANU/ZANLA control has been reinforced by the use of the ZNA and auxiliary and intelligence units as internal political enforcers and by military control over natural resources both in Zimbabwe and in the DRC. The relative effectiveness of the core of the ZNA, trained and at least partially integrated by the BMATT, is currently being undermined by the political and economic control exercised by a rapacious political elite that has turned most of Zimbabwe into an operational zone. The ZNA performed well in overseas postings in Mozambique, and it remains a formidable force. It has used this capability to engage in widespread purges of the opposition groups, including rival nationalists in ZIPRA, and the use of force internally has been largely political with the aim of regime protection. This protection has been purchased through increasing ownership and control by senior military figures of the means of production within the economy and made sustainable by access to diamonds. Thus economic security effectively fuels regime security. The eventual outcome of the integration process that began thirty years ago is therefore a capable but politically biased force that excludes those who were initially integrated.

ABBREVIATIONS

5B	Fifth Brigade—unit trained by North Koreans outside of the military
ANC	African National Council—nationalist party led by Muzorewa
APs	assembly points
BMATT	British Military Advisory Training Team
CIO	Central Intelligence Organisation
CMF	Commonwealth Monitoring Force
DRC	Democratic Republic of the Congo
FRELIMO	Mozambique Liberation Front—governing party in Mozambique
JOC	Joint Operational Command
MDC	Movement for Democratic Change—postwar opposition movement
PF	Patriotic Front—wartime alliance between ZANU and ZAPU
RENAMO	rebel party in Mozambique
RSF	Rhodesian Security Forces
RV	rendezvous points
ZANLA	Zimbabwe African National Army—military arm of ZANU
ZANU	Zimbabwe African National Union—African nationalist movement led by Mugabe, rival to ZAPU
ZANU-PF	Zimbabwe African National Union-Patriotic Front—postwar name of Mugabe's political party
ZAPU	Zimbabwe African People's Union—African nationalist movement led by Nkomo, rival to ZANU
ZIPRA	Zimbabwe People's Revolutionary Army—military arm of ZAPU
ZNA	Zimbabwean National Army
ZPM	Zimbabwe People's Militia—paramilitary group organized outside the military by Mugabe

NOTES

1. The literature on the Rhodesian war and its immediate aftermath is subject to extreme bias. As Bairstow points out, "Literature on Rhodesian counterinsurgent efforts generally fits into one of three categories: a large body written by black African nationalist (and often communist) revolutionaries, an even larger body of literature by Rhodesian ex-military and government authors, and a relatively small body of objective work written for academic purposes. The first leans so far toward the African insurgent that the authors downplay any Rhodesian government successes and exaggerate insurgent successes. The second holds up the Rhodesian soldier as the pinnacle of soldierly virtue and, at times, pines for the return of white-minority rule in southern Africa" (Bairstow 2012, 12).

2. However, note that when the Frontline States (Nigeria, Tanzania, and Zambia) moved

for the UN to censure Rhodesia through the UN Security Council, Britain used its veto to block this move.

3. The first leader of ZANU was Ndabaningi Sithole, its founder, who launched the movement in 1963. Following the government ban on the party he spent ten years in prison. A member of the Ndau people, he left when ZANU split along ethnic lines in 1975. Initially, neither party was founded on ethnic grounds; however, by the mid-1970s ZANU was predominantly a Shona party and ZAPU was associated with the Matabele, although neither was solely ethnically based. After internal struggles within ZANU in 1975, ethnicization accelerated, and by the time of the integration process both military wings were essentially divided along ethnic lines.

4. This is essentially a Maoist approach of forging a political war as much as fighting conventionally. This infiltration of communities by cadres tended to lead to political education of those communities and also make it more difficult for the government to separate insurgents from civilians.

5. See Wood (2009). Seeking to maximize their slim air assets and manpower reserves, the security forces developed what became known as Fire Force tactics. Along with the unconventional forces of "turned" guerrillas and groups like the Selous Scouts, these tactics are part of the mythology of Rhodesian counterinsurgency.

6. Moore (1992) titles his review of a book on African attitudes to the war at this time "Zimbabwean Peasants: Pissed On and Pissed Off."

7. Operation Quartz was apparently based on the assumption that if Mugabe was defeated in the elections, it would be necessary to carry out a strike against ZANU to prevent its forces from attempting a coup and taking over the country by force. ZIPRA forces had in fact already begun joint training exercises with the Rhodesian forces, and undoubtedly their leaders had been given an idea of what Op Quartz would entail. Although the full details of Operation Quartz have never been made public, some aspects of the plan have been revealed by former members of the security forces. It was divided into two parts: Operation Quartz, an overt strike against the terrorists; and Operation Hectic, a covert strike to kill Mugabe and his key personnel. This plan involved placing Rhodesian troops at strategic points from which they could simultaneously wipe out nationalist combatants at the assembly points and assassinate Mugabe and the other terrorist leaders at their campaign headquarters. The strike would be assisted by South African special forces and helicopters. The formations involved in Operation Quartz were in position three hours prior to the election result's being handed down, but the signal to initiate the plan was never sent. The reasons for this are still open to conjecture (Allport n.d.).

8. This higher estimate takes into account huge additions of conscripts (20,000) and guards (58,000), the evidence for which is difficult to verify. Even for the highest estimate, the regular army likely to be suitable for integration was about 19,800.

9. The Guard Force was an auxiliary unit set up effectively as a police support unit and recruited from local populations. Much air force equipment was on loan from South Africa in any case.

10. Many former special forces personnel served for several years within the South African military and police.

11. Not all white officers departed; some—e.g., Lieutenant Colonel Lionel Dyck, a former RAR commander—stayed, and according to Wood (2009) created a para-commando battalion of former RAR regulars and Selous Scouts.

12. These were to be numbered one through four; the now infamous Fifth Brigade was a latter addition. According to Chitiyo and Rupiya (2005, 338–76), "The new army was to be made up of three equal proportions of three battalions from the former Rhodesian army units, and nine battalions from former ZANLA and ZIPRA units."

13. On Operation SEED, Dennis (1992) notes that this apparently excellent idea foundered due to lack of commitment. By late 1980 it had become obvious that every PF soldier would have to be integrated and the process was further sped up to create three battalions per month, with a consequent increase in the size of the BMATT.

14. The Fifth Brigade was later supplemented by the Sixth Brigade. The other unit outside the structure was a para-battalion commanded by former RSF officers and made up initially of former RSF personnel.

15. The youth groups or militias have come to be known as "Green Bombers" and are particularly violent, engaging in intimidation, indoctrination, and even torture.

16. There were aspects of poetic justice here when several of those doing the destroying had their own homes destroyed. Overall, the social costs of this operation were catastrophic, leaving at least 300,000 homeless.

Chapter 5

MERGING MILITARIES: THE LEBANESE CASE

Florence Gaub

The Lebanese civil war is a classic case of the evolution of an internal conflict into something more complex: A conflict that started internally, with (mostly Christian) Phalangist and (mostly Sunni Muslim) Palestinian militias opposing each other, led to the emergence of other militias and to intervention by two neighboring countries. Although the war saw several national as well as international attempts to broker peace among the warring factions, it finally came to an end in an inter-Lebanese conference that took place in Saudi Arabia after fifteen years of conflict.

When the war ended in 1990, Lebanon's security sector was in shambles, with numerous militias, a shattered armed force, and two occupying powers on most of its territory. The issue of the disarmament, demobilization, and reintegration of the various armed groups on Lebanese territory was therefore crucial and found a prominent place in the final peace negotiations—long before DDR became fashionable in political scientists' circles. It was decided not only to merge a Lebanese Army split in two, but also to integrate former militia fighters into the armed forces.

ORIGINS

The Lebanese Army had long been regarded as one of the most integrated state institutions, although it was based on ethnic units. It was kept out of the civil war for a long time but was eventually dragged into it, ironically just at the end, when it split into Christian and Muslim units. After the war an amnesty was declared, and some Christians and Muslims on both sides were integrated into the new military. It was also decided that integration would proceed at the individual rather than the unit level.

The Historic Role of the Military

The Lebanese Army's role in Lebanese society has been largely symbolic. The last institution to be given up by the French colonizer, it has fought only one battle: its victorious one against Israel in 1948. Staying aloof from politics during the unrests of 1952 and 1958, it gained an image of multiethnic professionalism on which its popularity is based.

When the civil war broke out, the Lebanese Army was thus incapable of taking sides between the Maronite Christian Phalangists and the Palestinian militias—or, rather, Lebanese politicians refrained from using the armed forces in a biased way. If the Lebanese military wanted to retain its multiethnic image, it could not act against any part of society or the allies of any part of society. A second element that came into play was the fear of a possible disintegration of the armed forces upon deployment in a semisectarian scenario. Thus the passivity of the Lebanese Army during the civil war years was considered the price that had to be paid to protect the one institution that stood for the Lebanese model itself, namely, peaceful coexistence.

Yet the army suffered from this enforced paralysis. Drained by constant low-level desertion, frustrated by the never-ending conflict, and bloated by hasty reformist attempts, it was in a sorry state when it was dragged into a political quagmire that divided the country. When Lebanese parliamentarians and Syria (at that time the de facto occupying power) could not agree on a successor to President Amin Gemayel, Syria appointed the commander in chief of the armed forces, General Michel Aoun, first as prime minister and then as acting interim president. This occurred on September 22, 1988, only 15 minutes before the end of Gemayel's mandate. But according to Lebanon's unwritten National Pact, the president has to be a Sunni Muslim, and Aoun was a Maronite. Thus Salim Hoss, the previous prime minister, declared Aoun's appointment invalid. Because Aoun refused to resign, two Lebanese governments opposed each other: Hoss's in West Beirut and Aoun's in East Beirut.

While retaining his post as army commander, Aoun ordered his government to occupy the harbors, which were used by the militias to smuggle weapons and drugs. His goal, the recovery of Lebanese sovereignty over its entire territory, initially meant only the expulsion of the Syrian Army. Hence he declared a "War of Liberation" against Syria on March 14, 1989, and started the bloodiest episode in the Lebanese civil war as far as the army was concerned.

After the surviving parliamentarians of 1975 (no more elections had taken place since) had convened in the Saudi Arabian city of Ta'if in October 1989 and negotiated a peace treaty that put an end to Lebanon's civil war, Aoun rejected that treaty because it did not specify an exact timetable for Syrian withdrawal. When by far the biggest Christian militia, the Forces Libanaises, recognized the Ta'if Treaty and the new president Elias Hrawi, Aoun also declared war on it, fighting a war on two fronts against Syria and the Forces Libanaises. His offensive ended only when Syria launched massive air raids against the presidential palace (occupied by Aoun) and he fled to the French embassy.

This bloody episode eventually led the army to split into two. The existence of two prime ministers, and also two commanders in chief—the Hoss government had appointed General Sami Khatib (*Le Monde*, November 11, 1988), who was later replaced by General Emile Lahoud (*La Croix*, December 1, 1989)—affected the army's cohesion much more than any earlier interethnic strife. Violent clashes with both Syrian forces and the Forces Libanaises led to a gradual, but eventually massive, dropping out by the Sunni Muslims who had joined the Lebanese Army under Lahoud's command. Split into Eastern and Western Lebanese armies, the troops under Aoun's command decreased from 15,000 in early 1989 to between 12,000 and 13,000 in late June (McLaurin 1991, 553), and reached between 10,000 and 11,000 in January 1990 and between 3,000 and 5,000 in the autumn of that year. The Christian brigades thus opposed those who had decided to support the Ta'if agreement (Figuié

2000, 182). When Aoun finally recognized the futility of his situation and fled to France, he urged his men to merge with the Lebanese Army under Emile Lahoud's command.

It was no coincidence that Lahoud had been picked as the new commander in spite of his lack of command experience. The Hrawi-Hoss government saw clearly that the Shihabist wing within the army (named after the Lebanese Army's first commander in chief, Fouad Shihab) had to be strengthened if the army's unity was to be restored (*La Croix*, December 1, 1989). Shihabisme—meaning a strong sense of military professionalism and disregard, at least in theory, for religious sectarianism—stands at the center of Lebanese Army officers' professional identity. It stresses aloofness from sectarianism and politics and focuses on the special, supraethnic position of the armed forces in Lebanese society.

Lahoud knew the qualities of Aoun's men and, in the spirit of Shihabisme—again, military professionalism—avoided a brutal offensive against them. Those three thousand to five thousand men were the best motivated and best trained in the Lebanese Army, and Lahoud counted on them to rebuild the army after the war was over. After Aoun's departure, he followed the principle of amnesty and accepted into the newly reunited Lebanese Armed Forces almost all men who had followed Aoun (Rondé 1998, 122–25).

There was another reason Lahoud needed those men. They were mostly Christians, and without them it would have been nearly impossible to recreate a religious balance within the army (author's interview with President Amin Gemayel, Bikfaya, August 5, 2004). Those who had not excelled in the war against Syria were allowed to return to the Lebanese Army. However, rumor has it that 150 to 200 fighters were arrested by Syrian units and were subsequently executed or detained in Syria (Chartouni 1995, 42).

The reunification of the two armies happened remarkably quickly, helped by amnesty and Shihabisme. Because infighting and disintegration in the military is traumatic not only for the men but also for wider society, the official version of this event denies that the split occurred, the claim being that the war "never kept the Army's members from being united and communicating their brotherhood of arms, neither made them give up on their authentic principles and values or their desire to serve their homeland and their people which have always waited for a new rise for the Army's role" (Lebanon 2004). Three elements helped smooth the integration of the two Lebanese armies: First, a doctrine was adopted (the first one in the almost fifty years of the army's existence) that clearly identified Israel as the enemy and stressed the Arab identity of Lebanon. Considering that the war had erupted in part because Lebanese disagreed on the country's identity (and hence its enemies), this was an important step that helped orient the staff in one direction. The second element was the clear dedication to professional criteria for promotion, recruitment, and selection for assignments and training. During Aoun's command the army clearly suffered from his personalized style of posting men of his trust to some units while neglecting others. However, the new dedication to professionalism was sometimes at odds with the religious quota for officers, an aspect that is elaborated upon below. The third element that facilitated the merger was the internal army reform that mixed and matched all brigades and put an end to a system of composing brigades according to religious adherence (Lebanon 2003). Having accomplished the reunification of the two wings of the Lebanese Army, the institution accepted the integration of a limited number of militiamen into its ranks.

Militia organizations have a long tradition in Lebanon, dating back to French mandate times. Their existence and flourishing embody the distrust some factions of Lebanese society harbor toward the state and, most important, its security sector. The two oldest militias, those belonging to the Phalangists and the Syrian National Socialist Party, were founded as early as the 1930s. Due to a series of national and international developments, militias and arms both increased decidedly in the years leading up to the civil war, partly as a result of the Cairo Agreement of 1969, which gave Palestinian militias the right to conduct war against Israel from Lebanese territory and placed Palestinian refugee camps in Lebanon under the jurisdiction of the Palestine Liberation Army. The agreement, which was signed by the Lebanese Army commander Emile Boustani and by Yassir Arafat, effectively limited the Lebanese state monopoly on violence and led to the emergence of many more militias.

Within fifteen years of civil war, Lebanon had thus turned into a playground for a dozen larger and approximately forty smaller militias roughly clustered into two camps, the Islamo-Palestino-Progressists and the conservative Christians. In spite of this apparent division, infighting within the camps was more frequent than violence between them. Estimates of the number of these militias' members vary greatly; though some estimates of militiamen in prewar Lebanon are as high as 150,000, others believe that of these a maximum of 3,000 were full-time fighters. Nevertheless, for a small country like Lebanon (which had a population of 3.2 million in 1975), the number of men enlisted in militias (even if only part time) was impressive.

The Islamo-Palestino-Progressist cluster comprised Palestinian, Shiite, and politically leftist militias. Eight Palestinian militias alone claimed 22,900 men, of whom 7,000 were from Fatah. This cluster was allied with the National Movement and the Front of the Nationalist and Patriotic Parties, which comprised the militias of the Progressive Socialist Party (3,000 men); the Communist Party (1,000 men); the Syrian National Socialist Party (2,000 men); and several other, smaller groups. In addition to these there were a number of Shiite militias, such as Amal (several thousand men), the Youth of Ali, and the Soldiers of God. Hezbollah emerged only in 1982 (numbering up to 7,000 men), which indicates that parts of society were militarized only after conflict had broken out.

The Conservative Christian camp, later renamed the Forces Libanaises, was dominated by the Phalangists, who numbered almost 10,000 (some sources claim up to 25,000). In addition to that, the camp included the armed groups of the National Liberal party (2,000 men), the Zghorta Liberation Army (700 men), the Guardians of the Cedar, the Brigade of the Mountain, and the Lebanese Youth (Corm 1991, 23; Endrew 2000, 224; Hariri 1990, 150).

In addition to these there existed the South Lebanese Army, a militia occupying the South of Lebanon in conjunction with Israel and led by Major Saad Haddad, a Lebanese Army deserter. It numbered between 1,000 and 3,000 and collapsed with Israel's withdrawal in 2000. The Palestinian militias and the South Lebanese Army were never integrated into the Lebanese Army and are thus not discussed here.

When the Lebanese civil war came to an end in 1990, the country had approximately

36,000 men engaged in militias, although these numbers probably do not include part-time fighters (*Al-Hayat*, March 31, April 10, and April 11, 1991). The Lebanese Army for its part had grown from 10,000 in 1975 to 38,000 in 1990, but it had just survived severe clashes with the Syrian forces and the Christian Forces Libanaises and had undergone a political division that led to large-scale desertion (Sigaud 1988, 64). These scattered groups were subsequently supposed to be merged into one Lebanese Army.

The Emergence of Military Integration as an Issue

The idea to disband the militias and integrate them into the Lebanese Army arose rather early on in the conflict. Bashir Gemayel, the leader of the Forces Libanaises and very briefly president of Lebanon in 1982, had plans to incorporate all Lebanese militias (especially his own) into a strong and motivated army, while he wanted to disarm and expel Palestinian and "Iranian" militias (i.e., Hezbollah) (Karame 2009, 500). Because Gemayel was assassinated before he could even attempt this plan, the idea was taken up by his brother and successor, who integrated it into his peace plans. Likewise, the Syria-mediated Tripartite Accords of 1984—involving Forces Libanaises leader Elie Hobeika, Druze leader Walid Jumblatt, and Amal leader Nabih Berri—foresaw the same goal. But the persistence of the conflict, lackluster support from Syria as the occupying power, and failure to agree on key issues in the conflict put the idea of disbanding and integrating the militias to rest before it was ever implemented. It is important to note, however, that the idea generally received support at the time, although once its implementation neared, disagreement over the details showed that this support had been lukewarm from the start.

Ten years after Bashir Gemayel voiced the idea, the merger plan resurfaced shortly after the peace treaty that finally ended Lebanon's civil war. The Ta'if Treaty (officially bearing the name "Document of National Understanding"), which was drafted in 1989 in Ta'if, Saudi Arabia, by the surviving Lebanese parliamentarians, paved the way for an adjusted political system and the effective end of militia rule. Although the document makes a clear distinction between Lebanese and non-Lebanese militias (meaning Palestinians and Iranians, i.e., Hezbollah), it called on all of them to disarm within six months of the treaty's enactment (Malia 1992, 77). A ministerial declaration of 1991 restated this, declaring that all militias should be disbanded, disarmed, and reintegrated into Lebanese civil society. Shortly afterward, the council decided that "several" militiamen should be integrated into national administrative and military institutions. Candidates had roughly two months to hand in their applications. Soon Law Number 88 finally defined "several" as 6,000 men to be integrated in the first, and as it turned out only, round (Picard 1999, 6–7).

Although initially the government considered the integration of 20,000 men into the armed forces, this project never materialized. Defense Minister Michel Murr declared the number of men to be integrated should not exceed 10 percent of the armed forces' total strength, which was envisaged to reach a total of 40,000 (*Al-Hayat*, April 24 and April 26, 1991; Riad Takiedine's 1995 *The Army Revived*, in Arabic, 298; *Al-Raya*, May 13, 1991). The armed forces insisted on this rather low number, compared with the number of militiamen present in the country and willing to integrate. Recovering from their own division, and recalling the bad experience with a rapid intake of men in 1984, they feared not only difficulty in integrating these politically potentially compromised men but also the loss of their own professional identity.

The number of applications exceeded 6,000 by far. The Forces Libanaises alone handed in applications from more than 8,600 rank and file and also 650 officers, although it seems it was overestimating its share purposely. The Socialist Progressive Party of Druze leader Walid Jumblatt nominated 2,800 troops and 50 officers, and the Shi'a militia Amal roughly the same numbers (Picard 1999, 16). Because of the large response some of the candidates were integrated into the police and private security companies, the fire brigade, and customs. Even so, more than half of the interested parties were left out.

Of those who were accepted for integration into state institutions, 4,000 were designated to integrate into the army and the remaining 2,000 into other security institutions of the state. The integration of those who had been selected took place, however, in an unbalanced manner; 5,000 were Muslim (Picard 1999, 16). The Druze party managed to place 1,300 men, and the Shi'a party Amal (author's interview with President Amin Gemayel, Bikfaya, August 5, 2004) 2,800 (by far the biggest share), while only roughly 100 men of the Forces Libanaises were absorbed—and about 50 of these left the army after a few months because they felt mistreated (author's interviews with a former Forces Libanaises fighter, Beirut, April 21, 2004, and with Elizabeth Picard, Paris, June 12, 2004).

Supporters, Opponents, and Neutrals

Two groups opposed the merger: the armed forces themselves and the Forces Libanaises, the biggest Christian militia. The army, marked by a strong professionalism and pride in its institution and ideology in spite of the civil war, shunned those coming in with their own strong ideological or political agendas. In a sense this rejection expresses the occupational pride that the Lebanese Army had managed to instill into its men, differentiating sharply between the legal armed forces and the illegal militias. Interestingly, the same pride moved other army officers to perceive the army as the ideal place to reform these militiamen and mold them into good Lebanese citizens. The latter view prevailed in the end, possibly because the number of men to be integrated was very limited, especially from the Christian militias (Mansour 1993, 150).

There are many reasons for the comparatively low number of Christian militiamen who were integrated. The Lebanese Army maintains that the Forces Libanaises was never really interested in integration and depicts these numbers as representative of their disinterest, despite the fact that an internal report says otherwise.[1] According to the army only 500 men of the Forces Libanaises had applied, because their men mostly found new positions in business or emigrated to France (author's interview with Lebanese Army spokesman General Elias Farhat, Yarzé, April 21, 2004). The real reason for the rejection of Forces Libanaises fighters is probably political.

Although it was clear from the beginning that special Forces Libanaises units that had fought the army particularly hard under Aoun were never to be considered for integration, it is safe to say that almost the entire list of Forces Libanaises candidates, including those who had not been part of these special units, was rejected by the army. Several Lebanese officers confirmed that the war fought against the Forces Libanaises just months earlier was too fresh in the military's memory and obstructed a possible integration of the former enemy. In contrast to other Lebanese militias, the Forces Libanaises had fought an open war against the Lebanese Army, which effectively was about the question of which of the two was to be the "real" army of Lebanon. The other militias, such as Amal and the Druze, did not want to replace the Lebanese Army and were, for this reason, easier to integrate.

Although Christian militias did have less interest in integration than did the Muslim militias, there is no denying that at some point after Ta'if the Forces Libanaises was indeed interested in integrating a share of its men into the Lebanese Armed Forces (author's interview with President Amin Gemayel, Bikfaya, August 5, 2004; *L'Humanité* 1991). However, it was also clear that this integration had to happen according to its conditions. Aside from the request to integrate former "officers" without having them pass the army test or undergo the extended training program foreseen for militiamen at the officer level, the Maronite militia demanded that several of its units to be integrated should not be dissolved, but remain together—a move that would have been possible under the old system, under which the composition of units was based on religion and geography. However, the abolition of that system doomed the Forces Libanaises' hopes for a closed integration. Their men would be integrated and mixed into all the army's units, and the former Forces Libanaises structure thus would be broken up. Long before Ta'if, it was openly known that the Forces Libanaises planned the cantonization of Lebanon, in which it would replace the Lebanese Army in the Christian cantons, while the other militias would take over in the other regions (*Le Monde* 1984).

The Forces Libanaises considered itself a real competitor for the Lebanese Army. Therefore, it seems doubtful that its desire to be integrated into the army was free of hidden agendas. This is even more apparent when one considers the fact that the Forces Libanaises completely lost interest in integration when it was clear that the armed forces would get a new, deconfessionalized structure and that the Forces Libanaises would not be able to integrate as many men as it had planned. It used this as a reason, or pretext, for withdrawing its support for the Ta'if Treaty—which it had supported only grudgingly anyway—and for denouncing the "army of the Muslims." The Forces Libanaises, getting the impression that it had agreed to a treaty that was politically not profitable to it and that even thwarted its ambitions, restarted mobilization.

It is important to differentiate cause and effect here. Rather than failed integration into the army leading to the Forces Libanaises' rearmament, it is probably more likely that the Forces Libanaises never really abandoned the military option. Although its de facto exclusion from the army might have been the trigger for its decision to mobilize again, this was definitely not the only reason. Rather, the dominant role of Syria bothered the Forces Libanaises and persuaded it to reject involvement in government. However, it was crushed in 1994 when its leader, Samir Jaja, was convicted in a show trial and jailed. Shortly afterward the Forces Libanaises was disbanded. This, together with the conviction of Jaja and the ban on the Forces Libanaises as a militia and as a political organization (*L'Humanité* 1994), reinforced the belief among right-wing conservative Maronites that the formula of "no victor, no vanquished," which was considered the basis of post-Ta'if Lebanon, was not true in practice.

Compromises in the Process

The very biased integration of former militiamen in favor of smaller militias and lower-ranking, mostly Muslim men can be described as the compromise postconflict Lebanon found between militia integration and political reality. The Forces Libanaises was not ready for the commitment that integration ultimately signified. The dispersion of its men throughout the military would have broken its cohesion and made a desertion en bloc very unlikely. The Forces' concern was thus not that the Lebanese Army could be used against it, but rather that the Forces itself would not be able to manipulate the army to its advantage. Because the Forces was not really ready to commit to Ta'if, full integration was not in its interest. Rather, its acquiescence to the peace agreement in the first place can be understood as a move to position itself within Lebanese society as a peacemaker. For the other, mostly Muslim, militias, however, integration into the Lebanese Army proved to be an excellent opportunity to tip the confessional balance to their advantage and to occupy a crucial institution within the Lebanese state system. To this extent, the integration process reflects Lebanese sociopolitical stratification at the time.

CREATION

The new military was shaped by Lebanese supporters of the Ta'if agreements under the eyes of the Syrian government. Most of the integrated personnel were Muslim, but most of the new recruits were Christian, providing an ethnic balance. A 50/50 quota for officers was enforced.

Shaping the New Military

The face of the new, postwar Lebanese Army was essentially determined by the factions in Lebanon supporting the Ta'if agreement, as well as Syria as the de facto occupying power. Although Lebanon had a rather free hand in the selection of the people to be recruited—for example, the preservation of the confessional balance—it was clear that some militiamen, and members of the part of the Lebanese Army who had supported Aoun in his "War of Liberation" against Syria, were not eligible for employment. In an unspoken way, the Lebanese Army became a symbol of those groups in Lebanon that had united for the end of the war (civil society, the former parliamentarians, some militias) and those who saw no profit for themselves in supporting it (some militias, Palestinians). It is important to note that the support, or lack of it, for the agreement was not along religious lines. The newly elected government under Elias Hrawi consequently became the symbol of the signed compromise, which essentially affirmed Lebanon's Arab identity, its unity, and the end of confessionalism and the political dominance of Maronites in the Lebanese system.

Key groups opposed to the agreement, such as the (Maronite) Forces Libanaises, were thus effectively excluded from the new army, while the army sought to preserve, or establish, full confessional balance. In practice, this meant that while most integrated militia personnel turned out to be from Muslim militias, most of the newly recruited soldiers (89.8 percent) were Christians; this maintained the fragile confessional balance within the ranks (*Al-Jaysh*, October 1991, 23; November 1991, 28–32; May 1992, 6–10). The armed forces clearly preferred new recruits to men with a militia past. At the same time the former system of brigades that were largely homogeneous in religious terms was abolished, and full intermixture was established as a goal for the new, post–civil war army.

The reintegration of former militiamen into the Lebanese Army might have been biased and affected only a small number; it nonetheless expressed in a very tangible way the reconciliation that was so badly needed. Integrating was an important way for society to overcome internal divisions and start again, ending the militias that had dominated Lebanon for fifteen years. The reintegration process can therefore be considered a symbolic, yet vital, element in Lebanon's postconflict peace-building.

Although former Lebanese Army officers view the integration critically because they doubt the quality of the integrated men, today there is no visible difference between former militiamen and "normal" soldiers (*Libération*, April 14, 1995). In the media the integration was judged positively; it was described as "an operation of national and human integration that has largely surpassed expectations. The integration has been transformed into a social interaction in uniform, building bridges with groups that have been aloof from the army, which was aloof from their hearts."[2]

The way the integration finally took place (i.e., in low numbers, unbalanced in ethnic and in rank terms) indicates, once more, that its prime contribution to the postconflict stabilization process was symbolic. Nevertheless, it had a considerable impact on society at large.

One militia was famously excluded from integration or disbanding, namely, Hezbollah. Hezbollah, which had been founded in 1982 in a reaction to the Israeli invasion, differs from other Lebanese militias insofar as its enemy—Israel—has always been external; on the internal level, it mainly had clashes with the Shi'a militia Amal, not ethnically different Lebanese militias. This, along with Syrian support, led in the Ta'if agreement to Hezbollah's maintenance as an official resistance force against Israel, although the document called for the disbanding of all militias. The Lebanese Army itself sees its relationship with Hezbollah as pragmatic and complementary rather than competitive. A possible integration of the militia into the military, as is frequently demanded by some, would probably "not pose problems from an operational point of view, if and when such a decision is made" (*Daily Star*, April 20, 2006).

Personnel Selection and Rank Allocation

The militiamen who entered the armed forces after the war was over were selected according to their rank, while other concerns, such as military experience or human rights violations, were not taken into account. In a nutshell, the lower the rank and the less politically compromised or ideologically charged, the better the chances of selection for integration. Military experience, as in the case of the Forces Libanaises, which styled itself very much as a military institution, was an impediment because it meant that a man was opinionated and thus would be difficult to integrate. Also, the Forces Libanaises had objected to the mandatory training for officer-rank militiamen, arguing that their personnel did not need it. Overall, very few officers were selected.

The new Lebanese Army's main goal after the end of the civil war was to establish itself as the only legal force in the country. Having remained passive during the conflict, it was in a good position to claim this title, and it did not hesitate to impose its vision on the ensuing process. The armed forces had to shed their prewar "Christian" image due to Christian domination of the officer corps to style themselves as a trans- or supraconfessional military force. In practice this meant adopting an all-Lebanese discourse, while applying a quota to the officer corps.

The quota was not really new to the Lebanese Army; France had applied confessional quotas in the army's predecessors, the Troupes spéciales and the Troupes auxiliaires (McLaurin 1984, 83). Although it was not enacted under Shihab, who strongly believed in meritocratic values, the quota had provided the Lebanese Army with a Christian officer corps, which tarnished its supraconfessional reputation once the civil war broke out. In a sense, the army was not deployed precisely because of this Christian image. It is for this reason that a law enforcing a 50/50 quota for Muslims and Christians in the officer corps was introduced in 1978, and first enacted in 1980 with the graduation of 100 Christians and 98 Muslim officers (Jabbour 1989, 26). Because proper recruitment and promotion were effectively halted during the war, the quota truly came into being in 1990, once the Lebanese Army was essentially reborn. It is worth noting that the quota does not attempt to mirror Lebanese society, which does not comprise 50 percent Christians and 50 percent Muslims—rather, the exact composition of Lebanese society is guesswork because no official census has been conducted since 1932. Recent estimations give 59.7 percent Muslims and 39 percent Christians (Central Intelligence Agency 2010). The quota tries to establish an artificial balance between Lebanon's confessional groups.

In addition to the quota (which does not apply to the rank and file, which is about 60 percent Shi'a Muslim), it was decided that the commander of a company and his assistant must be of different faiths. Then, the commander of the brigade has to be of a different sect from the commander of the company. In practice this means that every brigade has six companies. If the commander of one company is Sunni, his assistant has to be a Christian, say a Greek Orthodox. In this case the commander of the brigade should also be a Christian, but not a Greek Orthodox (e.g., a Maronite). Of course this system cannot be applied rigorously everywhere, for structural and organizational reasons, but in principle it is followed.

The quota is frequently criticized. Because Christians apply in lower numbers than they used to, the quota effectively gives them an advantage over young Muslim men applying. As the army spokesman said, "When we have 100 posts to fill, and 200 Christians and 1,000 Muslims apply, we take 50 from each, that's how it is" (interview with Lebanese Army spokesman General Elias Farhat, Yarzé, April 21, 2004). Christians support the quota overall, not only because it gives them an advantage but also because, as a community, they suffer the most from the immigration of young men. For them, the quota is a tool to stop the brain drain and remain present in the Lebanese military. However, quotas are a double-edged sword, especially in an organization priding itself so much on meritocracy; though they create an artificial balance that calms ethnic fears and thus creates trust in the institution and its neutrality, they also formalize the ethnic factor, thus introducing potential division into the organization. More important, once a quota is created it is very difficult to abrogate.

Training

The new Lebanese Army needed training if its merger was to be a success. The training was conducted by the Lebanese Army itself, although training was aligned with Syrian courses following an extensive agreement with Syria (Habib 2002, 101). In order to anchor the notions of nation and fatherland in the consciousness of its men, the army doubled the intensive training courses aimed at the deconstruction of psychological barriers and the explanation of the reform's goals. The emphasis here lay on the ideological element and stressed the importance of the army's reconstruction. Commander in Chief Emile Lahoud and his men visited army bases all over the country, persuading the men of the benefits of the process (while the army command threatened discharge for opposing it). New training modules focused on the political and social importance of the armed forces as an agent of transconfessionalism and postconflict reconstruction, and effectively gained the men's support for reform. These modules proved rather successful, with little dropping out resulting from the reform.

Training itself, however, was a bone of contention when it came to the integration of former militiamen. The Forces Libanaises requested a training waiver for its officer-equivalent men, which was rejected (along with the demand that its men be regrouped in one single unit). In terms of military training abroad, before 1990 most officers and noncommissioned officers went to France, Belgium, or the United States. However, between 1991 and 1994, 42 percent of the officers went to Syria against 28 percent for the United States and 12 percent for France. Among NCOs 84 percent went to Syria, 4.5 percent to the United States, and 4.9 percent to France (Takiedine 1995, 309). Although it is traditional in the Lebanese Army to spend part of one's training abroad, the high numbers for Aleppo are striking. Obviously military training can possibly be used to forge political alliances.

OUTCOME

The Lebanese Army shows no interest in political involvement, but it retains considerable autonomy. Its military capabilities are limited, especially with regard to Hezbollah, which it regards as an ally against Israel rather than a potential internal foe, but its role as a symbol of integration in a divided country has been important.

Political Control

The newly merged Lebanese Army took the same position in Lebanon's social and political landscape that it had before the war broke out, namely, one of friendly ambiguity. Although, technically, the commander of the Lebanese Army is subordinate to the minister of defense, recommendations and guidance are effectively a two-way street. It is important to note that the Lebanese Army has shown no political ambition in the past and is generally at ease with the standard civilian control of the military. However, it also has refused to implement, or has even overturned, orders by the civilian government whenever it deemed Lebanon's stability or its own unity was at risk. Examples of such autonomous behavior include the refusal to intervene against anti-Syria demonstrators in 2005 and the actions taken against Fatah al-Islam in 2007. Though capable in a limited way of independent maneuvering, the Lebanese Army takes great care not to publicly contradict the government, and to stress the unity of the government and its institutions. Because the Lebanese government has not managed, since Syria's withdrawal, to develop a clear policy on Lebanese security interests, it arguably is sometimes difficult for the armed forces to act in this kind of setting (Nerguizian 2009, 11).

Military Capabilities

The Lebanese Army's performance since the end of the civil war has been mixed. Although it has succeeded in creating an internally balanced system in which different sects and "real" and "converted" military men coexist peacefully, it is also hampered in its actions precisely because of this. Hence the army played virtually no role during the conflict between Israel and Hezbollah. In a purely military sense, the Lebanese Army is not equipped, and not sufficiently backed politically, to counter an invasion by another army (especially not the Syrian or Israeli army). It is thus of limited use against foreign attacks.

Conversely, the armed forces conducted a four-month operation against a terrorist group hidden in a Palestinian refugee camp. However, the actions against this (primarily Sunni) group were reactive and took place only once the army had received the public's backing after a brutal attack on the military by this group. It appears that if the Lebanese Army intends to act against a particular group in the country, it needs large-scale public support (especially if that group is homogeneous). In this sense the armed forces are weakened by the country's scattered public opinion, which tends to form along sectarian lines.

However, the Lebanese Army is, at the same time, the one successful institution in the country that unites all ethnic groups. Not only do members of Lebanon's different sects coexist and cooperate peacefully in the military, but this fact is the pillar of the army's image throughout society. The army's fully integrated units and carefully balanced posts thus contribute greatly to the visibility of postconflict reconciliation, and give it a credibility other institutions in the country cannot. This means that any room for maneuvering by the armed forces (e.g., in intervening against terrorist groups, or in disobeying the government) is effectively due to this transconfessional image. Without this the Lebanese Army would be as limited in its actions as it was during the civil war—condemned to passivity.

It is for this reason that the Lebanese Army is usually perceived as making the resumption of large-scale violence less likely. The sense of security it inspires in the Lebanese population might be questioned by some because of its large Shi'a share (60 percent among rank and file) and a possible allegiance to Hezbollah, but the fact that it enjoys trust across all sectors of Lebanese society shows clearly that these qualms about its loyalty to Lebanon are not as widespread as some would like to believe. Thus the actual action scope of the Lebanese Army must be measured by Lebanese standards rather than by international or Western ones. Although this scope might seem limited compared with those of other armed forces, it is considerably bigger than it was at its lowest point, the times of civil war in which the army could not be deployed because it had a Christian image and thus was considered biased. The merger of the Lebanese Army, and its subsequent internal reform, were crucial to the military's repositioning within Lebanon's society and political system (table 5.1).

TABLE 5.1 Approval Rates of the Lebanese State and Its Institutions in Lebanese Society (percent)

Group	Agree with the statement "The Lebanese have confidence in the state and its institutions"	Agree with the statement "The Lebanese have confidence in their army"
Druze	62.5	75.1
Shi'a Muslims	43.7	80.6
Sunni Muslims	53.1	83.1
Catholics	46.9	73.0
Maronites	15.5	65.4
Orthodox	26.7	75.1
Lebanese	41.7	75.3

Source: Data from Azar and Mullet (2002, 741).

Sustainability

Although some studies suggest that the increase in an army's size actually contributes to the probability of relapse into civil war (Collier and Hoeffler 2006), the postconflict increase in size of the Lebanese Army is usually seen as a force for civil peace. When the civil war broke out in 1975, the Lebanese Army counted 15,000 men (International Institute for Strategic Studies 1975, 35), while the militias added up to many more—estimates range from roughly 40,000 to 150,000, although others argue that the number of full-time fighters never surpassed 3,000. Although exact numbers are not available, it is nonetheless clear that the militias, especially the Forces Libanaises and the Palestinian militias, were serious competitors with the army in the area of security. Although the army was certainly the best operational institution due to its structure, its training, and its equipment, one should not underestimate the effects of the militias' numerical superiority (Corm 1991, 23). The Lebanese concluded that a force superior to the militias in size, equipment, and legitimacy was necessary to prevent a relapse into violence. The large size of the Lebanese Army has to be understood in this context.

The increase in the Lebanese Army's size was thus, early on, considered an important step in regaining control of the security sector. Early in the conflict, military service was declared mandatory for all young males, and the army's staff was increased to 20,000 by 1980 (Kechichian 1995, 21). This number was further increased to 33,000 in late 1983 (McLaurin 1984, 106). As a deterrent to militias, the size of the Lebanese armed forces was increased to its current size of 59,100 men (57,000 army, 1,100 air force, 1,000 navy) (International Institute for Strategic Studies 2010, 260), which puts the ratio at 1,471 men per 100,000 inhabitants. In addition, there are 20,000 members of the Internal Security Forces, which puts the overall ratio at 1,969 troops per 100,000 inhabitants. By international postconflict standards, this is a reasonable number, situated between the 1,150 and 2,000 men recommended (Quinlivan 2003). When NATO entered Bosnia and Herzegovina in 1995, its forces were 2,260 soldiers per 100,000 inhabitants, whereas the number was 2,100 soldiers in the case of Kosovo (Quinlivan 1995, 60; Jones et al. 2005, 202). A stabilization force in a postconflict setting needs an appropriate size in order to function properly.

But this number is severely criticized for several reasons. On the economic side, a defense budget accounting for 13.5 percent of the overall government budget arouses criticism in a country with 9.2 percent unemployment, an estimated 28 percent of citizens living below the poverty line, and a per capita GDP of $13,100 (*Daily Star*, September 14, 2009; Central Intelligence Agency 2010). The military budget constitutes 5.1 percent of Lebanon's GDP, putting Lebanon in a tie with Yemen in terms of military burden. Most of the defense budget is spent on personnel and not on weapons, which are mostly donated by Western countries. The criticism thus targets mostly the large number of military men, especially in light of benefits that give the military staff access to funded groceries, housing, and mortgages.

More important, the repeated accession of former military members to the presidency (Emile Lahoud, now Michel Slimane) has contributed to the suspicion that the postwar Lebanese Army indeed pursues a political agenda that could eventually lead to its controlling the state. This view, however, does not take into account that the Lebanese Army is one of the few Middle Eastern armed forces that has never pushed itself to power. Depending on Lebanon's overall economic situation, the criticism might ebb. However, the reliance of the Lebanese economy on services such as banking and tourism (accounting for 76.2 percent) makes it particularly vulnerable in a volatile security environment such as that of the present day. Thus a deterrent military force contributes to Lebanese economic stability. As long as the military keeps its distance from politics, it will enjoy the support of much of the Lebanese population and thus be sustainable.

Alternate Outcomes

Of course, the postwar Lebanese Army could have taken a very different route. If it had not been for Shihabisme, there would not have been an ideology providing the reformers with a set of values to which most of the military could subscribe. Without an overarching, integrated ideology emphasizing the supraconfessional character of the military, it seems very probable that interethnic strife would have affected the armed forces not only in the aftermath of the conflict but also during it. Shihabisme needs to be understood not only as a work ethic created by General Shihab but also, and more important, as a nationalist ideology that put Lebanese identity above sectarianism and was a uniting force in post–civil war Lebanon.

The failure of any sect to win the Lebanese civil war eventually led to the victory of the one identity that was not part of the quarrel. This strong conviction led to the end of the civil war and subsequently to the end of the militias. Without the unity of almost all the other players, the strong rebuff of the Forces Libanaises' rearmament and challenge of the peace agreement would not have been feasible.

Thus the success of the post–civil war army is the result of a wider social development that was mirrored in an ideology prevalent in the armed forces since their inception. Of course, the capacity of Lebanese society to absorb former militiamen also constitutes an important part of the stabilization of postwar Lebanon, especially given the rather limited integration efforts of the government (Karame 2009).

HAS MILITARY INTEGRATION MADE THE RESUMPTION OF LARGE-SCALE VIOLENCE LESS LIKELY?

The merger of the Lebanese Army with former militias and with its own renegade parts has made the resumption of violence to some extent less likely. Although the military might have enough coercive capacity to provide security for the elites, it is unlikely to do so. The Lebanese Army's client, unlike those of other Middle Eastern armies, is society, and its actions must not jeopardize its multiethnic image. This symbolic value extends to the actual merger, which excluded many high-ranking and Christian militiamen and was ultimately too small to significantly affect the unemployment of potentially dangerous men. Trust in the Lebanese Army's multiconfessionalism is the currency the military deals in, and in this way it has contributed to postconflict stabilization.

The almost soothing effect of religious coexistence in the Lebanese Army has contributed to a feeling of security that has certainly helped the economy but that mostly has helped a society recovering from fifteen years of civil war by reestablishing trust among politicians, army personnel, and ultimately civil society. It is thus no wonder that military personnel remain the strongest bastion of belief in Lebanese coexistence, which is mirrored in the esteem society holds for the organization.

And yet the successful merger has not put an end to all of Lebanon's problems. The continued existence of Hezbollah as the only remaining militia, and its meddling with politics, constitute the most recent chapter in Lebanon's militia history. Because it managed to connect itself to the regional issue of Palestine and Israel, Hezbollah's existence is not just Lebanese anymore, so the problem of its continued existence is more difficult to resolve than was that of the Forces Libanaises. Yet one could argue that the integration process of 1991 reduced the number of militias to essentially two: The Forces Libanaises, not ready for the costly commitment to integration, was crushed by the large number of those who were; and Hezbollah remained because of support from Iran and Syria, and because its cause is considered to be just by many Lebanese. But domestic support is dwindling, especially since Hezbollah turned against its fellow Lebanese in 2008, and ultimately the "Party of God" might suffer the same fate as its Maronite counterpart. Only then can the integration process be considered accomplished, as at the moment Lebanese society is divided between those in favor of peace and a united armed force, and those against.

The Lebanese Army embodies the end of the civil war more than it acts as an actual military force. Built on Shihabisme, it has a multiethnic narrative that has provided the basis for reintegration and its reinvention as the ultimate symbol of coexistence. Its quota for the officer corps and its reluctance to act on controversial issues are characteristic of its largely symbolic, but nevertheless important, value to postconflict Lebanon. As an agent sandwiched between society and the political level, the Lebanese Army manages to impose itself as the prime mover of social exchange.

NOTES

1. Sayyed (1997) claims that of those men integrated, 15 percent were Christians and 85 percent were Muslims.

2. *La Revue du Liban* (1992): "C'est une opération d'intégration nationale et humaine ayant dépassé les prévisions. . . . L'intégration s'est transformée en une interaction sociale sous l'uniforme, ayant établi des ponts avec des groupes qui étaient éloignés de l'esprit de l'armée qui était éloignée de leur coeur."

Part II

AUTONOMOUS DEVELOPMENT

Chapter 6

FROM FAILED POWER SHARING IN RWANDA TO SUCCESSFUL TOP-DOWN MILITARY INTEGRATION

Stephen Burgess

For decades peacemakers have established arrangements to fashion integrated militaries out of previously warring government and rebel forces. In many cases successful military integration has proven to be one of the most important parts of the peace-building process. Four factors have proven significant in achieving success in integration (Burgess 2008):

- state strength,
- external involvement and assistance,
- the quality of contending forces, and
- high-quality management of the process.

Failure has occurred because one or more of those factors has been lacking. Strong states can withstand pressures from contending parties during the integration process, and high-quality forces have levels of professionalism conducive to integration. Most postconflict states are weak, and the quality of contending forces is often questionable; therefore, external involvement and assistance, and management of the process, take on added importance.[1] External supervision and training have expedited integration by ensuring that professional forces are developed and that spoilers do not block or reverse the process. The following ingredients have been proven to bring success in process management (Burgess 2008):

- The principles, values, and objectives of integration are reflected in the peace settlements.
- A strategy and resources are in place for the demobilization of combatants.
- Major parties have the political will to ensure that the rank and file comply with directives, and all parties are included in a flexible and adaptable military integration process.
- A realistic assessment has been made of the capacity of the integrated military to perform complex tasks.

In weak and failing states, external involvement and assistance are often necessary to overcome a lack of strategy and resources, political will, and state capacity. In many cases the United Nations and major powers have supervised the power-sharing transitions and military integration processes that enabled them to succeed.[2]

Given the complexity of military integration and the strains it puts on rebuilding states, integration is often not feasible. In some instances the internal situation is so volatile, and/or the government military and/or opposition militias are so corrupted, that integration is bound to fail. Alternatives to integration are to build a new national military with international assistance, or to preserve the existing military and absorb elements of nonstate militias. Rwanda tried an unsuccessful power-sharing-based process (Rwanda I); this was followed by successful top-down integration of ancien régime forces and rebels, featuring the use of traditional institutions for reeducation (Rwanda II). This chapter's focus on the top-down integration process and peace-building is different from that of others in this volume, in that success came after a military victory in a civil war.

The failed power-sharing agreement, part of the Arusha Peace Agreement of 1993, was based on a peacemaking and peace-building model that has been used for decades and is still being used in other postconflict countries, most recently in the DRC and in Burundi. From a theoretical perspective, the power-sharing design and military integration in Rwanda seem to have failed because of declining state strength, weak external involvement, and poor management of the political transition and integration processes. The ends of the military integration process were adequately designed in negotiations, but the ways and means were not. The peace agreement relied on the formation of a transitional government and the establishment of a joint military commission to supervise the disengagement, disarmament, demobilization, and military integration processes. The prospect of power sharing and military integration threatened vested interests inside Rwanda, and the processes were blocked by the former dictator, who exercised veto power in a weak multiparty government. Weak external involvement by the international community in the form of an inadequate UN peacekeeping mission and nonsupervision of the power-sharing and integration processes also contributed to failure and the degeneration to genocide in April 1994.

The primary importance of the Rwandan case lies in the subsequent successful top-down integration of ancien régime forces and rebels and the use of traditional institutions, which was executed by a revolutionary regime that sought to reengineer Rwanda from a country in which genocide had been committed into a united nation with no ethnic distinctions.[3] Once the Rwandan Patriotic Front and Army (RPF/RPA) took power, its leaders were determined to build a capable force that could defend the country from formidable guerrilla forces. The regime controlled the process so that recruits, including ancien régime soldiers from the FAR and rebel guerrillas, were integrated in waves over the span of a decade into the RPA and, after 1999, into the RDF. The high level of discipline, training, and esprit de corps of the RPA/RDF was essential in integrating thousands of ex-FAR soldiers and rebels while still creating one of the most professional and effective militaries in Africa. Of particular importance for the RPF regime was a process of indoctrination based on traditional practices, called *ingando* (Mgbako 2005; Ruhunga 2006).

THE FAILURE OF POWER SHARING AND INTEGRATION, 1993–94

This round of military integration was based on a negotiated settlement of an ongoing civil war (the Arusha Accords). It was similar to others discussed in this book, with quotas for personnel and demobilization, disarmament, and reintegration programs. However, it was resisted by the government and was then made irrelevant by the renewal of civil war and genocide.

Origins of Failed Power-Sharing-Based Integration

The origins of integration came when the RPF/RPA developed relatively high-quality guerrilla forces and tactics that forced the FAR onto the defensive and helped compel the Hutu regime led by President Juvénal Habyarimana to the bargaining table. There a new power-sharing agreement was negotiated and the issue of military integration arose. The RPF/RPA and moderate Hutu leaders in the government supported integration, while President Habyarimana and Hutu extremists in the FAR and the government strongly opposed the integration agreement. The compromises were made primarily by the government side.

According to the agreement, once the transitional government was put in place, the two sides would integrate their militaries into a single 19,000-soldier national army and 6,000-soldier *gendarmerie* (national police). Some 30,000 soldiers would have to be demobilized and reintegrated into society. A joint military commission would be established to implement the integration agreement. The officer corps was to be 50 percent FAR and 50 percent RPA, and the RPA was to comprise 40 percent of the integrated military. The preparations for the disengagement, demobilization, and integration of forces were scheduled for January to March 1994. From April to December 1994 there would be disengagement, demobilization, and the initiation of military integration. The integration of the armed forces would be completed between January and October 1995, and the new military would assume security responsibilities. A UN peacekeeping mission would withdraw at the end of the integration period (United Nations Security Council 1993). While a UN peacekeeping force (United Nations Assistance Mission for Rwanda, UNAMIR) was deployed to ensure that the process would not be derailed and the country revert to civil war, President Habyarimana and other powerful vested interests blocked power sharing and military integration and set in motion a process to wipe out all Tutsi and moderate Hutu in Rwanda.

From 1990 to 1993 the RPA featured high morale, rigorous training, and an educated officer corps and noncommissioned officers.[4] The guerrilla army developed a code of conduct in which fighters practiced frugality and self-discipline and interacted well with the population. Over time it developed the leadership and experience necessary to prevail in fighting what its leaders saw as a "protracted people's war" (Mills 2008, 73).

Initially, RPA guerrillas were motivated in the struggle by their belief in the Tutsi diaspora's "right to return" to Rwanda. However, the RPF's larger aim was to implement a vision of Rwanda in which politics were inclusive and society was free from sectarianism. This demanded vigilance and discipline against ethnic bias within the movement and created ownership and participation by both Hutu and Tutsi. At an early stage the RPA began to integrate ex-FAR soldiers and militiamen into the ranks. Finally, the RPA developed a remarkable ability to link battlefield intelligence with military operations, which enabled it to outmaneuver larger and better-equipped enemy forces.[5]

The FAR featured a core of 5,000 well-trained soldiers, who had the potential to integrate well as a professional military force with the RPA. However, the FAR experienced problems in maintaining effectiveness and professionalism, especially because of the rapid expansion of the force (to an estimated 30,000 soldiers by 1992) in response to the RPF/RPA invasion.[6] In addition, *interahamwe* militia gangs were formed to preserve Hutu power and terrorize the Rwandan Tutsi and moderate Hutu populations; their numbers grew to an estimated 30,000 to 50,000 militia gang members by 1994.[7]

The FAR was well armed and believed that it was fighting a struggle to preserve "Hutu power" and "Hutu land" from the RPF/RPA "Tutsi" invaders. It also received support from an estimated 600 to 1,100 French troops, who directed artillery support and managed logistics, flew helicopters, and provided airport security.[8] The FAR resisted the RPF/RPA during three years of heavy and costly fighting in a relatively small mountainous area between the RPF/RPA's Ugandan base and the Rwandan capital, Kigali. The RPA survived and gained the upper hand, because its leaders had had the foresight to change tactics to suit the circumstances.[9]

The RPA forced the Rwandan government to the negotiating table by advancing toward the capital, Kigali. The RPA expanded military operations, staged hit-and-run attacks on FAR contingents, and mobilized support within Rwanda. The tempo of the campaign ensured that the Habyarimana regime could not claim that the insurgency was a Ugandan ploy. Also, the Rwandan state had been weakened by the global recession, an International Monetary Fund–World Bank structural adjustment program, a declining dictatorship that was confronted by moderate democratic Hutu forces, and the RPF/RPA incursion.

External pressure led to the establishment of multiparty democracy, and President Habyarimana was forced to establish a coalition government with moderate Hutu parties that favored a negotiated settlement with the RPF. However, Habyarimana maintained veto power and control of the security forces. By July 1992 the RPF/RPA had forced the president and the coalition government to the negotiating table in Arusha, Tanzania.

In spite of the onset and continuation of negotiations, Habyarimana directed the FAR to keep fighting. When the Arusha negotiations stalled at the start of 1993 over the issue of military integration, the RPA, with an estimated 20,000 guerrillas, launched an offensive that ended on the outskirts of Kigali and cut the road from the capital to northern Rwanda—Habyarimana's home region (Mills 2008, 73). The offensive compelled the Rwandan government to succumb to RPF demands, which led to the Arusha agreement of August 1993.[10]

The Failure of Power-Sharing-Based Integration

In August 1993, the RPF/RPA and the government reached agreement at Arusha on a power-sharing arrangement, the integration of military forces, and multiparty elections. According to the agreement, the existing government would remain in office until a transitional government was set up, which was to be within thirty-seven days (in mid-September 1993) after the parties signed the accords. All registered political parties were eligible to participate in the transitional government and were allocated ministerial posts.

During the Arusha negotiations the moderate Hutu foreign minister, Boniface Ngulinzira, was willing to make concessions, which created the impression that the agreement favored the RPF/RPA.[11] Thus the Arusha agreement was flawed from the beginning, given that it relied on the good faith of Habyarimana and Hutu extremists and offered generous concessions to the RPF/RPA, especially in the integrated military. Also, the Arusha agreement was to have stripped powers from the president, transferring them to the transitional government. Subsequently, Habyarimana vetoed the delegation's decisions and refused to abide by the agreement.

The Coalition pour la Défense de la République (CDR), a Hutu extremist party, opposed the negotiations and was excluded from the process on RPF objections. Of twenty-one cabinet posts in the transitional government, the former ruling party of President Habyarimana was given five, including the all-important defense portfolio. The RPF got the same number, including the portfolio of the interior—in charge of policing—and the post of vice prime minister. In twenty-two months (June 1995) multiparty elections were to be held. Both the Rwandan government and the RPF agreed that Faustin Twagiramungu, the head of a moderate Hutu party, would become the prime minister of a broad-based interim government (Orth 2001, 83).

On October 5, 1993, the UN Security Council adopted Resolution 872, authorizing the UNAMIR and charging it with ensuring security during the transitional period. However, UNAMIR was not authorized to play a role in supervising the political transition and military integration processes; it was in Rwanda only to ensure that the cease-fire was observed. In late 1993, with the arrival of some 2,000 UNAMIR peacekeepers and attempts to implement the Arusha agreement, Hutu gang activities escalated, and hate radio broadcasts intensified. In spite of the lack of progress in implementing the Arusha agreement and the signs of escalating Hutu extremist violence, the Security Council and the UN Secretariat did nothing to adjust UNAMIR's mandate or rules of engagement.

Hutu extremists, aligned with Habyarimana, strongly opposed both sharing power with the RPF/RPA and the loss of control of the military. The Presidential Guard and other extremist soldiers wanted the president to resume the war with the RPF. However, Habyarimana was content with using delaying tactics by refusing to form the transitional government (Vaccaro 1996, 372–73). This pushed back preparations for the disengagement, demobilization, and integration of forces, which were scheduled to begin in January 1994.

As time passed, political squabbling was overshadowed by acts of violence and political assassinations by Hutu extremist insurgent forces, namely, the FAR-organized and FAR-directed *interahamwe* militias. In January 1994 prime minister–designate Twagiramungu accused the Rwandan Defense Ministry of giving military training to more than 1,000 members of the *interahamwe* (Orth 2001, 83). Violence escalated during early 1994 in an effort by the Hutu extremists to test the responsiveness and will of the international community, which thus far had done little in response to previous attacks. The military integration stalemate continued as violence increased, and the government and RPF/RPA prepared for renewed fighting. The launching of the genocide on April 7, 1994, destroyed any hope for political transition or military integration based on power sharing.[12]

CREATION: THE MILITARY VICTORY AND THE TOP-DOWN INTEGRATION OF ANCIEN RÉGIME FORCES

During the civil war the RPA had developed a set of techniques for integrating both enemy soldiers and militia personnel. After winning the civil war, the RPF had to rule a country in which it was a minority while facing a security threat from Hutu forces that had gone to the DRC. Military integration was used to create a security force that would both stabilize the country and make the government more legitimate.

The Emergence of Military Integration as an Issue

The process of top-down military integration and peace-building was launched after the RPF/RPA ended the genocide and secured a military victory in a civil war, which makes the Rwandan case different from the others in this volume. The origins of the top-down integration of ancien régime forces came in April 1994, as the RPF/RPA was confronted with the onset of genocide and the failure of power-sharing-based integration called for by the Arusha agreement. The issue of integration gained urgency in July 1994 because of the RPF's need to consolidate power and defend Rwanda from the Hutu forces that had fled the country and reorganized in the next-door DRC. Given the failed condition of Rwanda in July 1994, the RPF regime sought to ensure the security and defense of the country by forming a coherent national defense force, and it thus began the process of converting the RPA from a guerrilla army into a larger and more conventional force that could defend the country (Rehder 2008, 4–5).

Supporters, Opponents, and Neutrals

This conversion of the RPA was not an easy task; many of the rank and file of the RPA favored retribution against Hutu soldiers and militiamen and rejected the idea of reeducating them and integrating them into a new national army. However, Paul Kagame and other leaders of the RPF/RPA refused to compromise and insisted on integration, seeing it as the only way to secure victory and ensure peace-building in Rwanda. Integration involved reeducating both RPA soldiers and ex-FAR combatants. Obviously a big risk was taken in allowing so many ex-enemy soldiers to be admitted into the RPA. However, Paul Kagame and other RPF/RPA leaders were confident about the quality of the RPA and its ability to reeducate and absorb a large number of former enemies.

Shaping the New Military

In devising a strategy of military integration, Paul Kagame and the RPF/RPA leadership determined what the new military would look like. They planned to dramatically expand the RPA, incorporate ex-FAR soldiers and former Hutu guerrillas into the RPA ranks, and focus the military on national defense and not police duties.[13] Military integration was based on Protocol III of the 1993 Arusha Peace Agreement, which served as the basis for the principles, values, and objectives involved in the process. The strategy included an *ingando* reeducation/indoctrination process and intensive military training followed by rapid deployment to the battlefield in the eastern DRC and elsewhere. The strategy included fostering Rwandan ownership of the *ingando* program and the military integration process as part of nation building after the genocide. Nation building and military integration strategy had to be dynamic, dictated by the realities on the ground. A deliberate effort was made to establish *ingando*, an integrated military, and other national institutions. The strategy was to counteract the polarization of Rwandan society that had resulted from forty years of Hutu power and the genocide.

Personnel Selection

The criteria for selection for integration included military experience (e.g., FAR service) and a willingness to go through the *ingando* reeducation process. RPA officers chose rank-and-file ex-FAR soldiers and former guerrillas, and militiamen as well as ex-FAR officers for entrance into the new Rwandan army. Training was conducted by the RPA with little outside assistance. The problems encountered included resistance to *ingando* and integration by ex-FAR soldiers and former guerrillas, along with a lack of trust between RPA soldiers and ex-FAR soldiers and former guerrillas.

In July 1994, the RPF formed a government of national unity and ensured that the RPA had the monopoly of force and surveillance inside the country. The RPF regime began to engineer a top-down transformation of a country where an estimated tenth of the population (hundreds of thousands of Hutu) had participated in the genocide, as had large segments of the FAR. RPF leaders believed that massive political and social change was needed to move Rwanda away from being a society in which genocide had been committed and toward being a unified, nonethnic society. In the decade that followed, Paul Kagame and the regime forged a modernizing authoritarian state that proved capable of managing transformation and the military integration process (Edwards, Mills, and McNamee 2009).

After the genocide Rwanda was devastated and experienced a cumulative decline in GDP of 60 percent by the end of 1994.[14] The RPF regime had limited resources with which to build a larger and integrated military. Nevertheless, the RPF/RPA decided to try to move away as quickly as possible from reliance on external assistance, including with the military integration process. Accordingly, the regime placed an emphasis on the importance of local institutions in security-sector reform. The process worked on the psychology of the actors and was locally driven and owned, not imposed from outside. International partners were permitted only to complement local initiatives.[15]

In the short term the RPF regime relied partly on outside bodies for internal security. UNAMIR II and other UN agencies were present in Rwanda to provide peacekeeping and other forms of assistance. At the same time tensions arose between the Rwandan government and the UN, which had previously failed to respond to the genocide. In 1997 the RPF regime asked UNAMIR II to leave and UN agencies to reduce their roles (Rusagara 2008, 2).

Rank Allocation

Paul Kagame and RPF leaders understood that the divisive impact of the genocide and the independent way in which the RPF/RPA had secured victory would quickly erode the RPF/RPA's legitimacy, and they quickly began the process of integrating ex-FAR soldiers. The leadership's political will was important in ensuring that the RPA rank and file accepted ex-FAR soldiers and Hutu ex-guerrillas. In January 1995 several former FAR officers were given high positions in the new armed forces: "Colonel Marcel Gatsinzi became the Deputy Chief of Staff of the RPA, Colonel Balthazar Ndengeyinka became commander of the 305th Brigade, Lieutenant Colonel Laurent Munyakazi took command of the 99th battalion, and Lieutenant Colonel Emmanuel Habyarimana became a member of parliament and the director for training in the Ministry of Defence. Gatsinzi later became Director of Security and then Ministry of Defence in 2002" (Orth 2001, 99).

The integration of the ex-FAR soldiers and the militias took place in waves. Between 1995 and 1997, 10,500 ex-FAR officers and men were integrated into the RPA. In July 1994, after fleeing from Rwanda to the DRC, former FAR soldiers and the *interahamwe* militia gangs reorganized themselves in refugee camps containing more than a million Hutu and launched plans to return to power and complete the genocide. With the support of the Zairian dictator Mobutu Sese Seko and his military forces, the Hutu guerrillas posed a major threat to the RPF regime and launched a series of attacks, particularly in the northwest of the country from 1994 to 1996.

A key moment was in October 1996, when the RPA led an attack on "Hutu power" in the DRC and dismantled the refugee camps. Hundreds of thousands of Hutu refugees and tens of thousands of ex-FAR soldiers returned to Rwanda. This influx provided the basis for an expanded military integration process.

In August 1998, a new war began in the DRC, in which Rwanda was a major participant. In order to meet the new and expanded security needs of Rwanda, 39,200 ex-FAR soldiers and Hutu militia were integrated into the RPA between 1998 and 2002. The soldiers of the much-expanded and integrated RPA were jointly deployed into the DRC to prevent Hutu power from dominating the DRC government in Kinshasa and its eastern provinces (Rusagara 2008, 2).

Training

Disarmament, demobilization, resettlement, and reintegration programs that integrate a sensitization program and exit strategy, such as *ingando*, have been found to lead to sustainable stability and reconciliation.[16] In a study of interim stabilization of postconflict states, Nat J. Colletta and Robert Muggah found that Rwanda's *ingando* program (using so-called solidarity camps) was an example of a second-generation strategy for military integration and peace-building that had been overlooked in earlier attempts: "Other interim arrangements include *dialogue, sensitization programmes, and halfway-house arrangements*. This category is illustrated by the Rwandan *Ingando* process, through which former combatants were gathered in camps for 'problem-solving sessions' recounting the causes and taking ownership of the tragedy, exposing mutual myths and stereotypes, and endeavouring to rebuild trust after the deep trauma of the genocide in 1994" (Colletta and Muggah 2009; emphasis added).

In the national language, Kinyarwanda, the word *ingando* refers to a military encampment or assembly area where troops traditionally received their final briefing while readying for a military expedition. The briefing included instructions for the organization and/or reorganization of the armed forces and the allotment of missions and tasks. In such gatherings, soldiers were reminded to subject their individual interests to the common cause.

The RPF/RPA leadership created a process that suited its strategic purposes and that drew on its version of Rwandan history and culture. One ultimate aim was to erase the distinction between Hutu and Tutsi that had fueled the 1994 genocide. The stated objectives of *ingando* were to help the participants overcome mutual fear and suspicion caused by the 1994 genocide, and the temptation to carry out revenge attacks. It did so by encouraging conversation about the history of the Hutu-Tutsi conflict with the aim of healing the wounds of hatred.

Also, *ingando* was intended to make participants accept responsibility for harm done during the genocide, and to dispel negative perceptions Tutsi and Hutu had of each other. It was designed to encourage collective ownership of the genocide and agreement on future courses of action to prevent anything like it from happening again. *Ingando* camps were meant to integrate ex-combatants into society and the RPA, which entailed mixing ex-FAR and RPA officers and enlisted soldiers and giving them an opportunity to talk about the conflict (Ruhunga 2006, 50).

Critics of the RPF regime's authoritarianism have marshaled evidence demonstrating that the *ingando* process was primarily a form of political indoctrination and reeducation intended to meet the manpower needs of the RPA (Mgbako 2005). The solidarity camps were much like the reeducation and political indoctrination camps that the communist Chinese and other revolutionary forces used in top-down military integration processes, and the recipients had no choice but to submit. Like the communist Chinese, the RPF/RPA was faced with the problem of having too small an army to secure the country and counterrevolutionary forces (Fyfield 1982).

In 1994, a reeducation and training program was opened at the Gako integration center south of Kigali. At this camp, ex-FAR officers who had not been involved in the genocide were paired with their RPA counterparts and were exposed to the positive qualities that had brought victory to the RPA. The former US defense attaché to Rwanda, Rick Orth, was an eyewitness to this early phase of *ingando*: "One ex-FAR officer described the integration process as fairly simple due to the fact that everyone involved shared a common history, language (Kinyarwanda), and nationality. He regretted that the Habyarimana government had used divisive politics and acknowledged that the FAR had suffered on the battlefield due to the understanding that soldiers were treated as an expendable resource by the former government. In contrast, the officer outlined the strength of the RDF as an organization that values its people above all else" (Orth 2001, 99).

By 1997 ex-FAR soldiers who had graduated from *ingando* were integrated into the RPA and were helping to conduct a counterinsurgency campaign to defeat Hutu extremist forces in /DRC. From 1997 onward, waves of Hutu former guerrillas went through the *ingando* process.

The Rwandan Development and Reintegration Program Commission program supervised the integration of ex-combatants that preceded the three- to four-month phase of *ingando* sensitization. *Ingando* came to involve problem-solving workshops as a participatory conflict management strategy. These workshops, which had been designed to resolve protracted conflict such as Rwanda's (from the 1950s onward), encouraged the parties to analyze the conflict, its causes, their own attitudes toward each other, and the postconflict relationship.

Frank Rusagara, a former general officer in the RDF, outlined the four steps of *ingando*:

> Step 1: Ex-FAR soldiers, guerrillas, and Rwandan government forces (RDF) unburden themselves emotionally by talking about the conflict and its history. What the parties feel about the conflict and about each other is an important barrier that must first be revealed.
>
> Step 2: The former adversaries are jointly redeployed, which provides further opportunity for the participants to continue learning about the conflict, and facilitates bonding among troops through demystifying their differences and dispelling misperceptions they may harbor about each other. After their tour of duty in the DRC, or while they are on leave, the ex-combatants return to their communities. They are able influence their communities with their example of being fully integrated.
>
> Step 3: The government military (RDF) continuously facilitates exploratory dialogue through the office of the Civil-Military Coordination Office (CMCO) at the RDF Headquarters. The CMCO's part of the process is more analytical, and the participants are encouraged to analyze their conflict as a mutual problem. The process includes analyzing why the conflict began, and why each participant reacted to it the way he did, and coming to terms with the mutual losses and responsibilities of everyone involved. The CMCO ensures that no blame is apportioned. This stage can be emotional but is crucial and must be passed through, because in the end it facilitates a mutually beneficial solution.
>
> Step 4: Integrated ex-combatants meet and reevaluate the whole process. They may testify to having been convinced that there is a way out through working together, or to having developed doubts about the process; they may talk about contradictory reactions they received from their constituencies about the process. (Rusagara 2008, 4)

In Rwanda the disarmament, demobilization, resettlement, and reintegration process and *ingando* were generally found to promote stability and reconciliation, making the military integration process possible. *Ingando* was also found to facilitate military professionalism, which enhances combat effectiveness, civil–military relations, and societal reconciliation.[17] The policy of integrating the military and the militia into a coherent force served as a role model for the greater society that had been polarized and divided. With security in place, it was easier to build capacity in other sectors, especially the economy. Through justice and reconciliation processes, society generally came to accept that the ex-combatants had been victims of the conflict and were now partners in postconflict reconstruction (Rusagara 2008, 2).

OUTCOME

With regard to causal paths from military integration to successful peace-building, the top-down Rwandan integration process provided the basis for peace-building and the survival of the RPF regime in the face of strong external threats. Integration provided the basis for building a successful modernizing authoritarian regime. In contrast, the 1993 negotiated settlement and power-sharing-based military integration did not lead to successful peace-building. In fact, the terms of military integration in the Arusha agreement were one of the most important factors in turning the Habyarimana regime and the FAR against the peace agreement.

Top-down integration succeeded because the RPF/RPA scored a military victory rather than negotiating a settlement. Military victory enabled Kagame and the RPF to control the integration process and engineer a powerful military force. The power-sharing-based integration process had led to constant friction between the FAR and the RPA and eventually to the collapse of the entire peace process.

Military Capabilities

The integrated Rwandan military proved to be one of the most capable militaries in Africa. From 1994 to 1996 the RPA managed to defeat a strong Hutu extremist insurgency in the northwest of Rwanda. In October 1996, in response to provocations from Hutu extremist and DRC forces, the RPA invaded DRC, defeated those forces, and disbanded the refugee camps, encouraging hundreds of thousands of Rwandans to return home and escape the grip of the extremists.

The RPA then began a 2,000-kilometer march on Kinshasa, defeating DRC forces in a number of engagements and overthrowing Mobutu in May 1997. The integration of intelligence and operations with overall military effectiveness proved invaluable as a much smaller force defeated larger but less well-organized forces. In August 1998, the RPA attempted a daring and complex air raid on Kitona, DRC (to the west of Kinshasa), in order to link up with Congolese forces that were being trained by RPA units. The combined RPA and Congolese forces marched on Kinshasa and attempted to overthrow the regime of Laurent Kabila.

The RPA offensive was halted by Angolan and Zimbabwean forces, and Rwandan soldiers had to escape through hostile territory back to Rwandan lines. For the next four years Rwandan and allied forces managed to hold almost half of the DRC, as Rwanda sought a settlement that would guarantee the security of the country from attacks originating in the DRC and eventually disband the Hutu extremist forces of the Forces démocratiques de libération du Rwanda.

In August 2004, Rwanda sent 150 troops to protect military observers in Darfur as part of the African Union mission in the wake of the Sudanese government's genocide against the region's inhabitants. In 2005, Rwanda sent its first battalion of some 700 troops. By mid-2007, Rwanda had 2,000 troops in Darfur. With the establishment of the hybrid AU-UN mission (UNAMID) in late 2007, Rwandan contributions rose to more than 3,000 troops (Rehder 2008, 28–29). By April 2008, the 3,500 Rwandan troops made up almost half of the UNAMID force, while other African countries were still in the process of sending troops (Mills 2008, 78). In 2009, the UN chose a Rwandan general to be UNAMID force commander.

The Rwandans gained a reputation for being the most capable of the more than half dozen African contingents in Darfur. On many occasions Rwandan troops used force to protect civilians and to defend themselves against the Sudanese military, progovernment *janjaweed* militias and Darfur guerrillas. The US military took notice of the effectiveness of the Rwandan forces, first through airlifting Rwandan peacekeepers to Darfur and later through observing on the ground in both Rwanda and Darfur. As a result, in 2006 US European Command (EUCOM) elevated Rwanda to partnership status and involved the RDF in regional exercises.[18] The US Africa Command (AFRICOM) continued to build close military-to-military relations with the RDF.[19]

HAS MILITARY INTEGRATION MADE LARGE-SCALE VIOLENCE LESS LIKELY?

The Rwandan Patriotic Front/Army took over Rwanda in July 1994 and was faced with a large opposition force that waged guerrilla war in the northwest of the country until 2000 and with the former regime that was based among a million refugees across the border in DRC. The RPF/RPA had a choice—(1) to maintain the relatively small but cohesive RPA (30,000 troops) and wage counterinsurgency within Rwanda, or (2) dramatically expand the RPA to wage counterinsurgency and offensive operations against the former regime in DRC. Paul Kagame and his commanders decided for option 2, even though it entailed a high degree of risk—integrating Hutus who had been in the former regime's army and militias (10,500 between 1995 and 1997, and 39,200 between 1998 and 2002). If not handled skillfully, integration could have led to mutiny and a threat to the RPF regime itself.

All five of the causal paths outlined above were at work in this case:

1. *Costly commitment:* The Hutus who integrated were betraying the old regime and "Hutu power" and were taking the chance that the old regime would fail to defeat the RPF/RPA and return to power. Therefore, it was a costly commitment on the Hutu side, especially by former FAR officers, many of whom had rejected the genocide and decided to take a chance on a regime that talked about national unity. On the RPF/RPA side, integration risked the creation of a mutinous army and also dissension from Tutsi officers and soldiers who hated Hutus, especially after the genocide.

2. *Coercive capacity:* The new military provided security—defeating the insurgency in Rwanda and the former regime in DRC. The Hutus who integrated had the reassurance that the new military would not be used inside Rwanda against Hutus (which was largely the case). As a result, Hutu moderates participated in the government of national unity with Paul Kagame and the RPF (Kagame dominated the Government of National Unity). This made the spread of Hutu opposition guerrilla warfare outside of the northwest of Rwanda from 1994 to 2000 less likely.

3. *Employ fighters:* The integrated Hutus were used in RPA operations in northwest Rwanda and /DRC. However, Kagame and RPF leaders also wanted to

employ fighters who could have added to the regime's security challenges if they were left unemployed. The employment of fighters facilitated the new regime and the processes of national reconstruction and economic development.

4. *Example for society:* The integrated RPA (renamed the Rwandan Defense Force in 2002) became a symbol of national unity. The regime publicized the *ingando* process, the military integration process, and the significant numbers of Hutus serving in leadership positions as well as the achievements of the RPA/RDF in defending the nation and engaging in development projects.

5. *Trust:* Trust was created between Hutu and Tutsi leaders that enabled the new government to work until Kagame centralized power around himself and his Ugandan Tutsi elite in 2000–2003.

CONCLUSION

Power-sharing-based military integration failed because of low state strength and the veto power of President Habyarimana over a poorly devised political transition process. External involvement and assistance from the UN and the major powers were inadequate. Once the transition and military integration processes were blocked, the UN Security Council did not authorize a greater UNAMIR role in enforcing power sharing and integration. The quality of contending forces played a role in the failure, as the FAR degenerated from a relatively professional force in 1990 into a force intent on genocide by 1994.

The management of the process was poor, with little leverage available to force President Habyarimana to cooperate. Although the principles, values, and objectives of integration were reflected in the peace settlement, strategy and resources for the disarmament and demobilization of combatants were scanty or nonexistent. The ruling party did not have the political will to ensure that the rank and file complied with leadership directives. The blocking of the transition process meant that the RPA was not allowed to be included in military integration.

After the genocide, the RPF regime managed a top-down integration process. The *ingando* reeducation/indoctrination process built relations between Hutu recruits and RPA commanders. In regard to the quality of contending forces, the RPA integrated thousands of the more capable ex-FAR soldiers and Hutu former guerrillas, but only after they had gone through the *ingando* reeducation/indoctrination process. Integration into the RPA and the RDF was a continuous process that took place before, during, and after the cessation of hostilities. The regime devised a strategy for disarmament and demobilization of ex-FAR soldiers and Hutu former guerrillas in preparation for integration and established a well-managed and well-supplied process of demobilization. The RPF/RPA dominated the political process and had the political will to ensure that the rank and file complied with leadership directives, including those involving the integration of ex-FAR soldiers and Hutu former guerrillas into the RPA.

The RPF/RPA managed the military integration process but demonstrated flexibility and adaptability in accepting a large number of ex-FAR soldiers and naming a number of ex-FAR officers to prominent positions in the RPA. A realistic assessment was made of the capacity of the state and security forces to perform complex tasks, based upon the ability of the RPA to defeat the FAR in 1994. The assessment was validated when the RPA included thousands of ex-FAR soldiers in its 1996 assault on Hutu power camps in eastern Zaire and, in the first half of 1997, when the integrated forces marched and fought for more than 2,000 kilometers to unseat the Mobutu regime in Kinshasa in May 1997. Again the assessment was validated when the RDF—now with even more ex-FAR soldiers and Hutu former guerrillas, integrated from 1998 to 2002—fought, first against Hutu forces and Angolan and Zimbabwean armed forces in vast expanses of the DRC and then later in Darfur.

A comparison of the two processes clearly demonstrates the superiority of the top-down model for producing an effective military and a sustainable peace. Military integration, in power-sharing cases, can seriously undermine the peacemaking and peace-building processes. However, in fairness to the power-sharing model, in the Rwandan case political forces prevented the process from getting off the ground. If sufficient external supervision and management had been present, it might have been possible for an effective military to be developed and for genocide to be prevented.

ABBREVIATIONS

AFRICOM	US Africa Command
CDR	Coalition pour la Défense de la République—extremist Hutu party
CMCO	Civil-Military Coordination Office—central institution for *ingando* process
DRC	Democratic Republic of the Congo
EUCOM	US European Command
FAR	Forces Armées Rwandais—wartime government army; defeated
NRA	National Resistance Army—Ugandan rebel army
RDF	Rwanda Defence Force—postwar Rwandan military
RPF/RPA	Rwandan Patriotic Front and Army—victorious rebels
UNAMIR	UN Assistance Mission in Rwanda

ACKNOWLEDGMENT

The views expressed are those of the author and not necessarily those of the US Air War College or the US Department of Defense.

NOTES

1. The quality of the forces involved, indicated especially by their degree of professionalism, has determined whether they are capable of integrating, and the quality of their leadership has determined whether the leaders are suitable for officer training and development.

2. UN multidimensional peacekeeping operations with supervision of the military integration process helped to bring success in Mozambique, Namibia, the DRC, and Burundi, among other countries.

3. The author visited Rwanda in March 2008 and March 2010 and conversed with officials from the government and the RDF. In addition, he conversed with RDF officers at Air University in Alabama in 2008, 2009, and 2010.

There is great controversy about Rwandan history and politics. Critics of Paul Kagame and the RPF movement and regime assert that the RPF was as culpable as Hutu extremists in subverting the Arusha peace accords and the military integration process. They also claim that the RPF/RPA committed large-scale atrocities from 1994 onward that rivaled the genocide. There is also controversy about the Kagame/RPF regime, particularly regarding the degree of authoritarianism and coercion used in military integration, including the reeducation and indoctrination process (*ingando*). With regard to the violence that racked Rwanda from 1959 to 1962, some claim it was a "Hutu revolution" and others assert that it was an "anti-Tutsi pogrom."

4. Prunier (1998, 130); Mills (2008). The RPF/RPA's high morale was caused by the desire of Rwandan Tutsi refugees in Uganda to end three decades of exile and return to Rwanda. The Tutsi refugees who settled in camps in Uganda in the early 1960s were the elite of Rwandan society during the colonial era, and many of them sent their children to Ugandan schools and preached the goal of returning to Rwanda. The second generation of Rwandan Tutsi in Uganda finished their educations—many at university—and, from 1982 to 1986, thousands joined the Ugandan National Resistance Front and the NRA led by Yoweri Museveni—who became Ugandan president from 1986 to the present. From 1982 to 1986, thousands gained considerable military and guerrilla war experience. By 1986, 4,000 such refugees made up almost a third of the NRA, and they made up a higher proportion of the officer corps. They returned to Rwanda in 1990 determined to reclaim their place in society.

5. Rehder (2008). Rwandan Tutsis made up a large part of the intelligence branch of the NRA during the 1982–86 guerrilla war against the Ugandan government and played a large role in creating the synergy between intelligence and operations. Paul Kagame—today Rwanda's president—helped to command NRA intelligence.

6. United Nations (1996b, 224). The Rwandan military (FAR) was estimated to consist of 30,000 soldiers, and the security forces included an estimated 6,000–10,000 *gendarmerie*.

7. Mills (2008, 73). Estimates of the size of the FAR before the April 1994 genocide range widely, from as low as 12,000 to UN estimates of 30,000 to Greg Mills's estimate of 50,000. The *gendarmerie* and *interahamwe* militia gangs are sometimes counted along with the military in estimates of the size of security forces.

8. Estimates vary regarding the number of French troops in Rwanda, from 1990 until the genocide in April 1994, and their roles and missions. The Habyarimana regime bought weapons on the international market from Egypt, South Africa, and other countries.

9. Mills (2008, 73). After being defeated in October 1990, Paul Kagame changed RPA tactics to avoid direct battle with the FAR, and the RPA retreated to the Virunga volcano range in the northwest, where it regrouped and adopted guerrilla tactics.

10. Rehder (2008, 4). The RPA captured much of its ammunition and weaponry from FAR stocks and received funding through its diaspora in Uganda and elsewhere.

11. Ngulunzira had been in opposition to President Habyarimana before 1992.

12. The RPF/RPA was reorganizing its military positions in northern Rwanda, while President Habyarimana was also reinforcing his army. Arms were widely available among the civilian population, especially in the northwest, the bastion of Hutu extremism and Habyarimana's home region, due to increased militia activity.

13. From the ex-combatants and the demobilized soldiers, a new police force was created to take over the national policing duties from the military and to form local defense units in ex-combatants' areas of origin.

14. Rusagara (2008). Despite the collapse of the Rwandan economy, aid was not forthcoming from the international community until the end of 1996, when donors at the Geneva Conference for Rwanda pledged more than $600 million to be managed under a trust fund by the UN Development Program—which, along with the Office of the UN High Commissioner for Refugees, controlled most of the national budget, and thus compromised national priorities, including military integration.

15. Rusagara (2008, 2). In the justice sector, the RPF regime focused on the overwhelming caseload of genocide suspects by adopting a traditional mechanism of dispute resolution, the *gacaca* courts, which tried genocide suspects in their communities and sentenced convicts to hard labor.

16. Ruhunga (2006, 60) points out that "a former commander of the Hutu guerillas [Forces démocratiques de libération du Rwanda] Maj. Gen. Paul Rwarakabije [who] had been integrated in the RDF and was a commissioner in the Rwandan Reintegration, Development and Resettlement Commission (RDRC), provided an example of integration. He made testimony that *Ingando* helped to provide a way out of the conflict."

17. Ruhunga (2006, 56).

18. Discussions with EUCOM J-5 personnel, July 2007, Stuttgart, and via video teleconference, February 2008. Discussions with former EUCOM J-5 officer at Air War College, 2007–10.

19. Discussions with AFRICOM J-5 theater security cooperation personnel, February 2009, Garmisch, Germany, and via video teleconference, February 2010.

Chapter 7

FROM REBELS TO SOLDIERS: AN ANALYSIS OF THE PHILIPPINE POLICY OF INTEGRATING FORMER MORO NATIONAL LIBERATION FRONT COMBATANTS INTO THE ARMED FORCES

Rosalie Arcala Hall

In 1996, the Philippines embarked upon the unprecedented project of inserting former insurgent/combatants from the Moro National Liberation Front (MNLF) into the military. A total of 5,750 ex-combatants and their proxies were integrated into the Philippine army in line with the Final Peace Agreement signed by the national government and the MNLF. The processes underlying this integration/insertion project were controversial and had lasting effects on the composition and performance of the national army.

The agreement was a milestone in the ongoing three-decade conflict in Mindanao, southern Philippines, which has claimed thousands of lives and created widespread internal population displacement. The MNLF, the first of a series of armed movements anchored in the politics of differentiation (i.e., based on the idea that Muslims constitute a group distinct from the Christian-dominated Philippine nation), negotiated and accepted terms for a more robust regional autonomy. Although the terms of the autonomy itself (as opposed to separation) remain contested, the agreement was key to the transformation of the rebel group into a legitimate political player. But the MNLF was no longer the monolithic power it had been before 1978.[1] In the complex political landscape of Mindanao—in which a rival armed group, the Moro Islamic Liberation Front (MILF), politician warlords/strongmen and their private armies, and kidnap-for-ransom outfits vie for territorial control—the prospect for long-term peace through autonomy remains dim. The absorption of MLNF members into the Philippine Army must be considered in this context. Although conflict with the MNLF ceased once the agreement had been signed, the military's ground forces were still operating against other Islamic separatist and armed groups in Mindanao.

The integration program had little impact on the prospects for peace in Mindanao. Though a relatively successful undertaking, integration was embedded in a failed autonomy project that saw few advances in addressing poverty and minoritization. Its limited scope, the lack of an accompanying disarmament and demobilization program, the diversity of the recruits (ex-combatants and proxies), and the provision for mixed rather than separate units all served to strengthen the army's counterinsurgency capability but did little to neutralize the armed threat from MNLF and other nonstatutory armed groups. The program is nevertheless credited for its symbolic value: It introduced greater diversity into the ranks and engendered culturally sensitive military regulations and policies.

The Philippine MNLF integration project was introduced as part of a negotiated peace agreement resulting in the absorption of former fighters and their proxies into the existing national army. A more significant aspect of that peace agreement was enhanced autonomy for the Autonomous Region for Muslim Mindanao (ARMM), which was underpinned by the national government's fiscal support and the MNLF's transformation into a legitimate regional player.[2] The agreement, however, had serious flaws. Except for the MNLF integration aspect, for which there were quantitative targets, it failed to provide sufficient mechanisms to ensure effective local governance by the MNLF. The supposed power-sharing plan embedded in the agreement was made empty by the national government's failure to provide timely fiscal injections to make autonomy viable. In addition, the MNLF-led ARMM government was adjudged "corrupt and mismanaged" (Bertrand 2000, 37) and "wasteful and a poor performer," in part because it tended to "centralize" fiscal power in the hands of the executive (governor) (Gutierrez and Danguilan Vitug 1999, 193).

In 2001, Nur Misuari was removed from his position as ARMM governor following charges of terrorism. The support of the MNLF leadership was divided between him and the government-recognized Executive Council (Council of Fifteen), whose appointed chief is Muslimen Sema. These internal vicissitudes rendered the MNLF unable to steer the integration project.

The peace agreement itself reflected the uneven power distribution between the two key players. The Philippine government (and by extension the military) had more leverage in determining the outcome of the agreement, as the MNLF's cadre strength was greatly reduced owing to the emergence of a rival, the more Islamic-fundamentalist MILF. The Philippine Army saw the integration project as a means of augmenting its forces for counterinsurgency activities against the MILF and communist rebels in Mindanao and an opportunity to introduce diversity into the Christian-dominated ranks.

For the MNLF, the integration project was an important employment vehicle and key to sustaining political support from its members. With the Mindanao war at a stalemate, the integration project was an important carrot for the MNLF to offer its followers. At that point, the MNLF was a defeated force whose threat lay only in the potential exodus of members into more fundamentalist armed groups.

The military merger was a largely endogenous process involving the Philippine armed forces and the MNLF, with external actors playing a minimal role by providing financial support for the complementary reintegration process.[3] The recruitment process was undertaken by the MNLF, while the Philippine Army made the final selections based on relaxed criteria of admission (height, age, and education requirements being waived), and handled the training and placement of integrees. The integration process did not involve parallel demobilization or disarmament of the MNLF troops, which would have been unacceptable to the MNLF and potentially dangerous given the proliferation of other nonstatutory forces in Mindanao.

The MNLF's contribution to short-term or long-term peace in Mindanao is contested. One view holds that integration offers a credible security guarantee by reducing the armed group's numerical strength and its ability to mount renewed threats against the government (Simonsen 2007, 585). In their study of military integration cases, Glassmyer and Sambanis (2008) contend that the outcome is driven by economic incentives rather than security. For members of the MNLF, the financial benefits of being a soldier and the concomitant job security were very attractive. In Mindanao, where unemployment levels are high, ex-combatants are by no means the only ones interested. But designing an integration program to provide as many employment opportunities as possible creates serious fiscal pressures on the government. Moreover, integration is less useful fiscally than the public-sector investments needed in Mindanao.

As an economic vehicle, reintegration is a better alternative than integration—provided it is sufficiently funded.[4] However, the Philippine government did not put sufficient fiscal resources toward reintegration programs, despite an explicit provision in the Final Peace Agreement (20a) for "a special socio-economic and cultural program to cater to MNLF forces not absorbed in the armed forces and the police . . . to prepare them and their families for productive endeavors, provide for educational, technical skills and livelihood training, and give them priority for hiring in development projects."

In actuality, development projects that targeted ex-combatants and MNLF communities were funded mostly by foreign donors through a UN multidonor program called Act for Peace, which committed $500 million in support of the 1996 Final Peace Agreement (Cragin and Chalk 2003, 17). The program included livelihood assistance, vocational skills training, enterprise development, economic managerial training for ex-MNLF commanders, and the delivery of basic services (Rasul 2005, 71). There was also a reintegration program that specifically targeted ex-combatants and had 28,000 beneficiaries, which was funded by the US Agency for International Development.

In both programs, reintegration benefits were not given individually; ex-MNLF rebels had to be organized into cooperatives before they could gain access to reintegration benefits. No systematic study has been done of the impact of reintegration programs involving the MNLF. Program success varied widely across communities, according to a report from the US Agency for International Development (1999). Cragin and Chalk's (2003) study on the impact of development aid in Mindanao mentions positive reviews of aid targeting ex-MNLF combatants. Some authors (Lidasan 2006; Santos 2009) point out the weakness of the national government's effort to provide employment opportunities to ex-MNLF combatants who were left out of the army and police integrations.

Combined, the integration and reintegration programs' economic impact was limited. Neither program had a master list of veterans from which beneficiaries were selected or offered those beneficiaries money. The beneficiaries of both totaled more than thirty thousand, but this number equaled only 60 percent of the estimated MNLF strength in 1996. Whether the economic incentives dissuaded former MNLF combatants from joining the MILF is uncertain. MILF recruits tended not to be from economically deprived groups but from those with a strong Islamic-fundamentalist orientation. The greater danger lay in the potential recruitment of former MNLF combatants into armed kidnap-for-ransom gangs, of which there are many in MNLF areas in Central Mindanao and the Tawi-Tawi/Sulu island group.

Ferrer (1999) and Santos (2009) conclude that the actual integration (into the army and the police) did not significantly demobilize or disarm the MNLF. Those integrated were mostly the kin of ex-combatants and only a small fraction of the estimated MNLF strength. The integration program did not reduce the number of firearms in the MNLF's possession, as prospective applicants were ordered to procure their own to meet the army's "no gun, no integration" requirement. The guns-for-cash program increased demand for low-caliber weapons in the black market (Santos 2009).

From the Philippine government's perspective, the MNLF integration project was an opportunity to introduce cultural diversity into the Christian-dominated army. The MNLF has historically championed the distinctiveness of the Moro people and, as an armed movement, pressed for secession and, later, autonomy. Because the question of religious identity is at the core of the Mindanao conflict, having Muslim MNLF members in the force was part of the broader nation-building strategy of increasing Muslim visibility in the public sector.[5] At a symbolic level, the integration project was part and parcel of a broader government push to acknowledge Muslim distinctiveness, along with declaring Islamic holidays as national holidays and allowing public employees to observe Ramadan. The minority's symbolic presence inside the army is said to engender overall confidence that the institution is truly inclusive and that the rest of the Philippines, writ large, has become more accepting of Muslims.

ORIGINS

Both sides desired military integration, but their different motives resulted in disagreements. The MNLF wanted more people to be integrated and wanted them to be in Muslim units, which it could control; the army wanted fewer people to be integrated as individuals. These issues remain controversial, but the government's positions prevailed.

The Historic Role of the Military

The Philippine Army constitutes the core of the armed forces in terms of budget, personnel, and missions. The ground forces (infantry) have historically been involved in counterinsurgency operations on multiple fronts—against communist rebels, Islamic insurgents, and the offshoot nonstatutory armed groups that proliferate in Mindanao. After shifting in the late 1980s from a more combat-heavy counterinsurgency strategy to one concentrating on civil–military operations (CMOs), the army has carried out fundamental organizational reforms designed to enhance its CMO capacity. The entry of Muslims into the force through the integration program was expected to enhance its effectiveness in operations in Muslim communities. To the army, the Muslim recruits were critical to the strategy of winning hearts and minds. Thus, unlike military integration elsewhere such as South Africa (Williams 2005; Liebenberg 1997; Jackson and Kotze 2005), the Philippine integration project did not involve changing the army's orientation away from internal security operations; nor did it lead to a leaner military (Lingga 2009).

The Emergence of Military Integration as an Issue

The idea of Philippine military integration had its genesis in the 1976 Tripoli Agreement, which stated that "National Defense affairs shall be the concern of the central authority provided that the arrangements for *the joining of the forces of the Moro National Liberation Front with the Philippines Armed Forces* be discussed later" (paragraph 2, section 3; emphasis added). It was a key part of the negotiation agenda starting in 1986 (Rodil 2000, 18).[6]

Supporters, Opponents, and Neutrals

For the MNLF, military integration was beneficial because (1) it buttressed the MNLF-led regional autonomous government against rival traditional politicians and other nonstatutory armed groups (Azurin 1996); (2) it gave Chairman Misuari a way to discourage MNLF elements from joining the MILF or other criminally inclined groups by providing them with alternate employment (Ramos, as cited by Santos 2009; Makinano and Lubang 2000, 27; Gacis 2010); and (3) it served as a means of exercising political control. The MNLF, in trying to influence security affairs (national and local), saw military integration as part of its "dual track," the other being the creation of a separate Special Regional Security Force (SRSF). Of the two tracks, the MNLF was more invested in the SRSF, which the autonomous government could potentially control. Military integration was a smaller prize given the small number of ex-combatants who could be absorbed.

The idea of absorbing ex-rebels into the army was also received with equanimity by the military leadership—unsurprisingly, given its prior experience of working covertly or openly with insurgent factions and nonstatutory armed groups (e.g., vigilantes) during counterinsurgency operations. MNLF integration under the terms of a peace agreement lent formality and fiscal legitimacy to what was desirable for the military: additional manpower for counterinsurgency operations. But though not opposed to integration, the Philippine government intentionally did not press for the MNLF to disarm and demobilize; President Ramos understood that this would be unacceptable to Chairman Misuari, as it would erode his popular support. The armed forces addressed the problem of gun proliferation by implementing a modified guns-for-cash program in which integree recruits would have to turn in their weapons to be registered (Gacis 2010).

Compromises in the Process

The MNLF and the Armed Forces of the Philippines (AFP) both wanted the integration project. However, they differed about the appropriate number, composition, organization, and territorial assignments/postings of MNLF integrees. The MNLF wanted to retain control of its forces by having them integrated as separate units and posted only within the autonomous area, whereas the AFP wanted to integrate individuals into different units and insisted on the right to deploy the units anywhere they were needed.

The resolution to this disagreement is reflected in the slightly different terms of the 1995 Interim Peace Agreement and the 1996 Final Peace Agreement. In the interim agreement, detailed references are made to how during the transition phase the separate MNLF units should be assigned areas of responsibility by province and to assist the police. The final agreement is silent on these issues and assumes that there will be no separate MNLF forces after the transition phase. It concentrates instead on organizational concerns (e.g., how a "special" category of recruits could be accommodated with the least disruption to the military's established rhythm). The exact number of integrees was hotly negotiated, with the MNLF pressing for more than the AFP's proposed numbers based on the proportion of Muslims in the entire Philippine population (Rodil 2000, 113).[7] The government tried to sweeten the deal by promising future support to an indeterminately sized SRSF, which the MNLF could potentially control if it were to lead the autonomous government.[8] The final agreement was that 5,750 MNLF members were to be integrated into the army; in addition, 1,750 were to be integrated into the police.

CREATION

Various units of the MNLF were responsible for selecting individuals to be integrated. Special provisions were made for women. Training was basically the same as for the regular army, with some changes made to accommodate Muslim requirements.

Shaping the New Military

The Final Peace Agreement stipulated that (1) MNLF elements will be initially organized as distinct units during the transition phase but will be gradually integrated (as individuals) into regular AFP units deployed within the autonomous area; (2) the highest-ranking MNLF officer will be deputy commander of the AFP Southern Command, tasked to supervise MNLF integree units during the transition phase; and (3) a Joint Integration Board composed of AFP and MNLF members will oversee and troubleshoot matters pertaining to the recruitment, training, development, and deployment of the integrees.

The mechanisms and modalities of integration in turn are covered under Administrative Order 295, which sets a three-year time frame (beginning November 1996) and three distinct phases (processing; individual training; and on-the-job training, OJT) for the process. The processing phase involves screening candidate soldiers and officers from the list of names submitted by the MNLF.[9] Several things were agreed upon at this stage: (1) that the MNLF alone will supply the list; (2) that the AFP will waive the usual entry requirements (for age, height, and educational attainment); and (3) that everyone on the MNLF list must bring a weapon with him when reporting to the training site (no weapon, no integration).[10] The 5,750 positions were finally filled up in three recruitment cycles (during 1997, 1998, and 1999). There was a separate recruitment in 2008 to replace MNLF integration slots left empty by attrition. Recruits into the MNLF integration program were divided into regular (recruited from 1997 to 1999) and replacement (recruited in 2008) batches.

Personnel Selection

The 5,750 slots were divided into quotas for the various revolutionary committees (RCs), MNLF national units, satellite commands, and task forces. Each RC and military committee (MC) independently decided how it was going to fill its quota. Predictably, there was a great degree of variation in how the quotas were filled. Some slots were given to individual combatants, who could choose to use them personally or recommend a proxy (usually a family member). In the absence of a clear directive that the integration be limited to combatants, the process yielded a liberal interpretation, whereby those who contributed to the cause—whether in combat, on the diplomatic front, or in a civilian support group—living or dead, were eligible. There was a general belief that ex-combatants and the kin of ex-MNLF combatants and noncombatants alike deserved the slots as "blood payment." The final list (or lists) was compiled by the Committee on Integration and submitted to Misuari for final approval.

In determining to whom to give the slots, local commanders might have candidates draw lots, choose those who expressed interest, or make choices based on physical condition (Ayao 2009; Hadji Ebrahim 2009; Akbar 2009). Many opted out because of age and health problems.

It was important that those selected be able to provide their own guns in compliance with the army requirement (Akbar 2009). The gun issue was crucial, given that the MNLF did not plan to disarm, even with the peace agreement signed. The commanders agreed that the guns were owned by the units and could not be used for integration. A member of the MNLF Council of Fifteen was very candid in admitting that the majority of the guns entered into the BALIK-BARIL program were personal property (Salim 2009). In the end, the selection process privileged those with access to a gun over those who could not get one within the short period during which the list was finalized.

No women were included in the regular batches of MNLF integration trainees, although there was no official MNLF policy barring them. Given that integration was voluntary, MNLF leaders and commanders explained that the women combatants themselves and their female family members were dissuaded from applying (Salim 2009; Ayao 2009). In a policy shift, Muslim women were targeted for recruitment to compensate for attrition in 2008.[11] Twenty-eight were selected from among hundreds of applicants. The women were younger and more educated than previous integrees.

Interest in becoming part of the integration peaked in the second and third recruitment cycles. At first, few were interested because of suspicion about the government's motives in allowing ex-combatants into the army. The potential economic benefits of a steady government job were the primary motivation for many candidates. In Mindanao, where unemployment is high, any government job is much prized; the great demand has even led to the lucrative, illegal practice of selling government positions, including integration slots. Some claim that the names of Misuari supporters in the master list were replaced by those of supporters of the rival Council of Fifteen (Landasan 2009). Some who joined as proxies for their ex-combatant relatives did so because of family pressure (Gumampangan 2009; Abasama 2009). Having a family member in the armed forces or police is a source of prestige and an important asset, given the *rido* (clan wars) characteristic of Muslim communities in Mindanao.

Training

All candidate soldiers went through a six-month training course, and officer candidates did forty-eight weeks. The training centers for soldier candidates were in various camps in Mindanao (the bulk in Central Luzon), and officer candidates trained in Tanay, Rizal, and Capas, Tarlac. According to Lidasan (2006, 44), the integrees' training was very similar to the standard training for recruits, except that the integrees were allowed to use some of their training time for religious activities. To varying degrees, their religious needs were accommodated; they were given time off for Friday prayers, a prayer room and religious supplies such as veils and mats for the women, and a religion-specific (halal) diet.[12] As mentioned above, Muslim women were specially targeted for recruitment into the 474 slots (7 officers and 467 enlisted personnel) that made up the 2008 integree replacement batch. The men and women trained in separate facilities, but their training programs were the same.

Candidates from the regular batches went through on-the-job training (OJT), in which they were assigned to one of forty-seven rifle or ten engineering companies in brigades and battalions located within the ARMM's five provinces. The purpose of OJT, which lasted twenty-four weeks, was to give integrees the necessary field experience within the military's organizational setting, while maintaining their identity as a separate group (Lidasan 2006, 46). After this the integration process was completed, the separate units were dissolved, and individual integrees were reassigned to regular AFP units within the Southern Command. In the end, the entire process was extended by one year to end in 1999, owing to a delay in the MNLF's submission of the master list. A total of 5,990 people went to training, but only 5,191 (213 officers and 4,978 enlisted personnel) were fully integrated (Phillipines, Office of the President 2007).

The integration program was groundbreaking for the AFP as an institution, as it had to accommodate a fairly large intake of Muslim integrees. Although Muslims were already present in the army, the integrees constituted the single largest absorption of recruits from a minority group (4 percent of the entire AFP strength). Showing foresight, the AFP designed a training module sensitive to identity-based concerns by allowing the trainees to observe religious practices, assigning a sacred space to them (and later building mosques within the training camp), and observing religious requirements in food preparation.

After integration, the AFP also tried to help integrees to overcome their educational/literacy barriers. Its Paaral (i.e., education) Program allowed integree officers to go to college or university almost immediately after integration. It also offered remedial literacy programs to enlisted personnel to assist them in obtaining formal equivalency. Within each brigade or battalion, a system was designed to allow MNLF integrees the option to take extended Ramadan holidays (of up to two weeks) to visit their families; a sacred space or mosque was also dedicated for Friday prayers within the camp, and those stationed in the headquarters were able to take on lighter duties during the fasting season.[13] However, aspects of the AFP still do not accommodate religious customs (e.g., those pertaining to food preparation). Muslim integree subjects took to informally grouping among themselves to obtain and prepare their own food. Combat food packs/rations, however, tend not to be prepared in accord with Muslim dietary restrictions.

OUTCOME

The MNLF and the government/AFP continue to disagree about the integration program. Two MNLF leaders claim that they accepted the terms of integration because they were assured that the MNLF units would form separate units within the AFP and that they would be stationed within the area of autonomy (Salim 2009; Sema 2009). They contend that substantial changes, including the scattering of integrees throughout various units, and deployment outside the ARMM, violated the terms of the peace agreement. They also blame the government for not committing sufficient funds for the Special Regional Security Force. In a report, a former ARMM governor also argued for reassigning the MNLF integrees to the new ARMM territory, or at least to the Special Zone of Peace and Development areas (Hussin 2005).

The accusation that the government breached the agreement is a recurring theme in the MNLF leaders' official pronouncements. A professor at the University of the Philippines also regards the integration process as "half-baked, not fully implemented due to the absence of a separate MNLF unit and . . . used as a way to neutralize those who support Misuari" (Jundam 2009). Another professor argues that the MNLF leaders initially thought, and perhaps still think, of the Special Regional Security Force not as a police force but as a body in which integrees could serve as a homogenous unit within the ARMM army district command (Wadi 2009). This is a common interpretation; in the Final Peace Agreement, the reference to an SRSF is contained within the same proviso as integration into the Philippine military (20a), *not* within the proviso about MNLF elements joining the Philippine National Police (19a).

Confusion over the terms of the integration program is not confined to MNLF leaders; it is also experienced by some of the ex-combatant integrees (despite their having been in the service for more than ten years). It remains a puzzle to them why the integrees do not constitute a separate unit, and they remain hopeful that this arrangement will transpire in the future, reconciling them with former MNLF colleagues (Sali 2009; Pandian 2009). MNLF members (nonintegrees) are similarly convinced that the MNLF was given the bad end of the peace deal because the national government went on to unilaterally interpret the terms of the agreement. Their understanding, again, was that the integrees would make up a separate unit with their own command, and that they would be exclusively deployed in the area of autonomy (Hadji Ebrahim 2009; Ayao 2009).

A senior military officer rebuts the claims of MNLF leaders. In his opinion, the idea of a separate MNLF unit within the army is absurd from the standpoint of national security (Lucero 2008). The government simply would not have consented to such an arrangement because of the danger that it could be used by MNLF leaders to regroup. Additionally, the ARMM command, which covers units assigned to the ARMM area, is under the Southern Command. The Southern Command encompasses the entire Mindanao area, including the eastern regions threatened by communist insurgency. According to the AFP's organizational logic, the unified command lies at this level, not at the level of ARMM.

The AFP also cannot always arrange for integrees to remain in the ARMM area when their mother units are deployed elsewhere. However, individual commanders presented with such a request often allow it. In other words, the military, in practice, is already accommodating MNLF demands for integree deployment within the ARMM, although on a case-by-case basis.

The national government and the military on one hand, and the MNLF on the other, also disagree about the outcome of the integration process. The idea that the national government failed to deliver on its commitments is shared across the MNLF constituency—leaders, rank and file, and even the integrees themselves. This, along with other perceived government shortcomings in fulfilling the terms for Mindanao autonomy, fuels lingering distrust of the national government.

UNPACKING IDENTITY INSIDE THE ARMED FORCES

The MNLF integration was of great symbolic importance because it prompted formal and informal changes in the way the military operated. Although Muslims constitute but a tiny fraction of the total force, their concentration in units posted in Muslim-dominated areas of Mindanao called for a recognition of their unique needs. At the same time a national push was under way to recognize Muslim minorities by declaring Muslim holidays (e.g., Eid'l Fitri and Eid'l Adha) official public holidays. In the army, some guidelines were formulated to allow Muslim personnel to take vacation breaks during Ramadan (paralleling Christmas breaks for Christians). The army has also modified its benefits standards to allow for the possibility of multiple wives as benefit claimants, in deference to sharia law. At the informal level, some commanders have allowed their fasting Muslim personnel to take on lighter duties during the Ramadan season.

How have these attempts at accommodation affected the sense of identity of MNLF integrees? It is worth noting that they come from different ethnolinguistic backgrounds, consist of both ex-combatants and noncombatants, and vary in age at induction from fairly middle-aged (in their thirties and forties) to very young (early twenties). Moreover, though the majority of the integrees are Muslims, a handful of Christians were also included (not surprising, given that the MNLF was a secular movement). In the ten years since the program's completion, 474 integrees (9 percent of the total 5,191) have left the army, including those absent without leave. The relatively small number of exits points to the successful adaptation of integrees into military life.[14]

As to being labeled "integrees," there is a clear divide between those perceived as having made a "blood contribution" to the MNLF and others. Included in the first category are ex-combatants and those kin of former MNLF cadres seen as "deserving" of their positions. Those who purchased integration slots and have no clear MNLF connection are seen by ex-combatant integrees as lacking legitimacy. A decade after integration, the tinge of corruption associated with these cases remains a sore point among integrees and MNLF civilian leaders alike.

Another difference separates ex-combatants, who consider their MNLF credentials a basis for pride, and the generally younger noncombatants, who tend to resent the "integree" label, as they feel it connotes substandard credentials. Members of this younger set tend to question the label by arguing that they are no longer integrees but soldiers just like everyone else receiving the same pay and benefit entitlements—that they ceased to be integrees when they completed the training program. In their view, the persistent use of the word provides a basis for segregation and is suggestive of an inferior status within the organization. Integrees are also conscious of objective markers that separate them from the so-called regulars or organics. The Muslim integrees I interviewed were easily recognized from their uniform patches, which featured both their Muslim surnames and serial numbers of which the first three digits marked them as MNLF integrees. These markers plus their own knowledge of clan and ethnolinguistic origins (e.g., Tausog, Maranao, Maguindanawon) easily enable Muslim integrees to identify each other. Although Tagalog is widely spoken as a lingua franca inside the army, officers can often tell their soldiers' ethnolinguistic origins by their accents.

The waiver of the education requirement for the first batches of integrees is a sore point that continues to shape perceptions about them. Muslim soldiers who were admitted under the army's rigorous standards are quick to point out their difference from integrees. Officers in general acknowledge that integrees in their units do poorly when it comes to preparing papers and reports and have also been less interested than other enlisted personnel in further schooling. General Cayton (2009) admits that the rank-and-file integrees tend to advance in rank more slowly than their regular/organic counterparts; he thinks this is because the educational requirement was waived for them.

This pattern of accommodation, however, has limits. In a mixed-group environment, assertions of Muslim identity become more muted as the integrees come to terms with the army's ethos of work and discipline. Prayer time is considered important, and the physical limitations of fasting are recognized, at least outside combat operations. By and large, commanders accommodate the special needs of their Muslim personnel only at the garrison. Some practices in the army, such as Christian masses preceding official programs, make Muslim soldiers uneasy. Outward discrimination has become less overt, but misunderstanding persists, caused by ignorance and stereotypical assumptions about Muslims rather than by deep-seated prejudice. Some of these misunderstandings involve food (e.g., the requirements for halal food preparation and animal slaughtering), the taboo against shaking hands as a form of greeting, and the prohibition of drinking. Christian soldiers newly assigned to Mindanao units normally are unaware of these and other aspects of Islam, and require an informal cultural briefing to prepare them to interact not only with Muslim colleagues but also with the Muslim host population.

Military Capabilities

To the MNLF integrees who stayed on, the job of soldier trumps identity considerations. Their reliability as soldiers has been tested in the use of troops including integrees in the 2000 military campaign against the MILF (Buliok offensive). Previous assessments have also lauded the effectiveness of integrees in CMOs in Muslim communities (Depayso, cited by Santos 2009; Lidasan 2006). Civil–military operations are noncombat activities designed to "win hearts and minds." Mindanao units with integree elements were used in community dialogues and in the Salaam ("peace") program. Fighting enemies regardless of identity is part of their job (Abasama 2009). Engaging the MILF and Abu Sayyaf presents no moral quandaries for them. In addition, the integrees express pride in being soldiers and in what they have accomplished materially for their families as a result (Gumampangan 2009; Abasama 2009). Their attachment to their job is in no small measure due to the economic benefits they receive.

The integrees' adjustment had more to do with procedures inside the army than with identity concerns (Jacildo 2003). Integrees had to learn to be on call 24/7; seek official approval for and have limited vacation time; have a troop roster, orders from superiors, and battle plans; and focus on rank and troop safety during operations (rather than on charisma or gun ownership) as the basis for the commander–follower relationship. For ex-combatants, being in the army is much more difficult than being an insurgent, because of the demands for formality and documentation and the seeming inflexibility around familial considerations. They often cite long periods of separation from their families as a difficulty.

HAS MILITARY INTEGRATION MADE LARGE-SCALE VIOLENCE LESS LIKELY?

The conflict between the MNLF and the state is but one dimension of a complex landscape in Mindanao populated by rival Bangsamoro armed group MILF, clans, political warlords, and lawless elements. The MNLF integration project, which absorbed several thousands of ex-combatants and their proxies into the central government's standing army, was quantitatively limited. It made little difference to the quality of armed threat in the region. Moreover, with the MNLF neither disarmed nor demobilized, the integration did not completely eliminate MNLF's ability to mount an armed rebellion anew (as seen during the Jolo siege in 2001, where MNLF groups came in support of ousted Governor Misuari). The Philippine Army benefited from the integration bargain. The integration of mostly Muslim MNLF integrees enhanced its capability in nonkinetic activities in Mindanao while boosting its diversity credentials.

The Philippine integration case carried strong symbolic value of the intent of government and MNLF for a political settlement. It was premised neither on the need to enhance the coercive capability of the Philippine Army nor to undercut MNLF's armed muscle. It was primarily driven by economic considerations: to provide employment on the side of the MNLF and as a "cheap" side payment by the central government, which in the end reneged on the nonquantitative commitment to rehabilitation and a much larger Special Regional Force. The process, designed entirely by MNLF leaders and the army, reflected a bargain: The army required integrees to turn over a firearm, MNLF alone provided the list of eligible recruits, and the recruits were required to complete a training regimen similar to regular recruits with a proviso for religious peculiarities. It is a project that both elites found beneficial to their interest; the integration slots can be claimed by MNLF leaders as patronage resources akin to bureaucratic postings at the autonomous government, whereas the central government can claim money well spent for the army. When the integration project is compared with the other "unmet" provisions for autonomy in the Final Peace Agreement, it truly stands out as the only case where some level of success was met and can be equally claimed by both parties.

The outcome, mixed Christian-Muslim army units, provided unique operational settings from which trust was built between regular and integree army personnel. According to both those interviewed for this research and other studies (Depayso, as cited by Santos 2009; Lidasan 2006), integrees are good at their job and do not question their role as state agents. That the integration quota was met with minuscule attrition confirms this trust effect (notwithstanding dissatisfaction over the lack of dedicated special MNLF integree units). Regular army personnel and officers similarly commend integrees in their ranks for their service, especially in CMOs. Although many MNLF ex-combatants admitted to initial skepticism during the first round of recruitment, larger numbers subsequently became more interested in joining the army as positive stories got around. The later recruitment cycles became more fiercely competitive. Its social dividends also could not be underestimated. With younger Muslim men (and women) as proxies to MNLF ex-combatants joining the force, their families gain economic security which in Mindanao is hard to attain.

The visibility of Muslims in army ranks and the institution's solid reforms toward accommodating Muslim sensitivities in their training and operational design echo other government efforts to designate Muslim holidays as national holidays. They elevate the minoritization issue to a policy scale and make the general public in Mindanao and elsewhere aware of the value of political accommodation, even at a symbolic level. On one hand, they are seen as gestures of government sincerity, fundamentally different from previous programs whose purpose was the co-optation of rebel commanders and their men. The integration project, although modest and with inherent limitations, bolstered the central government's confidence that it can do business with MNLF (outside of running the autonomous government).

The project's "demonstration effect" on the rival group MILF is even more vital. The challenges and pitfalls of the MNLF integration process notwithstanding, political negotiations on "normalization" between the MILF and the central government are informed by lessons from previous experience. More cooperation and discussion between the MILF and the government was elicited by this policy preview.

CONCLUSION

The integration of MNLF elements into the Philippine army was part of but not central to a negotiated peace settlement that granted enhanced autonomy and local authority to the insurgent group. Although the idea of military integration had been articulated in negotiations since the 1970s, the government and the rebel group understood its merits differently. The MNLF saw integration, in conjunction with the desired separate units for integrees and exclusive Mindanao deployment, as a means of exerting continuing influence over its members (who are thankful to obtain government employment) and of gaining a foothold in the military organization. Its insistence on the separate SRSF under the control of the autonomous government follows the same logic of maintaining access to armed institutions.

For the government, integration provided additional manpower, enhanced the army's counterinsurgency capability, and created diversity in the ranks. It was fully aware of the symbolic importance of military integration to the task of nation building. The negotiations that yielded the integration policy also revealed the uneven political strengths of the national government and the MNLF. At that time, the MNLF was a numerical shadow of its former self, because it had lost a substantial number from defections to the MILF and criminally inclined armed groups operating in the same area. The looming security threat posed by the MILF informed each party's negotiation stance. Integration had an effect on the MILF; it shows that the government is serious about making peace with rebel groups and that the MNLF, by accepting peace, has earned legitimacy.

In the end the integration program favored the national government. Only a modest number of people (equal to roughly 4 percent of the total armed forces) were absorbed. The integrees were treated as an addition to the force, with recognizable CMO potential because they could liaise between the army and the Muslim host community. No adjustment was made to the military's focus on internal security/counterinsurgency or deployment patterns. The only concessions given to the MNLF were the waiver of the admission criteria and the independent selection of integree recruits. Integration proceeded on an individual basis rather than by units, allowing the military to break previous connections between MNLF commanders and their men. The training and deployment of the integrees were determined by the military, with the deputy commander of the Southern Command (the highest-ranking integrated MNLF officer) having no say in these matters.

In terms of contributing to lasting peace in Mindanao, the integration program had a very limited impact. Its potential for reducing the numerical strength of the insurgent movement was diluted by the hybrid composition of the integrees, who were both ex-combatants and proxies. Further, as integration was not accompanied by the MNLF's demobilization and disarmament, it has not made a dent in the pressing security concerns in the area. In fact, it has been suggested that the "no arms, no integration" policy increased the demand for low-caliber weapons in Mindanao, fueling further insecurity. At best, the integration project provided employment opportunities to ex-combatants and their male family members who would otherwise have been recruited by the MILF and other rival nonstatutory armed groups. A parallel, foreign-donor-funded reintegration program was conducted for the majority excluded from army integration, and this possibly served as a better livelihood vehicle; it certainly reached many more people.

Because military integration was secondary to more important elements of the peace agreement (e.g., ARMM autonomy, control over resource usage, education, and sharia law), the success in meeting program targets (the process being completed with only a slight delay in schedule and with an attrition rate of only 9 percent) did little to bring the Mindanao conflict to an end. Today the conflict persists with both new actors (like the terrorist group Abu Sayyaf) and old ones (clan wars and rivalries among political warlords with their own private militias). The MNLF integration program was mainly window dressing for largely failed efforts at much-needed structural reforms to address poverty and marginalization. The complexities and nuances of armed rebellion call for more serious political solutions, which unfortunately could not be realized given the MNLF's weakened position and the national government's failures to live up to its fiscal commitments. Integration has brought few peace dividends given its limited coverage and the focus in its design on meeting military goals.

The importance of the military integration lies in the symbolic realm. It introduced diversity into the army and prompted more culturally sensitive training practices and internal regulations. It has rendered the army more humane and more welcoming, particularly to Muslim communities in Mindanao. The success of the integration program was a worthy consolation in the face of the MNLF's failure of governance. Finally, integration was instructive for the MILF, which opted not to include the topic of military integration in recent negotiations with the national government; it chose instead to focus on more important matters, such as control over ancestral domains.

ABBREVIATIONS

AFP	Armed Forces of the Philippines
ARMM	Autonomous Region for Muslim Mindanao—a political unit created by a peace agreement
CMO	civil–military operations
MILF	Moro Islamic Liberation Front—rebel group that remains in opposition
MNLF	Moro National Liberation Front—rebels being integrated into the government army
OJT	on-the-job training
SPCPD	Southern Philippines Council for Peace and Development
SRSF	Special Regional Security Force

NOTES

1. The MILF, led by Hashim Salamat, separated from the MNLF in 1978. The MNLF also lost leaders who were co-opted by the government, such as the MNLF Reformists, the Sulu Magic Eight, and the forces under Colonel Ronnie Malaguiok. The MNLF's control over various commands was tenuous to begin with, and its boundaries with other nonstatutory armed groups blurry. The emergence in recent years of the Islamic radical group Abu Sayyaf and kidnap-for-ransom outfits can be traced to outfits or members formerly connected with the MNLF cause (San Juan 2007, 102–4).

2. The original ARMM, under Republic Act 6734 of 1986, covers four provinces: Sulu, Tawi-Tawi, Lanao del Sur, and Maguindanao. This act was superseded by Republic Act 9054 of 2000, which expanded the area of autonomy to Basilan province and Marawi City, as established by a referendum. As stipulated in the Final Peace Agreement, an interim structure called the Southern Philippines Council for Peace and Development (SPCPD) was created from 1996 to 1999 to formulate a development agenda for the Special Zone of Peace and Development. The SPCPD did not supersede the ARMM but rather functioned as a parallel body through which intensive aid for peace and development was channeled. The SPCPD could be seen as a test run for the MNLF in terms of governance. MNLF leader Nur Misuari was SPCPD chair from 1996 to 1999. He ran for ARMM governor in 1999 and served in this position until his arrest in 2001.

3. Reintegration is often presented as an option alongside military integration. It refers to social, political, and economic assistance given to ex-combatants transitioning to civilian life, and assistance given to the communities receiving them. Reintegration programs include skills training for livelihood, temporary financial assistance to cover immediate material needs, educational support, and job and medical referrals. Reintegration often comes bundled with disarmament and demobilization: the process of disarmament, demobilization, and reintegration (DDR) is carried out sequentially (see Rufer 2005; Knight and Ozerdem 2004).

4. Good examples are Sierra Leone, and South Africa (Peters 2007; Humphreys and Weinstein 2007; Williams 2005).

5. The notion that Muslim identity acts as a unifier is contested. Muslim adherents among the thirteen ethnolinguistic tribes in Mindanao, particularly those in the lower eco-

nomic strata, do not readily identify themselves as Muslims or Moros (see Blanchetti-Revelli 2003; Che Man 1990; McKenna 1998; Horvatich 2003). Muslim identity articulation is more prominent among elites whose upward mobility is restrained by minoritization arising from decades of national government neglect, assimilation attempts, and state-sponsored Christian migration into their ancestral homelands (see Tan 1993; Abinales 2000). To elevate Muslimness as a vector for national (Philippine) identity-building neglects such nuances.

6. In an interview, Yusuf Jikiri (2009) remarked that the MNLF's original intent was for integration to proceed using the criteria of proportional representation and with the MNLF as a separate unit (with a more complete merger to happen after trust was gained). He envisioned a time frame longer than the three years eventually adopted.

7. According to former undersecretary Gacis (2010), the MNLF wanted 15,000 absorbed. Agreeing to a number was the most difficult part of the negotiations.

8. The Final Peace Agreement states that the government will exert utmost effort in ensuring the integration of the remaining MNLF members into the SRSF. In paragraph 20a of the same document, a reference is made to a special socioeconomic program for MNLF forces not absorbed into the army, police, or SRSF, giving the impression that the SRSF is a separate entity from the military and police. Under the 2000 Organic Act for the Autonomous Region of Muslim Mindanao, the Philippine National Police Regional Command for the Autonomous Region was renamed SRSF. This police unit already had MNLF integree elements. This was a major disappointment for the MNLF, which expected a separate SRSF, with a potentially large number of ex-MNLF recruits, that an MNLF-controlled ARMM government could use. What emerged instead was an SRSF that is the same central police force as before, controlled by the national government.

9. It was also agreed that the AFP would take in 5,350 candidate soldiers and 160 officer candidates for training. The numbers were deliberately padded, as not all were expected to complete the course.

10. By contrast, there was no firearm requirement for police integration.

11. The women completed their training in January 15, 2009, and the men finished in July of the same year. Unlike the regular batches, the replacement batch are considered "enlisted" men and women. They are treated as regular/organics, and not integrees.

12. The experience of Muslims who were regular recruits stood in sharp contrast. Because there were few of them, they encountered many difficulties in trying to practice their religion. At this early date the army was not as sensitive to this issue in its training as it later became.

13. Those fasting are exempted from patrol duties, unless their unit has already been deployed for combat operations.

14. Many reasons can be found for the substantial number of cases of absence without leave. Some integrees dropped out early because their religious needs (prayer time, halal food, etc.) were not accommodated by their respective unit commanders; some were unable to cope with the rigid time demands of the army (which disallowed extended family visits); some had bad relations with fellow soldiers who regarded them as traitors; others were reneging on loans. The latter is a serious concern. Many recruits were at first drawn to the army because they could take out government loans of up to $2,000. Some skipped out of the service to avoid loan repayment.

Chapter 8

SOUTH AFRICA

Roy Licklider

> The term Security Sector Reform (SSR) had not yet been invented when South Africa embarked on it.
> —Gavin Cawthra, *Post-War Security Transitions*

South Africa has a fair claim to be considered the poster child for negotiated settlements, although its relatively low level of wartime violence makes calling it a civil war a stretch. Analysts spent several decades seized with the fear of a race war in a country with nuclear weapons; the eventual settlement caught almost all outsiders by surprise, and the country's ability to shift from white to black political dominance in a peaceful, democratic manner was, in retrospect, nothing less than astounding. Moreover, the settlement was largely a local product; certainly international pressure and support helped, but this was not a coerced settlement like those in, say, Zimbabwe, Bosnia, or Kosovo. For all its problems, South Africa is evidence—at least for the first twenty years—that negotiated settlements can work in very hard cases.

The period from 1991 to 1994 was perhaps the most violent in the history of the country; as Zartman (1995a, 167) noted, "a successful conclusion does not imply a friendly process." Nonetheless, the principals were able to agree to a Government of National Unity, an interim Constitution was approved by December 1993, and Nelson Mandela won South Africa's first democratic presidential election on April 27, 1994. The transition was solidified by security-sector reforms, including the merging of the apartheid-era national defense force with the forces of the "independent homelands" and the liberation movements.

ORIGINS

Military integration in South Africa involved eight different forces. They had different military cultures and did not even share a common language. Policy was set by senior personnel of the South African Defence Force (SADF) and the Umkhonto We Sizwe (Spear of the Nation, or MK) in extensive negotiations.

The Historic Role of the Military

Interestingly, the apartheid-era SADF was itself the product of military integration. By the end of the nineteenth century, there were two different European populations in South Africa: those of Dutch descent who spoke Afrikaans, and English colonists who spoke English. They fought in the South African War (1899–1902), and the British side finally won a very costly victory. In 1910, the Union of South Africa united the former Boer republics with the British colonies. As Rocky Williams (2006, 37–50) points out, the SADF was an amalgam of English and Boer units that had fought each other in the war, and it was intended to be apolitical. Jan Smuts, then minister of defense (and a former Boer general and future South African prime minister), said: "We want an organization that shall not be Boer or English, but a South African army. . . . Do your duty in a broad national spirit" (Williams 2006, 39).

The first senior staff officers' course at the new military college in 1912 included twenty-five British and twenty-five Boer officers who had fought against each other in the war. Interestingly, the new army had problems similar to those that came up during the merger of the SADF with other military groups, and it almost failed because of "rival political tensions" (Williams 2006, 41–42; cf. Seegers 1996, 10–25).

After 1948, most of the English senior personnel were purged (Williams 1994). In 1957, the force was renamed the South African Defence Force. The 1974 coup in Portugal resulted in independence for Angola and Mozambique, threatening the South African apartheid regime directly. South Africa declared a "Total Strategy," which linked economic reforms with a military buildup and the mobilization of the white community. The SADF launched destabilization activities, including military raids and invasions, on Namibia, Angola, Mozambique, Zimbabwe, Botswana, and Lesotho, with mixed success. During the final stages of apartheid, the SADF was employed in putting down disturbances in South Africa itself (Cawthra 1997, 27–53).

Historically, nonwhites had served in the South African armed forces; one authority traces their involvement as far back as the seventeenth century (Seegers 1996, 1–7). After the South African War they were kept in noncombat positions and almost always led by white officers. They were given limited combat roles when manpower became a serious issue; for example, both sides used them during the South African War, some units saw combat in World War I, and in the 1970s an incremental process led to the formation of ethnic combat units and even some limited racial integration within units (Grundy 1983; Peled 1998, 39–92). "In 1986 the SADF . . . was 76% white, 12% black, 11% coloured and 1% Indian. . . . In 1990 there were some ten black officers" (Mills and Wood 1993).

The rebel military, the MK, was formed in the 1960s as a result of the failure of peaceful protest. In general it showed considerable moral restraint in minimizing violence to civilians and remained under the control of the leaders of the African National Congress (ANC). Starting in the 1970s, the ANC's struggle for liberation can be described as having four pillars: mass mobilization and action, the political underground, the armed struggle, and the international campaign to isolate apartheid South Africa. The decision to resort to violence as part of the struggle against apartheid was guided by the notion that MK was not to operate outside the political sphere and separate from the liberation movement, but that, however unlikely it may have seemed at times, the principals should always strive to keep the avenues open for a negotiated settlement. This interpretation of the armed struggle was crucial in allowing the ANC and its allies to adapt and enter into negotiations with their political opponent, despite having taken up arms (Maharaj 2008, 13f). The MK was a guerrilla force; it did not confront the SADF in set-piece battles, so neither side could claim military victory, and it seems not to have been especially important for the ANC's political leadership (Mashike 2007, 604; useful histories of the SADF and MK, respectively, are given by Sass 1996 and Motumi 1996; cf. the magisterial Seegers 1996).

When President P. W. Botha was replaced by F. W. de Klerk in 1989, the white-minority government acknowledged the need for negotiations. Four factors played a crucial role in the "change of heart" of the old regime: The regime could no longer ensure the security and prosperity of the privileged minority; it could no longer effectively suppress the dispossessed majority; it faced international isolation in the form of sanctions from its former allies; and addressing these three factors was simply becoming unaffordable as the regime struggled to maintain the system. For example, by 1989, the government was spending 17.7 percent of its budget on the military, fueling inflation (Zartman 1995a, 148; Jervis 2005, 40).

The Emergence of Military Integration as an Issue

Negotiations between members of the apartheid regime and the liberation movement began as early as 1985 (Seegers 2012) and were formalized in December 1991 following the inauguration of the Convention for a Democratic South Africa (CODESA), which came to be dominated by two of the nineteen parties attending, the ANC and the government's National Party (Maharaj 2008, 26). It is common in civil war negotiations for security, during and after the transition, to be a major concern (Walter 2002, 104–5). Interestingly enough, the future South African military was not a major issue in the early negotiations, apparently because all sides wanted to keep their forces available if negotiations failed (Cawthra 2003, 36). Although a cease-fire was declared in 1990, the National Peace Accords of 1991 included some provisions about control of the police but nothing about the military.

However, in 1991 military leaders of the SADF and MK began a series of meetings and found common interests; each wanted to encourage control of the other side's radical allies, to exclude the other armed forces from negotiations, and to by-pass the designated but distrusted political mediator on its own side (Frankel 2000, 1–6; Shaw 1996, 18–19). They also shared an interest in amnesty (Seegers 2012). These meetings led to the formation of the Joint Military Coordinating Committee (JMCC) in 1993. Technically, this group had no direct power; it worked under the Sub Council on Defence of the Transitional Executive Council. In practice the committee, dominated by the SADF and MK, devised and implemented policies with remarkably little civilian involvement (Shaw 1996, 20–23). The ANC, in preparation for the discussions on integrating the formerly competing militaries, established a think tank on defense policy, the Military Research Group, which enabled it to become a major participant, but the party does not seem to have developed any firm ideas about what the new military would look like other than that it would be integrated (Mashike 2007, 606). The participants in the discussions on integration do not seem to have followed clear precedents (Esterhuyse 2012), although Zimbabwe and Namibia (Cawthra 2012a), and Germany after unification (Cilliers 2012), were mentioned in interviews.

Supporters, Opponents, and Neutrals

The new government, in accordance with Sections 224 and 236 of the Interim Constitution of 1993, faced the Herculean task of integrating all members of the eight separate military forces in South Africa at the time: the government's South African Defence Force; the MK; the forces of the four "homelands," Transkei, Bophutatswana, Venda, and Ciskei; the Azanian People's Liberation Army (APLA), the military arm of the Pan-Africanist Congress (PAC); and the Kwa Zulu Self Protection Force (KZSPF) of the Inkatha Freedom Party.

The easiest cases were the homeland forces that had been established as copies of the SADF; they were "translated" rather than "integrated." (For more extended analyses of these forces, see Mills and Wood 1993; and Reichardt and Cilliers 1996.) Nonetheless, even here there were difficulties. These fighters had been better paid than those in the SADF; easy promotion policies had produced top-heavy force structures; the forces' local orientation made fighters uneasy about being deployed nationally; and their personnel qualifications were often unclear. However, their small size, their similarity to the SADF, and considerable training made these problems surmountable. Moreover, by accepting their credentials, the SADF was able to create black senior officers who might be expected to be sympathetic to its interests in the new force (Frankel 2000, 49–56).

Integrating MK and the Pan-African group (known collectively as nonstatutory forces, or NSF) posed a more difficult set of challenges. They had been organized for guerilla warfare—to fight in small groups with minimal hierarchy—unlike large-scale, high-technology, hierarchical, and modern armed forces—so their skills and their attitudes toward command and control were very different. Many had given up chances for education to fight and now found themselves penalized for the choice. The SADF saw the problem largely in terms of training, but in fact the divisions ran much deeper than that. The NSF troops saw themselves as having won the war against the SADF and the homeland forces, so it was not obvious to them why they should adapt to the SADF model. SADF personnel, conversely, felt that they had never been defeated and resented the insertion of former enemies whom they regarded as unprepared. Language also proved to be a major barrier; the language of the SANDF was Afrikaans, which many of the NSF personnel did not speak and resisted learning because they thought of it as the language of oppression (Esterhuyse 2012).

The JMCC allowed officers from the apartheid-era national defense force and from the security forces of the independent homelands and the liberation movements to work together to create the national South African National Defence Force (SANDF). Because of their institutional capacity and experience in running complex planning processes, the SADF and MK officers dominated the process. The JMCC established a rotational chair position shared by the chief of the SADF, General George Meiring, and the MK chief of staff, Siphiwe Nyanda. Initially, the PAC's APLA was not involved in the formal negotiation process; it had to be dealt with as an add-on after the democratic government had been formed.

On the SADF side, three loose factions emerged in the debate over integration: a small reform-oriented group, centered mostly on the air force; a larger managerial-technocratic group within the South African army; and a so-called warrior element, which was determined to hold on to apartheid both for ideological reasons and because of bureaucratic self-interest. Members of this last faction were found across the board, but it had its roots in military intelligence (Frankel 2000).

By and large, the SADF wanted to simply absorb elements of MK and APLA into the existing force, arguing that the SADF was a professional conventional military and that the guerrilla armies were not. Negotiations were intense: "Since one side had not defeated the other, the self-satisfaction of the SADF, rooted in its superior numbers and technology, was more than matched by the arrogance of MK in the atmosphere of political victory" (Frankel 2000, 8).

Compromises in the Process

The SADF eventually agreed to establish a new military, to be composed of all veterans from all eight separate military forces at all levels, and to accept civilian control from the new government, with British military personnel as arbiters of the process. In return, the SADF's existing doctrine, personnel procedures, training structures, and equipment were accepted as the basis for the SANDF. A general amnesty was declared for all military forces for past acts, including human rights violations.

This simple summary, however, understates the intensity of the negotiations (more complete discussions of this process are Frankel 2000 and Shaw 1996). The SADF wanted to start the integration process before the national elections; the MK refused to do so. The two sides were unable to agree on a paramilitary force to police the elections. They differed on the question of when and how MK forces would be assembled and who would pay them and their dependents, not to mention the definition of a member of an armed force. The SADF tried to use its technical superiority in areas like logistics to its advantage, but MK succeeded in focusing the issues on political questions.

Over time, agreement began to emerge. The new military would be modern, which in practice meant that it would adopt the SADF model in many ways. It would be apolitical and subject to parliamentary control (a major change for SADF, but one that its leaders supported) and include a relatively small professional core, with a larger reserve. Some MK leaders would be given high-level positions, and its rank and file would be given training and fair opportunities for promotion. A Service Corps would be formed to ease the transition of demobilized soldiers, particularly those from MK. There would be a general amnesty. The integration process would start after the national elections of 1994.

The initial results of the negotiations suggest that the SADF had definitely done better than its opponent, but this impression is deceptive because the inevitable political victory of the African National Congress meant that many of the subsidiary agreements were simply overridden later (Frankel 2000, 20–42; Shaw 1996, 25–28). The SADF was compelled to accept the full integration of forces and such programs as affirmative action and the fast-tracking of members of the NSF. The NSF were compelled to accept a new SANDF initially led and very much controlled by members of the old SADF. The arbitration process needed a neutral umpire; and for this role the British Military Advisory and Training Team (BMATT) was acceptable to the SADF because it was a Western power and to the NSF forces because of its earlier work in Zimbabwe and Namibia (Seegers 1995, 47–48; Cawthra 1997, 204n6; Higgs 2000, 49; Hughes 2007). One of its tasks was to mediate disputes about rank and position. In the end, these compromises allowed a reasonably smooth integration process to take place.

CREATION

There was intensive debate about the mission of the new military. British advisers played a significant role as neutrals in personnel selection; it was particularly difficult to determine appropriate ranks for former rebels because their guerilla forces were quite unlike the structured organization of the SADF. Several MK personnel were given very senior positions and then sent to training courses; otherwise, the training was fairly standard.

Shaping the New Military

There was much debate about what the new military would look like, and in the beginning each side advocated a new SANDF based on its own model. The SADF wanted a new SANDF that would look a lot like the old SADF in terms of conventional and counterinsurgency forces. MK (as the strongest of the NSF) preferred a new SANDF that was based on the concepts of revolutionary warfare. In the early negotiations at CODESA, it was agreed that the new SANDF would be a "balanced, modern, affordable and technologically advanced military force, capable of executing its tasks effectively and efficiently," but this was not spelled out in a force design.

After a civil war, it is often unclear why a substantial military is really needed. There is also great pressure to reduce military spending to get a "peace dividend" to allow reconstruction and social and economic development. Both these factors suggest that the size of the military should be reduced, precisely as new people are being brought into the system. The result is likely to be personal and institutional uncertainty. This clearly happened in South Africa (Frankel 2000, 138–39). Drastic budget cuts (estimates placed them at 44 to 65 percent) were accompanied by disagreement about the military's future role.

Because the military remained dominated by SADF officers, it badly needed to establish its legitimacy. Therefore, while the details of military integration were essentially worked out in quiet negotiations, the military accepted an extensive and very public Defence Review process to discuss the goals and approaches of this new force. Civil society organizations and the general public participated in the debate in an unprecedented manner, along with the military, the secretariat of the Ministry of Defence, and the new Joint Parliamentary Standing Committee on Defence: "The breadth and depth of consultation exceeded any similar policy process in South African defence planning history and remained possibly the most consultative process on defence policy ever attempted by a modern democracy" (Williams 2003, 208; cf. Cawthra 1999).

The resulting Defence White Paper reflected the goal of preserving territorial integrity and sovereignty but gave the new force added responsibilities, in particular that of international peacekeeping (Williams 2003; Frankel 2000, 111–17; Winkates 2000, 459; Porter 2010, 18–31). However, it did not include a plan for carrying out these very diverse assignments, leaving problems that remain today. Not surprisingly, civil society's involvement in military policy lessened after this massive effort (Africa 2008; Schoeman 2007, 162–66).

Personnel Selection

Integration began with certification by the existing military organizations of individuals eligible for integration. These were to go to cantonments, where they would be assigned new positions by placement boards made up of representatives from the forces themselves, the new SANDF, and the BMATT. Non-SADF members would be given special bridge training as needed to bring them up to "international standards" and allow for future promotions; those who were rejected or chose not to join would be compensated.

In fact, however, the process was much less orderly than this description suggests. Once the process got started, everything was done at top speed; the SADF wanted to disarm the rebels, and MK wanted to get its people into the new SANDF because it had no money to support them. People at the working levels often did not understand the policies devised by their superiors and, as might have been expected, the hostilities among the different groups regularly produced explosive situations (Frankel 2000, 61–84, at 61–62): "Neither side at the grass roots, British military observers noted, fully understood the detailed mechanics of what had been decided in the JMCC. Neither had time to transmit what little was understood down through the military hierarchy to on-site staff. . . . Hence, for much of its early history integration was almost entirely haphazard, largely experimental, and a learning process for all participants."

The process was intended to incorporate all personnel of both the state and nonstate forces identified in the Constitution (with the belated inclusion of APLA and the KZSPF). Because the personnel records of the members of MK and APLA were incomplete, these groups were asked to create certified personnel registers listing the names of all their members. They faced a number of challenges: members used pseudonyms, combatants were sometimes reluctant to submit their names given the continued violence during the early 1990s, criteria for inclusion had to be determined, names were erroneously omitted during documentation on computers at integration centers, and many fighters had returned home when the fighting ended. Between 1994 and November 2002, a total of 42,020 names of former liberation movement forces were submitted for integration into the SANDF. Given the challenges mentioned above, these lists were incomplete; but in fact only about half those listed were accepted, which caused hard feelings (Mashike 2007, 607; cf. Gear 2002; Seegers 2012).

The placement process was fraught with conflict, but eventually it went fairly well, especially with officers. Cash was given to individuals from all groups who were not willing to accept the new regime or who were not accepted into SANDF, although the former rebels got much less than SADF veterans. Despite promises, relatively little was done to facilitate the return of demobilized veterans to society (Motumi and Mckenzie 1998, 194–203; Mashike 2008; Gear 2002).

Although there was no large-scale violence, tensions certainly remained during the process. Disputes centered on issues such as placement, salaries, and training, especially poor food and facilities in the training areas. The process started in 1994. There were disturbances, protests, and mutinies in several places; some people walked out; at least two white officers were apparently shot by former MK members; and a trade union movement within the enlisted ranks was attributed in part to unhappiness over the integration process. However, matters improved over time. Several hundred training instructors resigned from the army, and things seem to have fallen into place by 1995. All things considered, it seems to have gone fairly smoothly. (See Shaw 1994; Motumi 1996, 101–2; Motumi and Mckenzie 1998, 188–94; Mashike 2007, 612–13; Frankel 2000, 70–73; Cawthra 1997, 151; Cawthra 2003, 31–43; Williams 2002, 2005).

The process was facilitated by a 1991 decision by MK to retrain its members in conventional military skills and hierarchies, as opposed to guerrilla warfare, in preparation for the forthcoming integration. Training took place both within the country and outside; thousands of MK personnel were sent abroad for conventional command and staff training, mostly in the Communist bloc, to prepare them for the coming integration process. An informal civilian advisory group set up by the MK, the Military Research Group, suggested ideas that often entered into negotiations (Shaw 1994, 232–33; Shaw 1996, 15–17; Burgess 2008, 77).

The four homeland armies were all small and composed of SADF ethnic units, usually led by white South African officers. These groups played no significant role in the negotiations and were fairly easy to integrate into the new military. The Pan-Africanist Congress stayed out of the negotiations until the end but finally agreed to be integrated; the KZSPF Party militias were not brought into the process until 1996, and then only as new recruits (Motumi and Mckenzie 1998, 189). Interestingly enough, the PAC cadres, although fewer in numbers and with less combat experience, fared somewhat better in the integration process proportionately than those from MK (Frankel 2000, 75–76; Esterhuyse 2012; a history of this group is given by Lodge 1996).

Former MK and APLA combatants were subjected to an assessment process that focused on educational qualifications, experience, length of service, leadership qualities, evaluation reports from either APLA or MK, the results of preselection tests, age, seniority in their force, and military qualifications. There seems to have been no explicit effort to weed out human rights violators, perhaps because of the amnesty provisions of the peace agreement. Former members of the SADF and TVBC forces were subjected only to a verification process to confirm that they had earned their ranks through completing the necessary courses (Mashike 2008, 443–49).

Rank Allocation

Representation was a key concern; it was felt necessary to ensure equity in terms of race and gender representation within the new force. Given the imbalance in numbers—in 1994 nearly 70 percent of the total constituent force had been in the SADF—there was a need for rationalization strategies to meet this objective, such as the fast-tracking process for former liberation movement forces. Former SADF members were offered voluntary severance packages (LeRoux 2005).

Non-SADF officers were ranked by their organizations in order to go before placement boards that determined their new ranks. The criteria were command experience, operational experience, seniority, education, military training and qualifications, and length of service. Of the approximately 16,000 individuals from MK, the Pan-Africanist Congress, and KZSPF, about 1,770 (10 percent) became officers in the new force; another 500 officers transferred from the homeland armies. The former MK chief of staff became SANDF chief of staff; eight non-SADF officers were made generals and given general staff positions. Initially, their influence was limited because they were often sent away on training courses on the grounds that they were not sufficiently experienced, because they were not connected to informal networks within the SADF, and because Afrikaans remained the language of the South African military. The integration process was formally declared completed in 2003 (Williams 2002; Burgess 2008, 78-79).

Training

Clearly, training was central to this approach to integration. The SADF approach to training was largely technocratic. Of those individuals undergoing bridging training, those identified as having the potential to assume more senior, leadership positions were fast-tracked; they could complete the required military courses in only a few years, rather than fifteen or more. NSF veterans were simply put into existing courses; little attempt was made to provide them with information that SADF veterans would have as a matter of course (Esterhuyse 2012). The training was done by SADF personnel, with some supervision by the BMATT. Although the course content varied according to the particular requirements of these services, bridging training was generally designed as a two-stage process: a twenty- to twenty-five-week course of basic orientation, followed by formation training to "bring ex-NSF personnel up to the required standard for his/her specialization/arm of service" (South Africa 1995). Little attention was given to the more difficult task of aligning the values and attitudes of the state and nonstate elements that made up the new, integrated SANDF. This caused a number of problems.

One obvious issue affecting the entire integration process, including training, was racism, exemplified by a former APLA combatant's murder of several of his white colleagues in 1999 (Mashike 2008, 446). The problem was compounded by the perception that what was occurring was less a process of integrating all fighters into a new defense force and more one of integrating NSF into the SADF under a new name (Williams 2002; LeRoux 2005; Mashike 2007, 613).

The differences among the various armies that were to be integrated were also a problem—the conventional SADF was radically different from the guerrilla armies of MK and APLA. MK soldiers, for example, usually operated in small units with considerable tactical autonomy. Primarily concerned with political matters rather than conventional military ones, they could not easily make the transition to the professional, large-scale, high-technology environment of the SADF, whose officers tended to be focused on operational issues and somewhat anti-intellectual (Frankel 2000, 56; Esterhuyse 2012; cf. Perlmutter and Bennett 1980).

In the case of the former TVBC forces, personnel raised concerns about their status under the new political dispensation, morale was low, base facilities were lacking, and many were reluctant to submit to the verification process. Some complained that they felt marginalized by the integration process (Frankel 2000, 51). By and large, it was characterized by feelings of mutual suspicion.

OUTCOME

Perhaps the most startling change in the SANDF since the merger has been a demographic one. By November 2002, at the formal end of the integration process, 114,956 members from the various forces involved had been integrated into the new SANDF: 72 percent from the SADF, 10 percent from TVBC, 13 percent from MK, and 6 percent from the APLA (Mashike 2007, 606). Two thousand former KZSPF were added later (Cawthra 1997, 149). Between 1994 (just after the beginning of the integration process) and 2007, the total number of personnel was reduced to about 70,000. At the same time the proportion of Africans in the SANDF went from about 40 percent to almost 70 percent, while the white proportion dropped from 47 percent to 18 percent. However, these figures conceal important differences. Blacks dominate both the enlisted personnel (of whom only about 2 percent are white) and the highest ranks (brigadier general and up), where a majority are MK veterans; whites still occupy more than half the officer and noncommissioned officer positions, the so-called operational positions. That most lower-level officers and noncommissioned officers are white in part reflects major educational differences resulting from the apartheid educational system (Heinecken 2009, 29; Heinecken and Van Der Waag-Cowling 2011, 177–80).

Political Control

Traditionally, the SADF had been under tight civilian control; the discussion of a military coup to preserve apartheid and the militia group Afrikaner Volksfront seems to have been limited to members of the reserves (Cilliers 2012):

> Foreign embassies and their intelligence officers in South Africa would scurry around, trying to find the first signs of the inevitable military threat to the settlement process or worse, evidence of a military coup d'état that surely had to follow. Gradually, as the SADF became more transparent, observers grudgingly came to accept the obvious and only explanation for the support that it provided to the settlement process—that over decades the Afrikaner had developed an indigenous professional military culture within the SADF that could withstand the temptation to restore the status quo by force of arms. For years the SADF had preached that a political instead of a military solution was needed. Now its acceptance of change proved its belief in and support for political leadership. . . . It must . . . rank as the supreme irony that the SADF, the symbol of racial oppression and regional destabilisation, cemented the transition from white domination when, during the national election of April 1994, it stepped in to assist the independent Electoral Commission in its administration and ensured the success of the election which would inevitably bring the ANC to power. (Sass 1996, 119; reinforced by Esterhuyse 2012)

The MK had also been subject to political control, following the lead of the Eastern Europeans who had helped train it. As apartheid came under increasing pressure, the SADF had moved away from civilian control, particularly over the budget. Ironically, this may have facilitated integration, since the SADF was able to initiate and control the process after apartheid ended, excluding civilians on both sides.

The new Constitution clearly established civilian control, with the military reporting to the executive and the Ministry of Defense being responsible to Parliament, but the culture was much slower to change: Many former SADF leaders quietly resisted, despite often giving lip service to the concept. It was also difficult to create a new bureaucracy big enough and competent enough to control a large military, particularly under budgetary limitations. The first minister of defense, Joe Modise, was initially not perceived to be a strong supporter of the ministry when it came into conflict with the military. For similar reasons Parliament was not able to exert much control for some time, although the rise of organizations such as the Institute for Strategic Studies has added some expertise to the civilian side (Frankel 2000, 117–25), and the dismissal of the chief of the National Defence Force in 1998 solidified civilian control. The roles of women and trade unions in the armed forces were other hot issues in civil–military relations (Frankel 2000, 128–36), and civilian superiors did exert control in these areas.

Legally, the SANDF is under civilian control—subject to parliamentary oversight and guided by Parliament-approved policy and legislation. All decisions on budgets, acquisition, and deployments were vested in civil authority. However, in fact this control is fairly loose—not because the military is breaking the rules, but because most civilians, including most politicians, do not know or care enough for effective control to be exercised, and because turnover in Parliament and civilian leadership is so rapid (Jordaan 2004, 17–20; Esterhuyse 2012). Indeed, there is some feeling that the controls actually interfere with effective policymaking (Cilliers 2012). There is also some concern that the current military is becoming increasingly politicized, because it is closely linked to the ANC (Esterhuyse 2012).

Military Capabilities

Despite many negative perceptions and many real shortcomings, the SANDF is a functioning, integrated military. Only isolated incidents of internal conflicts have occurred within the SANDF. It has deployed internally in support of the police and civil authority. The SANDF can defend the territorial integrity of South Africa against any realistic threat. It can provide internal security and has done so when it has been called in to provide support to the police.

It has done reasonably well in peace support operations within Africa. South Africa has participated in no fewer than fourteen such endeavors under the United Nations and the African Union since 1999 (Heinecken and Ferreira 2012a, 20; cf. Heinecken and Ferreira 2012b, 2012c). Although its weak performance on its first attempt in Lesotho in 1998 "certainly did not mirror the robust SADF performance of earlier years" (Winkates 2000, 459; Frankel 2000, 150–89; Cawthra 2012a), the SANDF has done better in the Democratic Republic of the Congo (despite some problems) and especially Burundi and Darfur (Cawthra 2012a); it was even able to "deploy air assets" in the first two cases (Dempsey 2009, 135). It is generally seen as better than other African armies, though that is a fairly low bar (Cilliers 2012); many of the personnel in such operations are in technical rather than combat roles (Seegers 2012).

Sustainability

Affordability is a question. During apartheid as much as 20 percent of total government spending went to the military, not counting the secret and presumably very large Special Defence Account. In the new regime the SANDF was seen as an apartheid institution with no obvious mission and no domestic constituency, and thus was obviously in a weak position to compete for funds. Over the next few years military spending was drastically cut by between 44 and 65 percent (estimates vary because of continued secrecy), at the same time that integration greatly increased the number of soldiers. Procurement budgets were reduced disproportionately, raising the specter of a hollow military, personnel without equipment; the navy and air force were hit particularly hard.

This problem has been accentuated by questions about the appropriate role of the military. The decision to make the SANDF's major mission that of defending the South African homeland against foreign military attacks made sense in negotiations to end the conflict; but like many such decisions, it has caused other problems over time. This mission required a modern, mechanical military force, which privileged the skills possessed by SADF personnel, giving them an incentive not to oppose the settlement and reassuring the white community, while also mollifying the substantial South African arms industry. The ANC wanted to take the military out of domestic politics, and thus to reduce its threat to the black community (Cilliers 2012). This mission also fit in with the general image of a modern state as one with a modern military.

However, there is in fact no conventional military threat at hand to justify this selection. It also requires extensive investments in aircraft, armored vehicles, and ships; the major attempt to satisfy these needs was the Strategic Defence Procurement Package of 1998, which is still widely cited as an example of corrupt and inappropriate government spending. It also required skills that were concentrated among whites and were difficult for others to obtain in the apartheid educational system.

The military was later given new missions (disaster relief, border control, peacekeeping, and working with the police to control violence), despite some resistance within the services: "The insistence that the SANDF be primarily orientated towards external defence and conventional warfare, for fear of remilitarising society, has meant that it has not been configured, trained or equipped for what it actually does—rendering assistance to the police, border protection and peacekeeping" (Cawthra 2003, 53).

However, the South African government does not seem prepared to fund a military to carry out these very different tasks. One author claimed that at the end of 2008 only four of the twenty-four South African naval vessels were operational (Mbeki 2000, cited by Dempsey 2009, 135); Gavin Cawthra (2012a) noted that, when naval ships in port connected to the local electrical grid, there were power outages in Cape Town. One plausible alternative to the current military structure, given the lack of conventional external military threats, is a shift to a light infantry force structure for peacekeeping, crime control, and border protection (LeRoux 2007; cf. Baker and Jordaan 2010; Baker 2009). However, no one seems to expect this to happen (Cawthra 2012a; Cilliers 2012).

SANDF's problems are obviously serious. Nonetheless, given the major challenges of building a new military after the turbulent history of South Africa, it has performed well despite low expectations.

ALTERNATIVE OUTCOMES

The integration process did not start until late in the political negotiation process. Clearly, it would not have happened if these negotiations had broken down; the result would presumably have been mass violence and draconian military repression. The importance of Nelson Mandela in those negotiations has been widely noted; less often mentioned was the stroke that allowed Botha to be replaced by de Klerk at a critical time. They succeeded despite such dramatic episodes as a white mob invading and occupying the negotiation site in 1993.

This laid the political foundations for the new military, but there were deep divisions within the SADF and MK that could have caused major problems in the integration and training process and resulted in a political collapse. Many of the SADF personnel who were most opposed to integration were able to migrate to the private security sector (thanks to Austin Long for reminding me of this), and many of the most competent MK personnel found attractive civilian alternatives. It is also true that the professionalism of the personnel on both sides was absolutely essential.

Even so, the process itself was filled with risks. One critical question was how MK personnel could be assembled without either posing a threat to others or being at risk of being attacked themselves. Aboobaker Ismail was chief of ordnance for the MK, a negotiator in the peace process, and a major general in the SANDF. He says: "Both sides . . . recognised that should anything go wrong at these assembly points, the people would revolt against the ruling regime and could plunge the country into a bloodbath" (Ismail 2012, 77).

The SADF reportedly had plans for a massive repression if the process broke down, perhaps something similar to Operation Quartz discussed in chapter 4, on Zimbabwe. The worst case would have been an enormously bloody combat with a minority white army armed with modern weapons (including nuclear potential) repressing an increasingly organized and mobilized black majority. Of course this would only have happened if SADF personnel were willing to carry out such a policy, and there is good reason to doubt that. But it is good that the issue was not put to the test.

HAS MILITARY INTEGRATION MADE LARGE-SCALE VIOLENCE LESS LIKELY?

The integration process represented a costly commitment for both sides. It began fairly late, so its progress probably did not directly influence similar activities in other areas of society, but its failure certainly would have done so. It seems to be generally accepted that military integration was necessary to cement the peaceful transition to democracy (Heinecken 2012; Esterhuyse 2012; Cawthra 2012a) or at least that it was believed to have been necessary at the time.

The new military can protect elites from large-scale violence, which presumably made it possible for them to establish and develop new institutions in government and society. The new military cannot provide security against crime, a major problem in South Africa, but that is not its job.

The number of former combatants in the SANDF is relatively low, so it is hard to argue that military integration has reduced the likelihood of civil war by taking potential fighters off the streets; in fact, NSF veterans who were not integrated have caused substantial political problems (Esterhuyse 2012; Seegers 2012). The exit of SADF personnel who opposed integration into private security is an exit option that may not be available in other cases.

Despite its problems, the new military serves as an example of integration in South Africa: "The integration of the primarily White SADF with the other predominately Black African forces was one of the most far-reaching and complex undertakings in public reform in post-apartheid South Africa" (Heinecken and Van Der Waag-Cowling 2011, 170).

Blacks are happy that they do not have to fear the military, and they approve of its integration, show pride in the force on holidays and similar occasions, and tend to regard it as a reasonable career alternative. Whites, conversely, seem to regard it as a failing institution in which they have little future and that they do not wish to join or support; roughly three thousand South Africans are now serving in the British army (Esterhuyse 2012; Seegers 2012; Cilliers 2012).

The political process of compromise and transformation underlying the integration process contributed to the maintenance of peace. Appointing representatives of various parties to important positions was essential in making the peace work (Esterhuyse 2012). Speaking of the initial talks between the SANDF and MK leaders, one author asserts: "These series of talks granted considerable domestic and international legitimacy to the subsequent military. . . . Finally, the face-to-face interactions encouraged trust between the former enemies. This spirit of cooperation and relatively good will would facilitate future efforts at working out the more specific organizational details" (Porter 2010, 17).

On balance, then, the South African military integration has played an important, but not a dominant, role in its country's political transformation. The success of military integration has not guaranteed political success; its failure almost certainly would have caused political disaster.

ABBREVIATIONS

ANC	African National Congress—main resistance organization to apartheid, now the governing party of South Africa
APLA	Azanian People's Liberation Army—military arm of the PAC
BMATT	British Military Advisory and Training Team
CODESA	Convention for a Democratic South Africa—group through which transition processes from apartheid were negotiated, mainly between the ANC and the government
JMCC	Joint Military Coordinating Committee—group by which the terms of military integration were negotiated, mainly between the SADF and MK; technically subordinate to the Transitional Executive Council

KZSPF	KwaZulu Self Protection Force—military arm of the Inkatha Freedom Party
MK	Umkhonto We Sizwe, or Spear of the Nation—the ANC's military arm
NSF	nonstatutory forces—rebel military groups, the MK, and the APLA
PAC	Pan-Africanist Congress—resistance movement more radical than the ANC
SADF	South African Defence Force—apartheid government military
SANDF	South African National Defence Force—postapartheid government military
TVBC	Transkei, Venda, Bophuthastswana, and Ciskei Defence Forces—forces of nominally independent apartheid "homelands," closely linked to the SADF

Part III

INTERNATIONAL INVOLVEMENT

Chapter 9

HALF-BREWED: THE LUKEWARM RESULTS OF CREATING AN INTEGRATED MILITARY IN THE DEMOCRATIC REPUBLIC OF THE CONGO

Judith Verweijen

On May 17, 1997, the regime of one of Africa's longest-reigning autocrats, the self-proclaimed Field Marshal Mobutu of Zaire, was toppled by a heterogeneous coalition of forces known as the AFDL. Led by, among others, the former revolutionary leader Laurent-Désiré Kabila, this hastily recruited liberation army managed to conquer the country after an insurgency campaign lasting only seven months. Heavy backing from a regional coalition led by Rwanda, Uganda, and Angola was crucial to this victory.

Once the AFDL assumed power, Kabila took over the presidency, but he quickly started losing international and domestic legitimacy (Willame 1999; Lanotte 2003). While domestic support eroded due to his autocratic and erratic style of governance, he fell out of favor with his foreign allies because he failed to address their security concerns and attempted to gain more autonomy (International Crisis Group 1999b). On August 2, 1998, Uganda and Rwanda launched a new insurgency through a proxy group called the RCD with the intent of removing Kabila from power. Starting from the east, the RCD managed to rapidly conquer about a quarter of the Congo's territory. Not long afterward a second, a Uganda-backed rebel movement, the MLC, emerged in the northwestern part of the country. Largely due to the intervention of Zimbabwe and Angola, none of the rebel forces was able to conquer the capital. A military stalemate was reached, and the country effectively became divided into first three, and later six, zones of control, as the RCD splintered into several factions—RCD-G, RCD-K/ML, and RCD-N.[1]

What has come to be known as the Second Congo War was both the product and the engine of a set of interwoven conflict dynamics at the local, national, regional, and international levels. It drew in a large number of African states and foreign armed groups, leading to a regional war complex.[2] At the core of this complex were a number of competing, but at times also cooperating, transborder political–economic–military networks, made up of the main warring factions and their foreign allies. These were in turn linked to smaller-scale coalitions of rural militias and local economic and political actors. This resulted in ever-shifting webs of power relations among a multitude of "networks of profit, power and protection" operating at different scales.[3] Together, these networks drove and were driven by the development of a war economy, which fostered violent competition for territorial and economic control. Violence became the principal strategy to control production, fiscal functions, trade networks, land, natural resources, borders, and markets. The result was a coercion-based economy thriving on predominantly illicit exploitation and trade practices, such as pillage, wholesale looting, forced labor, and price-fixing.

In the eastern part of the country, these developments intensified and transformed long-standing local conflicts over political representation and access to land and other resources. These conflicts had traditionally been expressed in identity-based terms, mainly pitting "Rwandophones" against ethnic groups that portrayed themselves as "autochthonous."[4] The upsurge in violence unleashed by the war, in combination with the longer-term processes of the economic and social marginalization of rural areas and youth, led to the multiplication of local, ethnically recruited militias known as Mai Mai.[5] In order to recruit such militias, political and military entrepreneurs cultivated an image of the RCD-G as a Rwandan-controlled and Tutsi-dominated occupying force. The government in Kinshasa employed a similar strategy of inciting ethnic hatred in order to rally support against the RCD.[6]

Facing an enduring stalemate along the conventional front lines and disposing of a relatively weak army, Kabila started to rely on proxy forces to fight the insurgency in the east. In addition to using groups of Mai Mai, he also enlisted the support of foreign military groups like the remnants of the former Rwandan army and allied Hutu militia (*interahamwe*), which had fled to the DRC in the wake of the genocide. This extensive use of proxies entrenched a pattern of the use of armed groups by political, economic, and military leaders. In the short term, it led to the complete fragmentation of the political-military landscape, which severely complicated peace negotiations and the implementation of agreements.

After an endless series of meetings, a failed cease-fire agreement, and a change of leadership in Kinshasa, a peace deal was reached at the end of 2002 in Pretoria. The so-called Global and All-Inclusive Agreement (GIA) was signed by the main warring factions, civil society, and the unarmed political opposition. It stipulated the formation of a Transitional Government (TG), charged with reunifying the country, organizing elections, and setting up a restructured national army composed of the ex-belligerents' fighting forces.[7]

The TG, which was inaugurated in July 2003, consisted of one president and four vice presidents, drawn from the three main former warring parties (MLC, RCD-G, and the government) and the unarmed opposition.[8] This so-called one plus four formula epitomized the principle of power sharing. However, rather than signifying reconciliation, it came to symbolize the paralysis, entrenched factionalism, corruption, and distrust that characterized Congo's "transition."[9] This was succinctly expressed by the running gag "one plus four equals zero" that circulated widely during this period (Willame 2007).

It was in this context that the sensitive and difficult task of creating an integrated national army had to be accomplished. This chapter analyzes how this process fared by exploring (1) the origins, (2) the creation, and (3) the outcome of the development of the FARDC, as the new national army of the DRC was baptized.

ORIGINS

Historically, military groups in the DRC have been linked to particularistic political groups and have often incited rather than prevented violence within the society. The attempt to form an integrated, professional military was thus very ambitious and was not viewed favorably by many of the country's various powerful political factions.

The Historic Role of the Military

Although the debates during the peace talks had emphasized that the new integrated army should be "republican," no such army had ever existed in Congo/Zaire. From the colonial era onward, the military had been first and foremost an instrument of regime security and particularistic power strategies. As an important pillar of the projection of state and elite power, it had always played principally a domestic security role, combining policing functions with external defense. Furthermore, it had traditionally been as much an economic as an insecurity/security actor, by either protecting or actively engaging in a wide range of income-generating activities. Its practices in both these domains often amounted to brutal and predatory behavior, which severely strained relations with the population (Ebenga and N'Landu 2005).

As elsewhere on the continent, the Congolese postcolonial state was weakly institutionalized and fragmented. Rather than forming an obstacle to power projection, the resulting disorder and insecurity served as a political resource and instrument of control for dominant groups. One of the main sources of the production of insecurity was the state apparatus itself, in particular its security agencies (Schatzberg 1988). The Forces armées zaïroises (FAZ), as Mobutu's army was known from 1971 onward, was no exception to this. The FAZ was characterized by a number of tendencies that would continue in its successor forces: the personalization of power to the detriment of formal command chains and official rules; politicization; favoritism, resulting in ethnicization; and finally, factionalism fueled by the existence of a variety of parallel structures with overlapping commands and frequent rotations of office.[10] The inevitable results were deprofessionalization, deinstitutionalization, corruption, weak cohesion, and a lack of central command and control. Logically, this greatly undermined military performance. However, fearing a coup more than an invasion, Mobutu preferred to keep the army weak and divided while relying for his own security on a number of loyal elite units like his presidential guard. This allowed him to stay in power while neglecting the rest of the military, which was left underequipped, underpaid, and underfed (Young and Turner 1985).

This strategy of deliberate neglect was one of the main factors that pushed the military to engage in a wide range of illicit business activities, which fostered predatory behavior. As a result, the FAZ became a "free-floating source of insecurity," especially in the country's vast hinterland (Schatzberg 1988, 70). Paradoxically, this weakly performing army was at the same time a pillar of the Mobutist order. This was largely due to its brutality in responding to (perceived) threats to the regime. Rather than repressing in a systematic, totalizing manner, Mobutu's coercive apparatus proceeded by selective, ruthless displays of power. This strategy of deterrence was an important stabilizing factor and helped maintain coercive supremacy (Schatzberg 1988). It also allowed Mobutu to dispense with the construction of comprehensive administrative structures and the establishment of full territorial control. Instead, the projection of state power was mostly based on intermediary rule and followed an archipelago logic, with the direct exercise of power confined to certain political and economic core areas (Callaghy 1984). The military played an essential role in this order, as it provided both the insecurity and the repression required to reproduce the system.

Culturally, the role of the military was less pronounced. While it is believed to have contributed to the forging of Congolese nationhood in colonial times (Young 1965), the factionalism and ethnicization of the military in the postcolonial era prevented it from playing this role. Key positions in the FAZ, especially in the presidential guard, were almost exclusively filled by members of Mobutu's Ngbandi clan or other ethnic groups from northern Equateur and Province Orientale. Officers from other provinces were underrepresented and regularly eliminated by ethnic purges. This reduced the military's potential to serve as a symbol of national unity.[11] The FAZ chiefly represented and reflected Mobutu's power, not the nation.[12] This would not fundamentally change after Kabila took over power with liberation forces that had mostly been recruited in the east. The newly constituted army, the FAC, was perceived as heavily influenced by foreign forces and dominated by easterners. This diminished its potential to serve as a symbol of the nation. Moreover, important positions, especially in Kabila's elite presidential guard, gradually started to be filled exclusively with members from his home area in northern Katanga. This ethnic favoritism, and the ongoing power struggle between the various components that had made up the AFDL, generated important tensions and undermined cohesion, morale, and military strength (International Crisis Group 2000a). It was in part the resulting weakness of his military that forced Kabila to eventually accept a political and military power-sharing deal.

The Emergence of Military Integration as an Issue

Characterizing the war as principally an act of aggression by neighboring countries rather than a domestic rebellion, President Kabila categorically refused to meet face to face with the insurgents, let alone to contemplate power sharing. However, his African allies as well as the various mediators involved in the peace process insisted on including the insurgents in the negotiations and organizing a national dialogue.[13] After heavy diplomatic and increasing military pressure, Kabila finally agreed, which led to the signing of the Lusaka Cease-Fire Agreement in July 1999.[14] This agreement provided for the establishment of a UN peacekeeping mission called MONUC and the organization of inclusive peace talks in the form of an Inter-Congolese Dialogue (ICD), with the aim of sketching the outlines of a new, inclusive, political framework and a restructured national army comprising the signatories' fighting forces.

Supporters, Opponents, and Neutrals

The principle of power sharing was advocated by a wide range of domestic and international stakeholders in the peace process. Many African leaders were inspired by South Africa's successful political and military power-sharing deal, which had become an almost standard formula for peace negotiations on the continent. The Western powers, which were alarmed by Kabila's autocratic tendencies and cozy relations with outcast states like North Korea, saw it as a means to neutralize the increasingly discredited president. For Uganda, chief backer of the MLC, it offered a potential way to perpetuate its influence in the DRC in the long term, through its proxies. For the unarmed opposition and civil society, it was a window of opportunity to establish a different political order. Finally, for the insurgent groups, it offered access to state power and influence, which had been one of their leaders' main motivations for taking up arms in the first place.[15]

The only factions that were less enthusiastic about the idea of power sharing were, first, President Kabila and his foreign allies, Angola and Zimbabwe; and, second, Rwanda, which still hoped for a military victory and was not convinced that a regime issuing from power sharing would guarantee its economic and security interests. These doubts were shared by Angola and Zimbabwe, which had vested interests in the continuation of Kabila's rule, despite increasing dissatisfaction with the president's performance.[16] Consequently, they were less hostile to power sharing than Kabila himself, who at first outright rejected a plan that would significantly diminish his power.

Compromises in the Process

Kabila eventually signed the Lusaka agreement because of its explicit recognition of his position as head of state, which made him believe that he would be able to manipulate the inclusive talks of the ICD in his favor (International Crisis Group 1999a). However, his commitment to what was essentially an imposed agreement was weak. This lukewarm enthusiasm, in combination with the reluctance of some foreign powers to withdraw their troops, blocked implementation, and fighting continued over the next months. It was only after Kabila's assassination on January 16, 2001, and the subsequent power takeover by his son Joseph, that this stalemate could be broken. Because Joseph was more reconciliatory, more diplomatic, and less hostile to power sharing than his father, Joseph's presidency brought a new dynamic to the peace process. This culminated in the organization of the ICD in the South African resort Sun City in February 2002.

During the Sun City talks, no agreement could be reached on army restructuring, leading to a deadlock in the negotiations. The Kabila faction advocated merely absorbing "eligible fighters" from the former rebel forces into the existing government army, the FAC. This was to be mainly a technical process, allowing the ex-government side to retain a maximum of control. The leadership of the rebel factions and the unarmed opposition, conversely, advocated the merging of combatants from all factions into an entirely new structure based on a quota system. This would constitute a more comprehensive and more political process (Onana and Taylor 2008). Eventually, the second view prevailed, but agreement was not reached until the country was well into the transition. Therefore, the 2002 Pretoria accord that closed the second round of the ICD contained no details on the military integration process. It merely delegated the tasks of setting up a restructured army and drafting a new defense policy to the future Superior Defense Council.[17]

The vagueness of this deal was a double-edged sword. On one hand, it allowed the peace process to continue in spite of radically opposed views. On the other hand, it put a heavy burden on the TG, which had the onerous task of negotiating an issue that was at the heart of power relations in a climate of heavy distrust and ongoing violence. Rather than contributing to confidence building, the process of military integration turned into a divisive mechanism and a forum for the continuous power struggles both among and within factions.

In the background of this lurked the ex-belligerents' halfhearted commitment to military integration and the transition in general. Given that military power was the main determinant of their political and economic weight, most of them were extremely reluctant to dismantle their military structures. As a consequence, the transition was characterized by a close interaction between political and military processes. Violence, now concentrated in the east, was both manipulated from the top down and generated by local tensions. Often it acquired a momentum of its own and paralyzed the political process in Kinshasa, without ever entirely disrupting it. As a result, politics remained highly militarized and only partially replaced violence as a means of channeling power competition (International Crisis Group 2005).

Another reason why the ex-belligerents refused to fully dismantle their military structures was that these were crucial for retaining their zones of economic and territorial control. In spite of official national reunification, the country remained a patchwork of different economic fiefdoms in which illicit exploitation and tax systems persisted (UN Group of Experts 2005a, 2005b).

For some factions, concerns regarding the security of their constituencies also played a role. This was the case mainly with the RCD-G, which feared the persecution of Rwandophones, but it also applied to certain Mai Mai groups. The specter of electoral loss made these groups further drag their feet with regard to military integration. Because elections took place in a zero-sum-game political environment, they were anxious to keep open the option of a return to violence.

A final reason for the reluctance to progress with military integration was a general desire to prolong the transitional period as long as possible, because it offered tremendous opportunities for self-enrichment. The transition gave the signatories of the GIA access to well-paid state positions, lucrative business deals, and profitable state tenders, and it also allowed them to bring part of the networks of the war economy into the fold of the central state (De Goede 2007).[18]

In this context of engineered paralysis, disagreement on the military integration process was often as much a pretext as a reason for stalled progress with the transition. Disputes were usually not the result of differing visions on the armed forces but the product of power struggles among and within factions. The main standoff was between hard-liners in the entourage of President Kabila and a group of dissidents in the RCD-G, which was militarily the strongest of the former insurgent forces. The RCD-G was internally split between those committed to the transition and a Rwanda-supported faction that was opposed to the process, as it was determined to keep control over its stronghold in the province of North Kivu. Given the bleak electoral prospects of the RCD-G, which had become very unpopular during the rebellion, this latter group judged that the transition had little to offer them. It brought together a number of dissident politicians and military leaders, the most important of whom was General Laurent Nkunda, who had refused to take up his new position in the integrated military command. This group capitalized upon genuine fears about discrimination against Tutsi, a strategy that was facilitated by hawks in the Kabila faction who tried to galvanize support by adopting strong anti-Tutsi rhetoric. This playing of the ethnic card further poisoned the political climate and complicated the process of military integration (International Crisis Group 2005).

One of the more polarizing issues in the military integration process was control over the presidential guard. Because Joseph Kabila had little trust in the FAC, he relied principally on this Angolan-trained elite unit, which he expanded to an estimated 12,000 to 15,000 troops. However, the guard was neither regulated by a clear legal framework nor put under the control of the integrated army command.[19] Consequently, the other factions wanted to open this unit up to their own fighters and to drastically reduce its size. Not willing to compromise on this issue, the presidential camp deployed the guard to the most strategic sites in the country, while stationing five thousand of its troops in and around the capital.[20] This provoked deep suspicion among the vice presidents, who subsequently insisted on also maintaining sizable personal guards. The strongest of those was kept by Vice President Bemba, whose guard numbered around a thousand troops, whom he withheld from army integration. During and just after the electoral period at the end of the transition, this unit clashed on numerous occasions with the presidential guard in Kinshasa. This highlighted the dangers of the excruciatingly slow pace of military integration (Wolters and Boshoff 2006).

CREATION

International support for military integration was weak and divided. When the process was stalled, a compromise was suggested to integrate and train six brigades, which would presumably serve as models for further integration. High incentives for demobilization, disarmament, and reintegration (DDR) and pressures to meet quotas by the various groups resulted in recruiting many unqualified personnel, who were given only the minimum training by uncoordinated groups and then sent into the field with no support.

Shaping the New Military

The unresolved question of the presidential guard illustrates the increasing influence of the president and his entourage on the transition and the military integration process. Kabila kept direct control over military matters through the presidential military office, the so-called Maison Militaire, which constituted a parallel power network in the armed forces. The growing power of this network allowed him to marginalize the transitional structures charged with defense and security. Due to political infighting and distrust among its members, the Superior Defence Council became a lame duck. Because the vice president responsible for defense and security, Azarias Ruberwa, was from the RCD-G, decision making on defense issues became politicized to the extreme (Kibasomba 2005). The resulting slow pace of official decision making allowed the presidential camp to proceed unilaterally through unofficial channels. This was part of a wider trend of the centralization of power, which culminated in Kabila's electoral victory in 2006. The international actors involved in the transition facilitated these autocratic tendencies. Out of fear that the transitional process would be derailed, the CIAT, an ad hoc diplomatic mechanism regrouping donor countries and international organizations, was reluctant to put political pressure on Kabila in order to safeguard political space (International Crisis Group 2005).[21]

The reluctance to apply pressure was also visible in the military integration process. Donors adopted a classical technocratic approach, which focused mostly on training and infrastructure, while politically sensitive issues, like the unclear status of the presidential guard, were not placed prominently on the agenda (Melmot 2008). Furthermore, they failed to develop a common strategy or a comprehensive vision for army integration, in part because there was no lead donor (a good description of the problems with donor coordination in the army integration process is provided by Onana and Taylor 2008). Belgium initially played an influential role, as it organized a number of strategic workshops in Kinshasa in order to create a framework for army integration. It also immediately pledged funding and assistance for training and equipping. However, its leadership was contested by South Africa, Angola, and other entities that preferred African leadership of the integration process.

In 2005, the European Union established a small, in-country defense reform mission named EUSEC, with the explicit intention of taking up a coordinating role. This unilateral initiative led initially to competition with MONUC, which had also envisaged playing such a role through its security-sector reform unit. Furthermore, EUSEC's more outspoken political orientation was regarded with suspicion by some of the EU member states, which saw more advantages in old-school bilateral defense cooperation.[22] In the end, no agreement among donors could be reached, and bilateral, piecemeal approaches dominated. This was much to the liking of the TG, which had pushed hard for bilateral assistance, a time-tested strategy for reducing donor interference. The resulting laissez-faire approach meant that international actors ultimately had little impact on how the new military was shaped.

Another reason for this limited influence was that donors never committed sufficient funding in order to buy leverage. While they had a great appetite for funding the electoral process and the DDR of combatants, military integration clearly was not one of their priorities.[23] At the end of 2004, only $12 million had been pledged, versus $200 million for DDR, whereas the Congolese budget of that year did not even contain allocations for military integration, as the TG had expected it to be fully foreign financed. This gross underfinancing was an important reason for the slow takeoff of the process, as no funds were available to transport or canton troops (International Crisis Group 2006a).

Although an integrated General Staff and command of the country's eleven military regions had been set up by the end of 2003, troop deployment on the ground remained largely unchanged well into 2005. Predictably, shuffling only the higher command but not the troops led to major problems with authority, as some field-based officers refused to take orders from the new regional commands. This was the case with a group of ex-RCD-G officers in the province of South Kivu, for example. Their disobedience first sparked a mutiny and, in May 2004, an open confrontation in the provincial capital Bukavu, which fostered a split in the RCD-G's political leadership. These clashes motivated hard-liners around Kabila to push for a military solution to the ongoing power struggle with the RCD-G, which prompted him to send about ten thousand troops to North Kivu in December 2004. This offensive led to a standoff with RCD-G-controlled troops near Kanyabayonga, but failed to help extend central government authority over the province. These events illustrated both the fragility of the transition and the key role that military integration played in the process.[24]

The 2004 military confrontations temporarily froze the political process in Kinshasa, which blocked the development of a badly needed policy framework for military integration. This prompted South Africa to present an emergency plan, which obtained support from Angola and Belgium, both of which had earlier signed a bilateral defense cooperation agreement. Far from containing a wider strategic vision on military integration, the emergency plan was a short-term, practical measure aimed at facilitating the integration of the first six brigades of the new national army. This was to be accomplished through a so-called *brassage* (brewing) process, which consisted of the mixing of individual combatants into new units through a rudimentary forty-five-day training program in a Centre de brassage et de recyclage (CBR) or *brassage* center. It was only in mid-2005 that a more comprehensive policy for military integration was adopted in the form of the National Strategic Plan for Army Integration, which nevertheless still focused mostly on technical issues.[25]

Personnel Selection

Recruitment for the Integrated Brigades (IBs) was mainly carried out by the military leadership of each faction, with the SMI, the body responsible for the planning and execution of the integration process, communicating the needed numbers of combatants per brigade. The division key followed a quota system roughly based on the numbers of combatants that each faction had declared in Sun City, leading to the following division: 35 percent FAC, 17 percent MLC, 28 percent RCD-G, 8 percent Mai-Mai, and 12 percent other groups.[26] The last category included armed groups that had not signed the GIA, but that, under Article 45 of the 2004 Defense Law, were still eligible for integration.[27] Designed to ensure the IBs had a balanced composition, the imposed quota system soon created problems. Aware that military strength would be the key to the division of positions in the transitional institutions, the ex-belligerents had greatly exaggerated the number of their combatants. While the declared combined strength of their fighters totaled 220,000, observers estimated the actual number to be roughly half that size.[28] As a result, factions struggled to fill their allocated quotas, which undermined efforts at demobilization.

The 2005 Strategic Plan outlined a dual-track process that allowed individual combatants to choose between demobilization and integration into the FARDC. While those choosing demobilization would receive an initial cash payment of $110, followed by an additional $25 a month for the subsequent year, those opting for army integration would receive a meager $10 a month.[29] This discrepancy in financial benefits, caused by the asymmetry in available funding for DDR and army integration, prompted an overwhelming number of combatants or presumed combatants to demobilize. In reaction, commanders resorted to coercion in order to fill up their quotas, which rendered the principle of individual voluntary choice merely theoretical. A second strategy for producing the required numbers was fresh recruitment. As there was no individual identification of combatants nor any procedure to test for actual combat experience, civilians could easily enter the integration procedure. The virtual absence of admission criteria further facilitated such random recruitment. Although the established legislative framework barred children under eighteen and those deemed "physically or morally inapt" from entering the military, no mechanisms were created to apply these criteria.[30] Physical tests took place only in the CBRs and were largely symbolic. Moreover, there was no upper age limit, which led to the incorporation of many soldiers who were ripe for retirement.[31]

These lax recruitment criteria also worked against the safeguarding of human rights standards. No effort was made to filter out recruits with a dubious track record. On the contrary, the new national army became a refuge for those fearing prosecution, and there was even some recruitment in prisons.[32] One of the buzzwords of the "transition" was "inclusivity," and it was feared that excluding those suspected of gross human rights violations would generate an army of spoilers. Furthermore, the transition was dominated by ex-military leaders who did not have clean hands themselves, which further reduced incentives to address human rights issues. Additionally, there was little judicial or investigative capacity to screen many cases in a short time. Consequently, no attempts were made to bar suspected war criminals from obtaining superior ranks and positions in the FARDC. This short-term pacification strategy had perverse effects in the long term, as it gave off the signal that violence pays (Tull and Mehler 2005). Thus, it ultimately contributed to the continuing militarization of the east.

Rank Allocation

Unhindered by any human rights considerations, the selection of officers and the distribution of ranks and positions in the new army followed a mixture of political, performance, and patronage criteria. Initially, only the ranks of the former FAC were recognized, which caused resentment among the other factions. The task of resolving this was left to a ranks commission set up by the General Staff, to which each faction had to submit a list of officers. At the end of 2004, this commission promulgated an *ordre général*, which appointed the new army's officers. In general, a policy of generosity was applied, which resulted in an entirely lopsided military structure with a disproportionate number of officers.[33] Though most of the ranks of the factions with considerable military and political weight were recognized, this was not the case with the Mai Mai and some other smaller-scale armed groups. Predictably, this caused frustration and resentment, leading some of the officers of these groups to reject or drop out of the integration process. In many cases they viewed these events through an identity-based lens, seeing the new army as dominated by "Rwandophones." Another source of tension was the awarding of high ranks to officers with little or no military or even basic education. This was difficult to accept for those who had graduated from military academies and were used to a system of promotion based on certain educational criteria. Such well-educated officers often had little respect for their new superiors, which weakened cohesion and discipline (Eriksson Baaz and Stern 2010).

The issue of ranks was important, but the distribution of functions was even more so. Because the salary of even the most superior officers is negligible, positions, not ranks, are the main gateway to resources in the FARDC.[34] Though it had been agreed upon that the division of positions in the General Staff and the Regional Commands had to follow a strict system of balancing between the different factions, determining the actual distribution was a cumbersome process involving endless horse-trading.[35] The composition of the commanders of the IBs was based on a similar division formula, but also had to take into consideration the recommendations made by the CBR commanders at the end of the *brassage* period. While this system was meant to ensure that basic standards of competence were taken into account, power-mongering, favoritism, and political interference largely undermined this intention.

Training

According to the Emergency Plan, the first IB was to be trained by Belgium, the second by Angola, and the third jointly by Belgian, South African, and Congolese instructors. The remaining three brigades would be trained by the FARDC itself. In the absence of a coordinating mechanism, training approaches were not harmonized, and each IB was trained in a different military style.[36] By June 2004, Belgium had finished the training of the first brigade, which had taken ninety days instead of the planned forty-five. Due to the immense differences in the recruits' level of knowledge, skills, and capacity, the desired standards could not be easily reached. A lack of training matériel and infrastructure further complicated the training. These problems were even worse in the CBRs that were not under the supervision of foreign donors. Lacking basic facilities such as adequate shelter, sanitation, water, and electricity, the living conditions in these centers were appalling, which led to the outbreak of diseases that caused the deaths of dozens of soldiers. Furthermore, because salaries and supplies of food and medicine were grossly inadequate due to systematic embezzlement, combatants were pushed to prey on neighboring communities. Naturally, these harsh conditions undermined the effectiveness of the provided training and pushed scores of combatants to desert.[37] As the situation came close to qualifying as a humanitarian crisis, some donors swiftly approved funding for refurbishing the CBRs. However, this did not solve the problem of the overall underfunding of the military integration process. The absence of a budget for equipping and deploying the IBs was especially problematic, as it made training results difficult to sustain.

Another challenge to *brassage* was the wide range of ingredients out of which the new army was to be "brewed." Recruits had very different military backgrounds, which posed an obstacle to the creation of cohesion and unity. Whereas some had enjoyed extensive conventional military education, others were trained only in guerilla warfare, and some had no prior military experience at all. There were also acute differences between the larger insurgent forces, which had functioned more or less as conventional militaries, and the small-scale, rural militia like certain groups of Mai Mai. Furthermore, the heavy reliance of both government and insurgent forces on different sources of foreign military education and training had produced large variations in doctrines and styles. On top of that, linguistic and geographical differences divided the mostly ex-FAZ Lingalaphones from the west and the largely war-era Swahiliphone recruits from the east (Kibasomba 2005).

Perhaps surprisingly, given the bloodshed and atrocities that had characterized the Second War, vengeance and hatred for the enemy hardly played a role during *brassage*. Many combatants had not been mobilized along ideological lines but simply been recruited, either by force or voluntarily, into the faction that controlled the zone in which they happened to be at the outbreak of the hostilities.[38] This had facilitated instances of "cooperative conflict" (cf. Keen 2005), in which nominally opposed factions maintained cordial economic relations and established separate spheres of influence by mutual consent. However, several factions had capitalized upon and cultivated strong ethnic animosities, and in February 2006 a number of Tutsi soldiers were killed at the Kitona CBR in the western province of Bas-Congo. Although this was an isolated incident, it did foster a fear of participating in *brassage* or leaving the Kivus among Rwandophones (Amnesty International 2007).

Although ethnic tensions influenced the social dynamics during the integration process, no provisions had been made for addressing this issue. *Brassage* was approached as a purely technical process, being little more than the juxtaposition of the former belligerents. Little attention was paid to the social or ideological aspects of creating a new, republican army. The delays in the development of a strategic vision for the armed forces meant that the soldiers of the new national army entered a doctrinal and ideological void. And an army that was still in search of an identity could not successfully transmit its professional values to its soldiers nor elucidate their role in society (Kibasomba 2005). In combination with the short period of training, this further hampered the dissolution of former allegiances and identity-based loyalties.

OUTCOME

Given the enormous problems that have already been discussed, it is not surprising that the new army, like the old, is deeply divided, ineffective, and predatory. Most of the integrated brigades have stayed together, but their performance had not been notably different from that of other units, and no more have been added.

Political Control

Reluctant to give up their independent military power and zones of economic and administrative control, the ex-belligerents tried to have it both ways. They formally committed to the army integration process, sending some of their fighters to *brassage* and profiting from the power and money that important military positions offered. Consequently, the militarily weaker factions—like MLC, RCD-N, and RCD-ML— eventually gave up most of their separate military structures or kept them dormant. This formal commitment to the integration process is illustrated by the fact that, technically, the integrated military has remained largely intact and violence has not continued along old factional lines. The largest share of the troops of the original eighteen IBs remain in the FARDC today, although individual commanders and groups of combatants have regularly deserted, with some later reintegrating.

However, factions which agreed to dismantle their military structures did not necessarily abstain from militarized power politics. The ex-belligerents adopted two main strategies to offset the potential loss of influence caused by army integration: First, they tried to maintain economic and political control by building up power bases within the political and administrative institutions—for example, by entrenching themselves locally or provincially in unelected administrative positions or by forging alliances with factions that were likely to have good electoral results.[39] Second, they attempted to maintain military spheres of influence by building up client networks both within and outside the military. For example, several influential commanders managed to withhold their brigades from the *brassage* process, while others liaised with the armed groups that had refused or dropped out of the integration process. Consequently, the war-era networks that had been absorbed into the military remained partly intact as parallel power structures *within* the army, although many would gradually dissipate or transform. The end product was a "semi-integrated" or "half-brewed" military, in which several unintegrated units and separate spheres of influence intermingled with the integrated command and brigades. Logically, these convoluted power structures weakened the influence of the central command.

The main power network that undermined the General Staff's grip over the military was the Maison Militaire, which was led by the president and had a preponderant influence over a number of military agencies and units, notably the presidential guard and the military intelligence services. But the Maison also maintained parallel chains of command, supply, and logistics in other parts of the armed forces, occasionally even giving orders directly to the Regional Commands or brigade commanders in the field, while bypassing the General Staff (International Crisis Group 2005). Though influential, the presidential circle did not fully control the whole military apparatus. Rather, it had established a number of islands of control in the form of loyal client networks, which made it powerful enough to prevent the military from becoming a threat to the regime. However, the presidential patronage network was but one among the many poles of power. Because the power networks that had been absorbed into the military were relatively autonomous, the new military was subject to strong centrifugal tendencies. As a result, the president relied chiefly on his presidential guard to stay in power. It was also the strength of this guard that allowed him to gradually concentrate power and dominate the transition.

The consolidation of presidential power was an important stabilizing factor and kept the fragile transition on track. However, it also curbed the potential for the development of civilian oversight and control of the defense sector. This potential was not very large to begin with. First, there was no tradition of democracy or civilian control of the military that could be built on. In preceding regimes, the Ministry of Defence, let alone parliamentary committees on security, had largely had symbolic functions (Ebenga and N'Landu 2005; Young and Turner 1985). Second, many factions in the transition greatly benefited from the lack of transparency and civilian control over the defense sector, as it allowed them to manipulate the integration process and embezzle funds. As actors without a military background had become marginalized in the political domain, there was no critical mass to push for more control. In this way the militarization of the civilian sphere precluded more civilian control of the military.

Of the main ex-belligerents, only the dissident faction of the RCD-G rejected the transitional framework entirely and opted for continuing armed struggle. The military component of this group gradually came to be dominated by General Nkunda, who eventually launched his own politico-military movement in North Kivu called CNDP. The latter was at first largely recruited from unintegrated or badly integrated ex-RCD-G troops and covertly supported by Rwanda. Nkunda justified his creation as necessary for the protection of the Tutsi community against a discriminatory government army and the still sizable Hutu-dominated FDLR (Stearns 2008). The ethnic orientation of the CNDP exacerbated intercommunal tensions and contributed, together with unresolved local disputes over access to land and the distribution of power, to the remobilization of groups of Mai Mai.

These local militias and other small-scale armed groups also mushroomed in other parts of the east.[40] Many such groups were led by commanders who felt they had nothing to gain from army integration, as they had neither the qualifications nor the influence to obtain positions of importance. Furthermore, they preferred to stay in their local strongholds, where they could exercise political and economic influence and protect local communities. Some of these militias had been abandoned by their former leaders, who were now integrated into the FARDC; but more often, these integrated leaders used their newly obtained position to sustain local armed constituencies. Both domestic and foreign armed groups continued to be used by political, economic, and military entrepreneurs, including FARDC commanders, for furthering particularistic power projects. As a result, the complex pattern of violence in the east that had emerged during the wars largely remained intact. Rather than putting an end to this violence, the new national army became a crucial factor in sustaining it (UN Group of Experts 2010).

Military Capabilities

In December 2007, after the elections, the government launched massive operations against the CNDP, which had become one of its biggest military threats. This offensive, which involved more than 20,000 troops fighting against an about 4,000-strong rebel movement, ended in a resounding defeat for the FARDC. This ruthlessly exposed the structural weaknesses of the new army in terms of logistics, administration, operational effectiveness, cohesion, and command and control (Boshoff and Hoebeke 2008). Furthermore, it showed that the military was not able to deploy to all corners of the country and that independent fiefdoms remained (UN Group of Experts 2005a). These included border zones, implying that the FARDC could not effectively fulfill its external defense functions. This was evidenced by the regular incursions of militaries or armed groups from neighboring states, which were, however, also the result of a general governmental incapacity and reluctance to establish effective border control. Finally, the 2007 debacle painfully showed that the size of the new army said little about its strength. Although the 2005 Strategic Plan had established the size of the army at 120,000, by the end of 2007, an estimated 164,000 soldiers were on the FARDC's payroll, while a further 70,000 to 90,000 combatants still had to be either demobilized or integrated (World Bank 2007).

Sustainability

Although the real number of troops was likely to be much lower given the large number of "ghosts" or fictitious soldiers, the army still appeared to be bloated. International experts recommended a military of 70,000 as more manageable and more sustainable, both financially and from the point of view of implementing reforms.[41] Of the $200 million 2007 defense budget, no less than $190 million was spent on salaries, leaving next to nothing for administration, logistics, and equipment.

However, not much of this money arrived at the soldier in the field, as most of it was embezzled at various stages along the command chain.[42] Aside from irregular and insufficient pay, soldiers also suffered from a total lack of institutional support. After finishing the *brassage* process, most of the IBs were sent straight to the front lines. Once deployed, they were abandoned, receiving almost no supplies and little logistical and material support. In most areas of deployment, there were no barracks and brigades lacked equipment, means of transportation, medical care, and other basic necessities (Onana and Taylor 2008). Furthermore, their low incomes and the absence of arrangements for military families led many soldiers to bring their wives and children along on deployment, which was not only dangerous but also undermines military functioning.

HAS MILITARY INTEGRATION MADE THE RESUMPTION OF LARGE-SCALE VIOLENCE LESS LIKELY?

The consequences of these abominable service conditions were manifold. Aside from undermining morale and operational effectiveness, they reinforced the Congolese military's habit of extralegal income generation, which traditionally had been accompanied by abusive and predatory behavior. With the basic incentives for behavioral improvement missing, the brief training in the CBRs eventually had little impact on the military's conduct. Furthermore, neither the FARDC nor foreign donors had planned any follow-up training or monitoring of IBs. Additionally, much of the brigade leadership was weak, and troops in the field were badly supervised, which further lowered the threshold for abuses. As a result, the FARDC, like its predecessors, has become an important source of insecurity for the population, although patterns are mixed and some units appear to behave better than others (Verweijen 2013).

Aside from its own behavior, the military contributes to this insecurity by sustaining the presence of armed groups. It does so in a variety of ways. First, many FARDC officers maintain economic ties with these groups, and thus they trade arms, ammunition, and natural resources. This leads to treason and a reluctance to attack these groups, even when ordered to so by the military hierarchy. Second, the FARDC's low operational capacities are a sort of life insurance for armed groups, as they render military defeat less likely. Often, operations merely disperse the targeted group, without the FARDC being able to hold the cleared areas afterward. Moreover, its structural weaknesses often prompt the FARDC to seek temporary coalitions with armed groups or use them as proxies in military operations.[43] Third, the FARDC's perceived lack of neutrality and weak fighting capabilities stimulate the mobilization of armed groups that intend to protect local communities under a real or perceived threat. Lacking trust in the protective capacities of the new army, such groups find a justification to take up arms under the pressure of both unresolved intercommunity tensions and the manipulations of political-military entrepreneurs.

The features sketched above do not only contribute directly to armed group proliferation; they are also essential factors in sustaining what might be termed "revolving door" military integration. As no clear end date was set for army integration, rebel fighters continued to be absorbed into the army, even after the last IB (the Eighth) left its CBR at the beginning of 2008. The knowledge that the door to the army would always stay wide open became an incentive for discontented groups and leaders to desert or to remain outside the army so they could later renegotiate their integration on better terms. This allowed dissident groups to turn threats of desertion or the rejection of army integration into bargaining chips. Furthermore, for those who desert, this policy provided incentives to produce violence, as this would guarantee them a better negotiating position in the next round of (attempted) integration. Rather than punishing dissidents and deserters, ongoing military integration rewarded them for it. As a result, some parts of groups alternately deserted and reintegrated, which led to significant volatility (Eriksson Baaz and Verweijen 2013).

Continuing armed group activity in the east was thus both a cause and a partial consequence of the stagnating army integration process (Hoebeke, Boshoff, and Vlassenroot 2008). Ongoing military operations, while indeed rendering army integration difficult, started to serve as a pretext for the slow progress in this field, masking a lack of commitment. At the same time, military operations offered multiple opportunities for the embezzlement of funds and justified the lagging civilian control over and the lack of transparency in the defense sector. This, in turn, facilitated the "leaking" of arms and ammunition to armed groups, which were an essential part of the militarized networks that underpinned parallel power structures in the army. In this manner, the military's imperfect integration and the existence of armed groups became mutually conditional.

The DRC presents a puzzling case of a peace settlement made up of a double political and military power-sharing deal, of which the political component more or less held, the military component was half-accomplished, and the "peace" was never fully achieved. During the "transition," a legal-rational institutional framework inspired by the Organisation for Economic Cooperation and Development was erected, which culminated in the inauguration of an elected government in February 2007. The long-term stability of this framework, as well as the extent to which it has promoted real changes in governance, would seem to have been doubtful. However, more than ten years after the signing of the final peace agreement, it still stands. Paradoxically, this might be in part precisely due to the half-achieved nature of army integration and the ongoing violence. By allowing the former warring factions to maintain sufficient military, economic, and political control, they remained committed to the "transitional" project. Yet at the same time, the half-brewed character of the new military enabled wartime power networks to remain partially intact and allowed for the manipulation of the military for private ends, thus blocking a transformation of the war economy. In sum, faltering military integration at once contributed to the relative "success" of the political deal *and* the failure to end the violence. In the following subsection, this paradox is further explained, by means of an analysis of the causal pathways identified in this research project.

Costly Commitment to Increase Trust

It appears that in the DRC, military integration worked to a limited extent as a mechanism to ensure an ongoing commitment to the transition. Ultimately, the transition created a type of self-sustaining momentum that made many groups believe they could not jump off the bandwagon anymore, as there appeared to be no better option than profiting from the rents and positions offered by the transitional institutional framework. Even most of the groups that did continue to resort to violence did so not to challenge the new national institutions but to perpetuate their control over local strongholds. This momentum, however, was very incremental, and it was caused more by the benefits of being part of the transitional institutions than by a commitment to military integration per se.

As mentioned, the ex-belligerents committed only halfway to military integration, thus hoping to reduce the costs of reversing this process, should the outcome of the elections not be in their favor. They also tried to hedge against a potential loss of influence in other ways, for example, by carving out spheres of influence in the administrative apparatus or by liaising with local militias which controlled local strongholds. This eventually made them give up most of their old military structures. As a result, the costs of a reversion to violence became too high. This is evidenced by the fact that some factions gained little from the elections but did not resort to violence afterward. Hence, in spite of the ex-belligerents' halfhearted commitment, military integration did hamper a return to violence and therefore helped sustain the transition.

Coercive Capacity to Provide Security

Paradoxically, the fact that the FARDC never developed a strong coercive capacity worked partly as a reassurance for the leadership of the different factions. A strong FARDC would have posed a direct threat to these elites' political and economic power projects, including to the transitional regime itself. Furthermore, a weakly institutionalized military allowed them to manipulate parallel power networks in the army, which were key to their own security. But the reassurance of a weak and fragmented army was only partial, because competing power networks within the military could always turn against one's own. This made it necessary not only to cultivate a powerful patronage network within the military but also to ensure a military backup outside it, by arms caches, a reserve of fighters that could easily be remobilized, or alliances with armed groups.

Employment of Possible Fighters

Although the FARDC offered only moderate reassurance as regards physical security, it appeared to be more reliable as a source of economic security to faction leaderships. Limited transparency and weak control allowed for the massive embezzlement of salaries, operational funds, military equipment, and money destined for the *brassage* process. Furthermore, all factions benefited from the diverse income-generating practices in which the military engaged. These benefits were sufficient to sustain commitment to army integration, at least for the leadership. For the rank-and-file, the process was less advantageous, and many initially opted for demobilization. But despite the promises of the DDR package, the military still proved attractive enough for tens of thousands of former combatants to enroll. However, their integration into the army did little to improve security within society at large, because it merely swelled the ranks of an army living off the population's back in an often abusive manner. In sum, the FARDC failed to work as a mechanism for improving security by means of offering employment to former fighters.

Symbol of National Unity

The bad behavior of the new military, along with the fact that its ranks were replenished with suspected war criminals, reduced its potential to assume a positive symbolic function and act as a signifier of national unity. Furthermore, the groups among the ex-belligerents that had a more pronounced ideational profile, particularly the most radical factions of the RCD-G and the Mai Mai, dropped out of the military. This reduced the military's potential to contribute to reconciliation, while also undermining its image as a neutral, unified national institution. Fierce internal power struggles, sometimes even leading to open clashes between different army units, further reduced the FARDC's chances to become a symbol of reunification. Instead, its perceived lack of neutrality appears to have turned it into a permanent reminder of the main fault lines within society—at least in the east, specifically the Rwandophone/autochthone divide. This impression was reinforced by the rampant distrust within its own ranks, especially among its officer corps.

Trust Inspired by Negotiations

As was already explained above, the negotiations for fleshing out the military integration deal did not work as a trust-building mechanism but proved highly divisive. It has been theorized that the increased interaction and joint problem solving entailed by the negotiation of power-sharing deals may lead to the creation of new rules, perceptions, and norms that foster reciprocities and trust. This allows for overcoming the two major stumbling blocks in negotiations: the problems of uncertainty created by unreliable and insufficient information and the perceived incredibility of commitment.[44] But nothing of the kind happened in the talks surrounding the military power-sharing deal in the DRC.[45] By contrast, these only seemed to reinforce uncertainty, as they painfully exposed both the lack of reliable information and the lack of a genuine commitment. They showed, for example, that none of the ex-belligerents was willing to be transparent about the real numbers and locations of their combatants and arms stocks. Furthermore, the ongoing violence constantly exposed the discrepancies between promises made at the negotiating table and realities on the ground. The presence of international mediators did little to overcome these problems, as they made few efforts to improve the transparency of military processes. Furthermore, they ultimately had little impact on parties' levels of commitment, in particular because they lacked the means to put military pressure on dissident factions. In this respect, it is important to note that MONUC's mandate was limited to promoting disarmament on a voluntary basis.

Alternative Outcomes

Some of the main features of the organization and functioning of the FARDC are remarkably similar to those of its prewar predecessors. This is not surprising if one takes into account the important continuities with the prewar order, in particular the nature and workings of the state (Vlassenroot and Raeymaekers 2004). The DRC is still a political order where power is projected to a large extent through Big Man networks and via "indirect rule" or accommodation processes with intermediaries (Tull 2005). State institutions, including the military, continue to function predominantly as vehicles for the extraction of resources, the maintenance of basic order, and the power projection of elites at both the national and the local levels, with public service provision being irregular and highly insufficient (Trefon 2009b). Furthermore, similar to the immediate prewar situation (but different from the heydays of centralized rule under Mobutu), the center has only a relatively weak grip over local power networks, especially where these are inscribed in powerful transborder networks (Vlassenroot and Raeymakers 2004). In such a context, it is difficult to develop a strongly institutionalized, coherent, and operationally effective military. It is not only undesired by elites, to whose interests it would pose an immediate danger, but it is also hampered by the center's lack of strength to bring under control the militarized networks in the east, both inside and outside the FARDC.

Research on transitions from autocracy has emphasized the importance of preexisting institutions for both the dynamics and outcome of the process (Bratton and Van de Walle 1994). Similarly, the new institutions created after civil wars are not purely the product of the contingent interactions of the key players in the transitional process but are strongly shaped by the structures of the social order in which this process takes place (Paris 2004). This principle of "structured contingency" also applies to military integration processes and their outcomes.

In this light, the strong continuities with the prewar order rendered a radical break with the DRC's military history unlikely. These continuities also reduced the chances for international actors to create very different outcomes. Donors might have contributed to generating a momentum that would have resulted in a less chaotic, more solid integration process, which would have yielded better initial results. However, in the face of the unchanged political-economic order, it is doubtful whether such an outcome would have been sustainable. It might have made a difference in the degree of institutionalization of the military, but could possibly not have fundamentally altered its nature. The track record of externally driven institutional engineering in Africa is not very impressive (Englebert and Tull 2008), especially not in the DRC (Trefon 2009a).

Most analyses of the DRC military integration process ascribe its lukewarm results to the following flaws: the lack of international commitment, weak coordination among donors, the failure to adopt a more political approach, and, finally, the DRC government's "lack of political will" (International Crisis Group 2006b; Melmot 2008; Hoebeke, Boshoff, and Vlassenroot 2008). All these explanations focus more on agency than on structure and tend to ignore the intricate nature of conflict dynamics and power politics in the DRC.

More resources and a more robust international involvement could have certainly jump-started the military integration process, given that they would have made available the logistical and financial means for rapidly transporting, cantoning, training, and equipping troops. However, "shock therapy"—in the form of the immediate dismantlement of factions' military structures—would have likely undermined their commitment to the transition, as it would have taken away their guarantees for physical and economic security and the bases of their political power.

Furthermore, although more intrusive involvement might have boosted the FARDC's operational effectiveness, the question is again how sustainable its results would have been. The history of Mobutu's army, the FAZ, is a telling example. It was only through extensive foreign tutoring by what were called "godfathers" at the time that the FAZ disposed of a number of operationally effective units (Young and Turner 1985). However, as soon as foreign tutors withdrew, the gains made were rapidly reversed. Aside from the issue of sustainability, high levels of foreign financing open up the dangers of further reducing accountability toward domestic constituencies and of entrenching existing power structures by enabling dominant actors to profit most from the incoming money flows (cf. Moss, Pettersson, and van de Walle 2006).

Concerning donor coordination, it is often believed that this would have produced more intensive defense transformation by allowing for concerted pressure and the tackling of more politicized agendas. However, it is doubtful that concerted pressure per se, if not complemented with additional resources, would have produced far-reaching and enduring change, especially in light of the unchanged nature of the state apparatus as a whole. And even if it would have made Kinshasa more committed to army integration, the weakness of the political center and the fragmented nature of the FARDC would have made it difficult to effectively implement top-down reforms. In this respect, it is often overlooked that Kinshasa's engagement in army reform is not only shaped by its "political will," but also by the "political space" available for imposing reforms. Though the political will to advance with military integration and army reform was indeed only limited, the political space to do so was equally circumscribed. It is not certain whether donors would have been able to create more such space because this would have required addressing a whole range of issues going beyond the narrow scope of army reform, like conflict dynamics in the east, regional interference, and the workings of the DRC's political-economic order in general.

Finally, it is an open question whether strong pressure on the ex-belligerents would not have triggered a relapse into war. As we have seen, it was precisely the half-achieved nature of the military integration process that enabled the transition to remain on track. It is only through the creation of a loose-knit and opaque military that the ex-belligerents could be induced to erect the shaky legal-rational skeleton of an OECD type of state. Arguably, it is also precisely the half-brewed nature of the resulting military that can make this skeleton eventually collapse.

ABBREVIATIONS

ADF	Allied Democratic Forces—rebel Ugandan group
AFDL	Alliance des forces démocratiques pour la libération du Congo-Zaïre—the most important government party, led by Kabila
ANC	Armée nationale congolaise
CBR	Centre de brassage et de recyclage—centers where integrated brigades were trained
CIAT	Comité international d'accompagnement de la transition—committee of donor countries and international organizations
CNDP	Congrès national pour la défense du peuple—breakaway group from RCP-G led by General Nkundu
DDR	disarmament, demobilization, and reintegration
DRC	Democratic Republic of the Congo

EUSEC	Mission de conseil et d'assistance de l'Union européenne en matière de réforme du secteur de la sécurité en RD Congo—defense reform commission set up by the European Union
FAB	Forces armées burundaises—Burundi rebel group
FAC	Forces armées congolaises—Congolese army
FAR	Forces armées rwandaises—former Rwandan army
FARDC	Forces armées de la République démocratique du Congo—Congolese army
FAZ	Forces armées zaïroises—former Congolese army of Mobutu
FDD	Forces pour la défense de la démocratie'Burundian rebel group
FDLR	Forces démocratiques pour la libération du Rwanda—Rwandan rebel group
FLEC	*Frente* de libertaçao do enclavo de Cabinda—Angolan rebel group
FNL	Forces nationales de libération—Burundian rebel group
GIA	Global and All-Inclusive Agreement—peace settlement
IB	Integrated Brigades
ICD	Inter-Congolese Dialogue—peace negotiations
LRA	Lord Resistance Army—Ugandan rebel group
MDRP	Multi-Country Demobilization and Reintegration Program of World Bank
MLC	Mouvement pour la libération du Congo—rebel group backed by Uganda
MONUC	Mission de l'organisation des Nations Unies en RD Congo—United Nations Mission to the DRC
RCD	Rassemblement congolais pour la démocratie—group organized by Uganda and Rwanda to remove Kabila, splintered after failing to do so
RCD-G	RCD-Goma—largest of the splinter groups
RCD-K/ML	RCD-Kisangani/Mouvement de Libération—one of the smaller splinter groups
RCD-N	RCD-National—one of the smaller splinter groups
SMI	Structure militaire d'intégration—planning body for integration
TG	Transitional Government
UNITA	União nacional para a independência total de Angola—Angolan rebel group
WNBF	West Nile Bank Front—Ugandan rebel group

ACKNOWLEDGMENT

This chapter is primarily based on literature research, but it contains additional information gathered in the course of thirteen months of PhD thesis fieldwork on civilian–military interaction carried out in North and South Kivu between January 2010 and February 2012.

INTERNATIONAL INVOLVEMENT

NOTES

1. Due to competing interests between Uganda and Rwanda as well as among local actors, the RCD fell apart into the larger Rwanda-backed RCD-Goma (RCD-G) and two smaller, Uganda-backed factions called RCD-National (RCD-N) and RCD-Kisangani/Mouvement de libération (RCD-K/ML). The district of Ituri in the northeast was not under control of any of these factions, but occupied by a variety of armed groups. Due to the proliferation of Mai Mai militia, control was also fragmented in other zones in the east.

2. Kabila was supported by the militaries of Angola, Zimbabwe, Namibia, Sudan, and Chad, while his opponents were aided by the Rwandan, Ugandan, and Burundian armies. The most important foreign rebel groups fighting on Congolese territory were the remnants of the former Rwandan army (FAR) and allied militia (*interahamwe*); the Burundian FDD and FNL; the Ugandan ADF, WNBF, and LRA; and the Angolan UNITA (Lanotte 2003; Reyntjens 2009).

3. This description is used by Vlassenroot (2004, 23) to describe the parallel governance structures that emerged during the Second War. These structures fostered "an independent process of politico-military control, redistribution of economic resources and rights to wealth" and were "essentially non-liberal, (or informal, transboundary) and socially exclusive (ethnic, clan-based) in nature."

4. Rwandophones refers to speakers of Kinyarwanda, in this context referring to both Hutu and Tutsi of Rwandan and Burundian descent living in eastern DRC. The presence of these populations, who migrated to the Kivus at different points in time, has generated tensions that have occasionally erupted into violence since the colonial era (Willame 1997).

5. The Mai Mai phenomenon has its historical roots in rural tribal militia that emerged in the colonial era and in the local militia that mobilized during the Mulele rebellion at the beginning of the 1960s. These militia resurfaced in the Kivus at the beginning of the 1990s, when strong tensions around access to power and resources, often expressed in identity-based terms, led to open violence (Vlassenroot and Van Acker 2001; Vlassenroot 2003).

6. Although Congolese Tutsi and Hutu (at times assimilated to Tutsi under the "Rwandophone" label) played an important role in its leadership and rank and file, the RCD-G was by no means a monoethnic movement. RCD-G was seen as a Tutsi-controlled movement because of its image as a proxy for the Tutsi-dominated Rwandan government rather than its composition (Tull 2005).

7. The signatories to the GIA were the government, the MLC, RCD-G, RCD-K/ML, RCD-N, the unarmed opposition representing twenty-eight different political parties, the somewhat ill-defined category of "civil society," and, finally, representatives of the Mai Mai. However, the latter were a heterogeneous and loose coalition, and not all groups felt represented by the delegation that signed the agreement. The armed groups from the Ituri district were excluded from the deal altogether. Groups from both of these categories continued fighting in the course of the "transition."

8. The TG was composed of President Joseph Kabila and the following vice presidents: Abdoulaye Yerodia Ndombasi (representing the former government), Azarias Ruberwa (RCD-G), Jean-Pierre Bemba (MLC), and Arthur Zahidi N'Goma (unarmed political opposition).

9. "Transition" is put in quotation marks here because it did not lead to a fundamental transformation of the political-economic order (Raeymaekers 2007).

10. These features are by no means unique to the Zairian army but can be found, in varying degrees, in many postcolonial armed forces in Sub-Saharan Africa (Howe 2001).

11. Although his power was ultimately based on the military, Mobutu did not draw upon it for legitimizing his rule. Instead, he tried to anchor his legitimacy in an eclectic mix of "ideologies," including Mobutisme and "authenticity," which emphasized the "rediscovery" of Zaire's supposed precolonial cultural roots, combining this with revolutionary rhetoric and a personality cult. In order to diffuse these legitimizing ideas and foster national unity, he mostly relied on the country's single political party, the Mouvement populaire de la révolution, and also on state propaganda and mass gatherings (Schatzberg 1988; Young and Turner 1985; Callaghy 1984).

12. According to Honoré N'Gbanda Zambo (1998), Mobutu's security adviser, the FAZ was nationally and internationally known as "Mobutu's army."

13. During a meeting in South Africa in January 1999, Angola, Zimbabwe, and Namibia agreed to include the insurgents in a potential cease-fire agreement. This idea was also propagated by Blaise Compaoré, then president of the Organisation for African Unity, and Fredrick Chiluba, president of Zambia and principal negotiator. Another leader who played an important role in preparing the ground for inclusive talks and a power-sharing deal was South African president Thabo Mbeki (Willame 2007).

14. The Lusaka agreement was signed on July 10, 1999, by Angola, Namibia, Rwanda, Uganda, Zimbabwe, and the Government of the DRC. The MLC signed on August 1, and the RCD, which was involved in internal leadership struggles, signed only a month later (International Crisis Group 1999a).

15. Tull and Mehler (2005) argue that the RCD and MLC insurgencies were, at least as far as their domestic leadership was concerned, partly a product of blocked political aspirations of sidelined elites, who aimed to insert themselves into the state system, rather than subvert it.

16. Though Angola mostly feared that a new regime would support the rebel group UNITA, Zimbabwe wanted to safeguard the business contracts it had signed with the Kabila government as well as the repayment of its investments in the AFDL campaign. Rwanda declared itself opposed to any form of peace agreement as long as Rwandan Hutu rebels remained active on Congolese soil, but it also preferred to maintain the status quo of war due to the economic benefits accruing from its military presence in the DRC (Lanotte 2003).

17. According to Articles VId and VIe of the GIA, the Superior Defense Council was to be composed of the president; all four vice presidents; the ministers of defense, the interior, and foreign affairs; and, finally, the chiefs of staff of the FARDC and the navy, air force, and ground forces; *Accord Global et Inclusif*, 2002, VId–VIh.

18. E.g., in 2005, the costs of the functioning of the presidency were budgeted at $17 million, and vice presidents' cabinets were granted $1.8 million each. Senators and deputies earned $1,000 a month, and had each been given a 4x4 vehicle (Willame 2007). Even though the budget for the *espace presidentiel* was already disproportionate to other budgeted posts (e.g., it was twice that of the Ministry of the Interior), several of its members grossly overspent their allocations, which led to enormous budget deficits (International Crisis Group 2006a).

19. Article 140 of the 2004 Law on Army and Defense refers to the presidential guard, but it provides no further details as regards it size, role, or functioning.

20. The TG promised several times that the presidential guard would be downsized and integrated, but eventually only one 800-strong battalion was integrated at the Kibomango site in Kinshasa in September 2006, while troop strength was never reduced (Amnesty International 2007).

21. The Comité international d'accompagnement de la transition (CIAT) was mandated by the Pretoria agreement to politically support the transition. It included fifteen states and international organizations and was chaired by the special representative of the secretary-general of MONUC. Its membership consisted of the African Union, Angola, Belgium, Canada, China, the European Commission, the European Union, France, Gabon, Russia, South Africa, the United Kingdom, the United States, Zambia, and MONUC (De Goede and Van der Borgh 2008).

22. EUSEC set up a politically sensitive project to separate the chain of salary payment from the chain of command and to establish a realistic payroll. Furthermore, the EU had been more outspoken in its criticism of the TG than individual EU member states (International Crisis Group 2006b).

23. Hesitation to fund the military integration process was in part related to the fact that most funding destined for security-sector reform could not be classified as official development assistance. However, the rampant corruption and lack of transparency in the DRC's defense sector, along with the military's abusive behavior, also played an important role (Onana and Taylor 2008; International Crisis Group 2006b).

24. The influence of certain elements from the RCD-G over the southern part of North Kivu, its wartime stronghold, remained largely intact during the "transition." Governor Eugène Serufuli, aided by the commander of the Eighth Military Region (North Kivu), both from the RCD-G, ran a parallel administrative and military organization consisting of roughly 10,000 to 15,000 local defense forces. Serufuli's entourage occupied key positions in the local administration and economy and had strong links to Rwanda (International Crisis Group 2005; Wolters 2004).

25. The plan provided for the establishment of eighteen infantry brigades of 4,200 troops each before mid-2006. A second stage was the creation of two to three brigades of a rapid reaction force, followed by the development of a main defense force capable of taking over from MONUC (International Crisis Group 2006b).

26. The eventual composition of the IBs would differ substantially among the brigades for both pragmatic and political reasons. As the funding and the logistical means to transport combatants over long distances were lacking, many of them wound up going to a CBR nearby. Additionally, many factions refused to send their combatants far away, fearing a loss of local influence. As a result, the principles of geographic distribution and faction balancing were flexibly applied.

27. Article 45 of the 2004 Defence Law stipulates that the Ituri armed groups signatory to the 2004 Act of Engagement as well as "all armed groups non-signatory to peace agreements" are still eligible for army integration.

28. Whereas at Sun City, the ex-belligerents had declared that the combined strength of their forces totaled 220,000, this number had mysteriously increased to 340,000 when the first

payroll of the new army was drawn up. The Superior Defense Council eventually brought it down to 240,000, but doubts about the real number of combatants persisted. This prompted South Africa to initiate a census in 2005, the results of which suggested that between 30 and 55 percent of the registered combatants were fictitious, leading to an estimated loss on salaries of $4 million a month. These numbers were rejected by the TG, which subsequently ordered the FARDC high command to organize its own census. The latter established the rate of ghost soldiers at 30 percent (Amnesty International 2007).

29. In 2006, soldiers were given a pay raise to $24 a month. However, given that salaries were systematically embezzled, this hardly translated into improvement for the rank and file.

30. Large numbers of child soldiers initially entered the integration procedure, but most of them were removed in later stages of the process by alert child protection actors (Amnesty International 2007).

31. The admission of elderly soldiers should also be seen in light of the absence of functioning pensions and welfare systems in the military. It would have been difficult to lay off long-serving aged or ill soldiers without any provisions for their survival. For the same reasons, the wives and orphans of deceased soldiers were sometimes allowed to be put on the payroll. In 2007 the number of soldiers to be retired was estimated at 60,000 (International Crisis Group 2006b).

32. Amnesty International (2007) reports such an incident for the central prison of Beni in North Kivu. In March 2005 the deputy regional commander ordered fifty prisoners to be released for army integration, fifteen of whom were suspected of having committed serious crimes. This was against instructions of the Tenth Military Region to release only military personnel detained on charges of less serious crimes.

33. An estimated 70 percent of the FARDC are officers and noncommissioned officers; only 30 percent are privates.

34. Senior army officers received around $50 a month; this was later increased to $60 to $70.

35. In mid-2003 the negotiations about the division of the regional commands ended in an impasse, which was only broken after intervention by the CIAT. The latter proposed a new division formula, while exerting heavy pressure on the ex-belligerents to compromise (De Goede and Van der Borgh 2008).

36. This diversity of approaches increased when first China and later the United States also started providing training to the FARDC under bilateral military cooperation agreements. Furthermore, in 2007, MONUC launched a short-term training program in order to retrain a number of integrated battalions, using still another training concept.

37. The conditions in the Mushaki and Luberizi CBRs were so appalling that 2,500 of the 6,000 integrated troops deserted between March and August 2005 (Boshoff 2005).

38. This lack of sharp ideological mobilization was mirrored in the political sphere, where, during the "transition," politicians regularly migrated from one faction to another—a phenomenon described by Obotela Rashidi (2004) as "political transhumance."

39. This was for example the case with the RCD-K/ML. Many of the officers of its armed wing never integrated into the FARDC but obtained powerful positions in civilian security and administrative structures in their former stronghold, the northern part of North Kivu. This allowed them to retain a high level of political and economic control over this zone.

INTERNATIONAL INVOLVEMENT   **162**

Furthermore, their leader, Antipas Mbusa Nyamwisi, decided just before the presidential elections not to run but to ally with Kabila, with whom he had been collaborating since Sun City. This assured the RCD-K/ML of three ministerial positions in the first elected government (Bucyalimwe 2007).

40. Local militia and armed groups maintained an often highly violent presence in the northern part of Katanga, South Kivu, North Kivu, and Province Orientale, most importantly but not exclusively in the Ituri district.

41. Mobutu's FAZ counted 60,000 to 70,000 at its high point (International Crisis Group 2006b).

42. The pay situation of rank and file would gradually somewhat improve due to EUSEC's payment chain project.

43. The various forms of cooperation between the FARDC and armed groups have been extensively described in reports by the UN Group of Experts (2009, 2010).

44. Rothchild (2005) presents an overview of theories of the long-term and short-term effects of power sharing.

45. Because the details of military power sharing had not been enshrined in the final peace agreement, and no concrete strategy was presented at the start of the "transition," negotiations about the shaping of the military integration process were ongoing and continued during the implementation phase.

Chapter 10

MERGING MILITARIES: MOZAMBIQUE

Andrea Bartoli and Martha Mutisi

Mozambique has been at peace now for longer than its previous civil war lasted. The peace period is remarkable because the country experienced a long war of independence before the internal struggle. Having been a Portuguese colony for centuries, Mozambique emerged as a political player through a military struggle with first the colonial power, and later civil war. The historical and cultural role of the military in the country has been marked by these prolonged armed conflicts. The very flag of independent Mozambique maintains the rifle as part of its essential symbolism.

Since October 4, 1992, the country has maintained a multiparty democracy while adjusting to very challenging conditions regionally. Its GDP has risen sharply, and while the HIV/AIDS epidemic has reduced average life expectancy, general living conditions have improved dramatically since the end of the war. This improvement would not have been possible without the merging of the military forces that fought a very bloody and violent civil war for more than sixteen years that caused extensive destruction and more than a million deaths, along with the internal and external displacement of 4.5 million people. The merging of the military made the state stronger, credible, and stable. It contributed enormously to the creation of the infrastructure that led to the country's postwar economic development, most significantly making possible one of the first major demobilization, disarmament, and reintegration programs in the world.

ORIGINS

A colony of Portugal since the late fifteenth century, Mozambique experienced a long independence war that ended with the military coup in Lisbon in the early 1970s. It can be argued that the prolonged and brutal war in Mozambique contributed to the decision of the Portuguese military to overthrow the government, thereby granting independence to the colonies. Frente de Libertação de Moçambique (Frelimo) was recognized as the legitimate authority in the transition and has led the new independent Mozambique since 1975. Its model for power distribution and state formation was the Soviet Union, and its centralization policies and its methods of social control led to the emergence of a resistance that was shrewdly used by neighboring Rhodesia and South Africa to oppose the government. Resistência Nacional Moçambicana (Renamo) was born from dissatisfaction with policies that abolished private property, disempowered traditional chiefs, and imposed relocations, among other measures.

Mozambique's success is due in large part to the interruption of a long cycle of violence and the introduction of direct negotiations between the Frelimo government and Renamo. Through the talks, Renamo became a political party, abandoned its military struggle, and became part of new political institutions. Frelimo, too, underwent a profound transformation. The shifts were encouraged by macro-level political change (especially the end of the Cold War and the subsequent global and regional realignment of power structures) that permitted a reimagining of Mozambican conflict and movement toward reconciliation. Still, the process defied the conventional wisdom of postconflict peace-building and set the stage for transformations based on important endogenous processes.

Since their inception (although at different moments in history), both Frelimo and Renamo had been military forces. They were used to thinking strategically, and the military aspects of their actions were always very relevant. Although the internal dynamics of the political–military relationship varied between the two, the military component has been key to the identity, operation, and trajectory of both movements.

The ability of the Mozambicans to end the civil war rested on the dynamic interplay of several critical components, one of which was the military assessment that neither opponent could win. Though Frelimo controlled the cities and the legitimacy of the state, Renamo was capable of controlling the countryside and challenging intermediate areas. Both sides finally calculated that neither could win, and that the war should therefore be ended.

The transformation of the conflict must be understood also in terms of the openness of Frelimo to political change, the redefinition of the Mozambican polity, the active facilitation of third parties, the state of the Mozambican security system, the importance of effective leadership, and the role of traditional and tribal mechanisms of redress. The peace process was in many ways constitutional in nature and contributed greatly to the redefinition of Mozambique as a state. In this context it is clear that the merging of the military forces was a crucial element in the state formation process.

An overarching theme of the Mozambique peace process is the idea of direct ownership of the process by the people it sought to serve. This is also true for the process of military integration. Conflict transformation in Mozambique happened in the streets and villages as well as in other countries. In many ways the process is still continuing. Although the risks of political instability are still palpable, especially concerning the issue of developing a truly inclusive and representative polity, the transformation in Mozambique remains a critically important case study of the successful merging of military forces.

The process that led to direct negotiations between Frelimo and Renamo to end the war in Mozambique was long and tortuous. Although contacts were established following Joaquim Chissano's appointment as president in 1986, it was only in July 1990 that the parties were able to send delegations to Rome for the first direct talks. These were facilitated by the Community of Sant'Egidio in Rome and observed by two of its members (the founder, Andrea Riccardi, and Don Matteo Maria Zuppi), a representative of the Italian government (Mario Raffaelli), and a Mozambican Roman Catholic archbishop (Jaime Gonçalves). These four were subsequently nominated as formal mediators for the peace process. The success of the first encounter was captured by the Joint Communiqué. In it the parties professed "to be compatriots and members of the great Mozambican family"; they

> expressed interest and willingness to do everything possible to conduct a constructive search for a lasting peace for their country and their people.... Taking into account the higher interest of the Mozambican nation, the two parties agreed that they must set aside what divides them and focus, as a matter of priority, on what unites them, in order to establish a common working basis so that, in a spirit of mutual understanding, they can engage in dialogue, in which they discuss their different points of view. The two delegations affirmed their readiness to dedicate themselves fully, in a spirit of mutual respect and understanding, to the search for a working basis from which to end the war and create the necessary political, economic and social conditions for building a lasting peace and normalizing the life of all Mozambican citizens. (US Institute of Peace 1992b)

The Emergence of Military Integration as an Issue

Although the military dimension was not explicitly mentioned in the Joint Communiqué, it had been included in the informal conversations that led to the meeting in Rome and was frequently addressed informally in the conversations of the observers with the parties. Frelimo's initial inclination was to quickly fix the conflict by achieving a cease-fire in a few months' time and address the other pending issues such as democratization, political reform, and devolution of power afterward. Renamo, which considered its military capacity its only leverage in the negotiation process, strongly opposed this vision and succeeded in postponing the cease-fire signature to the end. In 1990, officials of the US State Department attempted to negotiate a Christmas truce, which failed because no political settlement had been reached.

The idea of a military merger as a final outcome emerged almost immediately in the negotiation process. Renamo was a military power, and it was clear from the informal preliminary conversations that an agreement could be reached only if Frelimo was willing to create a new unified military force. Frelimo on its own wanted a small, agile military force under tight political control. The peace process was in many ways constitutional, which set the stage for the new postwar Mozambique. The redesigning of the military caused intense political struggle. All those involved agreed that their goal was not a Costa Rican option without a military but rather a unified military force that could support the new Mozambique's claims to independence, unity, and peace.

The peace process in Mozambique occurred while the United Nations was undergoing an intense transformation, with the Department of Political Affairs (DPA) transitioning from primarily organizing conferences, as it had done throughout the Cold War, to analyzing and responding to immediate political problems. Although the department attempted to assert itself as a significant player in many contentious issues, it was frequently resisted by officers working on the ground—as in the case of Aldo Ajello, special representative of the secretary-general. Some other departments of the United Nations also resisted the transformation of the DPA into a more relevant player. This was evident especially with the Department of Peacekeeping Operations (DPKO), which was younger, very operational, and traditionally had more resources and power than other actors within the United Nations, especially under the leadership of Kofi Annan. In the end, tensions within the UN system were actually healthy, helping the DPA, DPKO, and many other sectors grow dialectically.

Supporters, Opponents, and Neutrals

For different reasons, both Frelimo and Renamo endorsed the notion of merging the militaries. Renamo wanted to demonstrate that the new Mozambique was not controlled by Frelimo and that Renamo could contribute on an equal footing. By including Renamo in the state formation process without surrendering its own influence, Frelimo felt it could prove that the state was being reformed according to plan and still retain control.

The overall negotiating process was conceived in phases. First, there were preliminary informal conversations, with the observers and mediators. These set the tone for the negotiations and established the objectives of both delegations. The second phase was the discussion within the formal mediation process that led to the approval, on October 4, 1992, of Protocol IV of the General Peace Agreement, to be applied simultaneously with the other outstanding protocols (US Institute of Peace 1992a). These negotiations were delicate, complicated, and somewhat long. There were several unknowns, in particular the reaction of the militaries themselves to integration. The third phase was the direct negotiation between the military leaders that followed the approval of the political framework. This negotiation was unexpectedly pragmatic and effective. At all these stages the UN offered some input. The final stage was implementation, and this was executed within the context of the United Nations Organization for Mozambique (ONUMOZ).

Those who were involved in the operations stressed that the military merger—like the demobilization, disarmament, and reintegration (DDR) activities—was highly political, with collective and personal dimensions.[1] Fighters who did not know much besides how to fight had to decide whether to trust the new military force and enlist or choose civilian life with all the uncertainties of the postconflict conditions. Security was part of the calculation, as were economic incentives and the general outlook of the country. DDR operations end up being a sort of de facto indicator of faith in the future of the country. If fighters trust their own country's capacity for peace and development, they will tend to demobilize faster and better, reintegrating into productive civilian life. DDR—at least in the case of Mozambique—was not a process controlled by commanding officers. Rather, it was the expression of a generalized mood and individual choices. The UN has been grappling with these issues since the Mozambique case, which was one of the first instances of DDR and, partially because of the quality of the agreement reached in Rome, one of the most successful to date.

In general, the leaders of both Frelimo and Renamo were all in favor of merging the military forces. The main point of contention was the structure of the state, which initially was patterned by Frelimo after the Soviet model and then was transformed into a multiparty democracy. Once the fundamental transition was under way, the question became one of managing continuity (Frelimo's priority) versus imposing structural changes (Renamo's priority). The conflict in Mozambique was never a secessionist one. Although Renamo controlled many of the rural areas and Frelimo controlled many of the urban ones, neither side had ever called for separate governments.

During the first, informal phase the primary actors were the heads of the two delegations, Frelimo's Armando Guebuza, minister of transport and communications, and Renamo's Raul Domingos, head of the Department of External Relations. Each had direct access to his group's leader (Joaquim Chissano and Alfonso Dhlakama, respectively) and to its military structure (the head of the Frelimo military force was Guebuza's son-in-law; Domingos was himself proud of his military rank of general and had contributed to Renamo's strategic reasoning).

The role of the observers and mediators should not be underestimated, especially in this first phase. On countless occasions, contentious issues such as the merging of the militaries were informally explored with their participation. It was clear from the outset that the observers and mediators, especially the Italians, did not have a specific outcome in mind and were not defending preconceived interests. Rather, they were using their openness to ask questions that led to new explorations. This engagement was particularly important for Renamo, which did not have established political positions and grew in understanding and clarity through this process. Through these interactions, the participants decided to put the issue of the military on the formal agenda.

The second phase was the actual examination of military issues (not only the merger). Once again Guebuza, Domingos, and the mediators and observers were the participants. However, during these exchanges the parties' leaders (including military leaders) became more involved; though they were not themselves present in Rome, the discussions reflected some degree of exchange with Chissano, Dhlakama, and the military leadership of both Frelimo and Renamo. During this second phase major issues—especially in the informal conversations—were the internal tensions between the political and military leadership within each group (especially for Frelimo) and concern about how militaries of both parties would meet and negotiate directly. One major breakthrough was the Agreement on a Partial Ceasefire signed on December 1, 1990.[2] This agreement, which was negotiated by Guebuza and Domingos and received the green light from Dhlakama, allowed Zimbabwe to remove its 7,000-strong troop contingent in central and southern Mozambique and brought three Renamo military commanders to Maputo—which until then would have been unimaginable.

A major concern was whether combatants would comply with the agreement. It was important that the compromise that ended the war be seen as honorable and sensible enough to be acceptable to those who had suffered the most. Western observers, including Stephen Stedman at the US Institute of Peace in July 1992, worried about fragmented armed groups unwilling to accept an unsatisfactory agreement.

The protagonists of the third phase, in which military leaders from both sides negotiated the implementation of Protocol IV, were Tobias Dai (Frelimo) and Hermínio Morais (Renamo). The meetings were more frequent in the last year of the long (more than twenty-seven months) process that was subsequently concluded in Rome; they were not conclusive but were extremely important because both sides expressed a high degree of pragmatism and flexibility, which eliminated all fears of military resistance to the political agreement and set the stage for effective implementation. It was also very useful to have the actual procedural elements of the merger discussed and agreed upon. One particularly important agreement was that Renamo fighters would retain their former military ranks even if they had never had proper training. Also crucial was the fact that Renamo was guaranteed a set of positions in the newly designed military force; this reassured the rank and file as well as the officers that the new structure was balanced and truly "Mozambican." Dhlakama himself sealed the legitimacy of this settlement in a live radio broadcast to all of Mozambique at the concluding October 4, 1992, ceremony, when he ordered his men to stop fighting. And that is what has actually happened! It was a political speech but also a military order, imparted by the commander in chief himself.

The fourth phase focused on implementation, including the interaction between internal and external actors, which was increasingly important. The difficulties were clear at the first encounter between Aldo Ajello, in his role as special representative of the UN secretary-general, and the representatives of the parties. The timetable negotiated in Rome had been extremely ambitious. But because the UN did not have substantial military forces on the ground, immediate action was not feasible. The schedule had to be renegotiated to give external actors a chance to quickly instruct Renamo personnel who were to handle key tasks in the transition. This improvised training, organized by Dhlakama with the help of foreign scholars, was endorsed by Ajello as a way to facilitate the transformation of Renamo into a political party.

Mozambique's military merger put stress on a United Nations system led by a special representative of the secretary-general who did not want to have much done outside the UN framework. This created tension both with Frelimo leadership and with UN headquarters in New York, which felt unprepared for the tasks. These issues were worked through in the Joint Commission for the Formation of the Mozambican Defence Force (CCFADM), which had been created by Protocol IV. It was composed of representatives of the government and Renamo, who were assisted by representatives of the United States, the United Kingdom, France, and Portugal; and it was chaired by the United Nations. In the end, the new UN chair proved to be extremely effective.

Compromises in the Process

The most significant compromise involved the very nature of the transition. Was the new military going to be the old Mozambican military force with some inclusions from Renamo, or a completely new force, as Renamo was advocating? The tension between Frelimo's desire for continuity and Renamo's desire for transformation informed the whole negotiation process and involved other issues as well, such as multiparty representation and elections. However, Frelimo had already started significant internal movements to change structurally. With the end of the Soviet bloc and the transformation of South Africa, the whole southern Africa region was politically in flux, and Frelimo's response was pragmatic, although it was not possible after a violent war and years of enmity to easily carry out an institutional makeover. Renamo was entitled to equal representation in the proposed new force of thirty thousand, to consist of fifteen thousand Frelimo and fifteen thousand Renamo troops. It also achieved a significant presence at the level of military leadership, even in areas (such as the navy) where it had not previously been present.

The war had exhausted both the population and the parties. In particular, life in both armies had been very hard, with low pay, terrible lodging, and scarcities of both food and equipment. At the end of the war, most people wanted to end the hostilities and return to civilian life.

CREATION

The new military was shaped by the two competing organizations. Fewer people than expected wanted to join, so the total force was considerably less than anticipated; but in the long run this probably made it more sustainable. Training was provided by outside powers and stressed the need for political allegiance as well as more conventional military skills.

Shaping the New Military

Procedurally, one of the most important steps in the merging of the militaries was the creation of the CCFADM. It was responsible for overseeing the process of forming the new military, the Forcas Armadas da Defesa de Moçambique (FADM). FADM was headed by a high command and placed under the authority of the new Ministry of Defence. The name of the new force took some time to negotiate, for ideological and reputational reasons (Renamo had suggested Forças de Defesa de Moçambique, FDM).

The fundamental decisions regarding what the new military would look like were made by the parties themselves. This was characteristic of the whole process. The input of external actors was not controlling but proved to be useful. In fact, the success of the Mozambican negotiations can be ascribed to the capacity of the mediators to enhance Mozambicans' ownership of the process. Both Frelimo and Renamo were proud of their participation and kept control of the process with the assistance of the mediators. The lack of dominant external actors allowed the parties to explore their concerns in a way that stressed cooperative security. The representatives of the government and of Renamo initially alternated chairing CCFADM. This arrangement made decision making almost impossible because each chair had the incentive to forestall delicate decisions until he was in the leading position. After some inconclusive meetings, it was agreed that the chairmanship of the commission would be held by a representative of the United Nations.

Merging the militaries necessitated not only trust but also an institutional commitment to let the security of each party depend on the compliance of the other with the new structure. The Mozambicans themselves, through the cooperation of Frelimo and Renamo, determined what the new military would look like. Within the parameters negotiated in Rome and spelled out in Protocol IV, the leadership of the two military forces agreed on procedures and implementation. This centrality of the Mozambican leadership led to the new military's subsequent flexibility to address challenges (unexpected by some) such as the choice of civilian life by many ex-combatants and the consequent lack of military personnel.

Personnel Selection

Frelimo and Renamo made the first personnel selections. As noted above, the militaries of both parties were ready to demobilize. The number of those who wanted to enlist in the new army was relatively low, and in many cases the process was not that of selection among many candidates but rather of strongly encouraging people to take part in the new regime. It must also be noted that ONUMOZ had an interest in demobilizing and offering alternatives to those enrolled in the military. It offered a very generous long-term package that allowed each demobilized soldier to receive up to two years of compensation. Every two months those who were demobilized could redeem vouchers through any bank in their hometown. This was a very clever innovation that induced early reintegration in ex-combatants' areas of origin. It encouraged compliance and supported the redirecting of productive energies to new opportunities, especially in agriculture. Many ex-combatants were given land and were able to start new self-sustainable lives with their families. Only a minority enrolled in the new armed forces.

Both forces were smaller than they had claimed. Although this was somewhat expected for Renamo, which tended to inflate its numbers at the negotiating table to maximize its influence, the political leadership of Frelimo was surprised to discover that the numbers for its own armed forces were not realistic either. As a result neither side was able to produce its fifteen thousand soldiers, and because of this the new force was smaller than anticipated; the final count was about ten thousand, with about seven thousand coming from the government and the other three thousand from Renamo. The current force is the same size and is certainly sustainable in terms of size.

Rank Allocation

Military experience was relevant for those who were offered officers' positions. As mentioned above, Renamo's ranks were recognized by Frelimo and made operative in the new force. However, Renamo and Frelimo officers were clearly different. When the British started the training of the unified infantry, the parties selected the officers who were to become trainers themselves. Notwithstanding the fact that they were in many ways high-quality officers, the Renamo representatives were in no way formed militarily. Yet after just a few months of joint training, the officers from the two groups were fundamentally indistinguishable.

No screening for human rights violations took place; this was consistent with the general tone of the peace agreement and led to a de facto amnesty. Some years before the talks, the Frelimo government had unsuccessfully offered a unilateral amnesty, which was viewed by Renamo as a patronizing attempt to neutralize it.

Both Frelimo and Renamo participated in the selection of the officers and the noncommissioned officers, but, as stipulated in Protocol IV, the FADM command structure was strictly apolitical and would receive directives and orders only through the appropriate chains of command. FADM was commanded by two generals of equal rank, one appointed by each party, and command decisions were valid only when signed by these two generals. FADM had a single logistics service for all three branches; the Logistics and Infrastructure Command was established under the authority of the FADM High Command. Appointments to the FADM High Command, as well as to the commands of the three branches of FADM and the Logistics Command, were proposed by CCFADM.

Training

The need for training was identified in Protocol IV, and training was made one of the responsibilities of CCFADM. The United Kingdom provided training for the army, while Portugal focused on the navy and special forces. The French played a marginal role in training a de-mining team. Although China had offered military support to Frelimo during the civil war, Sino-Mozambican relations after the war were mainly diplomatic and economic. China was not involved in military training following the end of the civil war.

The focus of training—especially as conducted by the United Kingdom for the army—was on the need for allegiance to the new political institutions and the legitimate government. In an interview, Aldo Ajello recalled the transformation that occurred in the trainees, who after a few sessions were ready to participate in the new structure and were respectful of the new chains of command. This is another proof of the benefits of military pragmatism vis-à-vis political and ideological rigidities.

The major problem during training was the lack of formal military experience among the Renamo members, who had learned military exercises mainly as rebel forces. The level of illiteracy was very high, and while discipline was not necessarily a problem, the challenges of fitting into a formal structure were evident. All training before and during the merger was done under the auspices of ONUMOZ and took advantage of the lessons learned by the DPKO, at that time led by Kofi Annan, with whom Ajello had a very good working relationship.

When we analyze the negotiating process, we observe that the very effective and successful discussion of military integration by military leaders increased trust enormously between the elites of the two sides. It was not necessarily a surprise that the military leaders could negotiate directly; but the speed and quality of their interactions, as well as the effectiveness of the merger process, was most certainly unexpected. After the agreement on the military protocol, the negotiators moved on to other issues with much greater confidence that agreement could be reached, and this made renewed civil war less likely. The actual existence of the military was not irrelevant, because the military component was actively participating in the process. Moreover, the synchronization of political and military negotiations strengthened the general mutual security discourse that emerged through the direct negotiation formula. Mozambicans were in charge, and their agreement on the military demonstrated that even the most fraught issues could be addressed and resolved successfully.

The ability of the military to reimagine itself as an integrated institution in a divided country became a confirmation of emerging national unity. The existence of the military was not in itself as important as the fact that it was indeed an integrated institution—the first in the new country. Its actual military capabilities remain irrelevant. What was relevant is that the military institution was integrated, inclusive, and subject to legitimate political control. Political elites trusted their counterparts in the emergent political system more because of the successful negotiation of the merging militaries, and were able to move on to other issues with some confidence that agreement could be reached, which made renewed civil war less likely. The negotiation of the military merger created the first experience of trust among the key players within the forces, and this trust was strengthened even further during the shared training. Confidence and trust were transmitted from civilians to the military and vice versa in a process of mutual reinforcement. The military was willing to support the elites, who advocated cooperation with former enemies and therefore strengthened the civilian investment in the peace process.

The negotiations stipulated not only that the new military would employ former combatants (who would otherwise be unemployed because of their lack of skills and the weak economy) but that they would be included at the highest possible rank. This type of recognition was as important as actual monetary compensation in making it difficult for elites on either side to mobilize forces for renewed violence.

OUTCOME

The new military seems firmly under the control of the civilian government—even though, because there is no formal power sharing, this means under Frelimo control. It appears to be sustainable both in terms of numbers and expenditures.

Political Control

Frelimo chose a strategy of no formal power sharing at the level of government, and while Renamo has a significant presence in parliament and in regional assemblies, the actual control of the military force lies with the state, especially the president. Significantly, the current (reelected) president of Mozambique is Armando Guebuza, who negotiated the General Peace Agreement for Frelimo in Rome. The shift in allegiance from the divided armies to the unified one occurred relatively early in the transition. UN observers noticed that immediately after the training offered by the British to the army officers, the troops participating in the unified structure were openly pledging allegiance to the proper line of command.

Some authors have suggested that the new military was fairly independent because its civilian superiors, in both the executive and the legislative branches, were either unable or unwilling to exert much control over it. However, the first minister of defense after the agreement was signed was Aguiar Mazula, a Frelimo leader who had participated in the Rome negotiations. Creating budget and promotion policies was a prerogative of his ministry and, though the capacity of a state such as Mozambique in terms of command and control is often doubtful, it is fair to acknowledge that the bureaucracy has been fundamentally in the hands of Frelimo.

Military Capabilities

The force performs all fundamental functions of a military, not only remaining in existence without its members killing one another in large numbers but also contributing to the development of Mozambique as a unified, independent, and peaceful country. Unlike other war-torn countries, such as Angola, Mozambique's actual military expenditures were relatively low even in the midst of the civil war, never reaching more than 3.5 percent of GDP. It is now less than 1 percent. Mozambique's GDP has grown significantly since the end of the war, which accounts for some of this reduction in percentage.

The unified military force is deployed in all areas of the country and is not bound by ethnic or local divisions. On the contrary, its deployment strengthens state institutions and legitimacy.

We do not have enough knowledge of other cases to compare Mozambique's military force's ability to defend itself against foreign attacks from other states. However, this force was not expecting a foreign attack at the time of writing. Rather, it has contributed greatly to the development of the Southern African Development Community (SADC), a regional economic organization that deals with significant security issues.

In terms of internal use, the state is now unified, independent, and at peace. The president of Mozambique is the negotiator of the General Peace Agreement for the Frelimo side, Armando Guebuza. At no time during the period when the agreement was signed and implemented by ONUMOZ (1992–94), did military force need to be used on a massive scale. It seems plausible that force would have been used against any group in response to legitimate orders.

The resumption of large-scale violence became less likely in Mozambique after the peace agreement negotiated by the Mozambicans themselves in Rome was welcomed by the population at large. The response to the signing of the peace agreement was very positive. People danced in the streets for days, and many refugees and internally displaced people started going back home without waiting for the UN to organize their repatriation. Both the war and the peace process were truly Mozambican. Because the Mozambicans themselves truly owned the military merger, it contributed to the larger phenomenon of unification that has made the resumption of large-scale violence less likely. However, not all grievances have been addressed, and some have mentioned the example of Kenya as a model for what happens when needs are unsatisfied.

Sustainability

How big should a military force be? What would be the right size? The arbitrary number negotiated in Rome was never met. Renamo fell considerably short of fifteen thousand troops, and even the government was unable to mobilize that number. In the end no one complained, and all involved accepted the existence of a smaller, sustainable military force.

The state can probably maintain this military force indefinitely, economically and politically. It is relatively small, not too costly, and now not at war. Its participation in the region is of paramount importance for the country, and it is therefore highly improbable that Mozambique will embrace a military-less solution (à la Costa Rica).

Mozambique is also part of a still-unstable region where its military presence can be an asset. The country must be part of all continental, regional, and subregional organizations, and the presence of a credible military force helps counteract regional instability. Some observers have advocated transforming SADC into a more robust structure, as the Economic Community of West African States has done. The unified military force could contribute more to international peacekeeping operations if training and equipment were up to standard. It could be argued that the modernization of the new military force should be part of the larger goal of strengthening state institutions in the country and its projected power internationally.

Alternate Outcomes

Several possible circumstances could have produced a different outcome. The issue of power fragmentation was directly raised by Stephen Stedman in July 1992:

He anticipated that Dhlakama would not be capable of controlling Renamo and predicted a proliferation of local feuds that would make Mozambique ungovernable. The hypothesis was plausible but fortunately was proven wrong by Renamo's ability not only to fulfill its obligations during the negotiations but also to contribute to the implementation of the peace agreement. Strengthening Renamo was central for Ajello and ONUMOZ. Throughout the process, on matters of any importance the special representative of the UN secretary-general communicated with Renamo only through Dhlakama. When other interlocutors to Renamo suggested themselves, they were consistently turned down by the internationals. Dhlakama was given direct access to the funds made available by the international community to facilitate the transformation of Renamo into a political party. The more the UN recognized Dhlakama, the more he was able to control Renamo and do what was essential to move the peace process forward.

The United Nations' perception of the General Peace Agreement has been very positive. Secretary-General Boutros Boutros-Ghali spoke of the "original mix of institutional and noninstitutional approaches" and of an "Italian formula":

> For many years, the [Community of Sant'Egidio] worked with utmost discretion . . . in order to bring both parties in contact with each other. It did not keep those contacts for itself. It was very effective when it came to involving others who could contribute to a solution. The Community let her technique of informal discretion converge with the official work of governments and of intergovernmental organizations. Since this experiment, the expression "Italian formula" has been coined for this unique combination of government work and non-governmental peace efforts. Respect for both parties in conflict and for those who work in this area are indispensable for the success of similar initiatives.[3]

However, mismanagement by the United Nations was a second cause of possible disaster. At this time it was greatly expanding its peacekeeping missions, and, while its internal capacity was growing, the expectations implied in the Rome agreements (negotiated in the presence of a UN envoy) could not be met. The UN did not have access to personnel to fill the positions of ONUMOZ, and the soldiers arrived seven months after they were expected. Peace was maintained less through the presence of the Blue Helmets than by the restraint of the warring parties, together with the UN's creative capacity to project a greater presence than it had actually provided. The first few months on the ground were extremely difficult for the UN mission because the DPKO resisted the bolder role of the Department of Political Affairs. Fortunately, this particular relationship was resolved by the forceful intervention of Boutros-Ghali himself.

Weapons were a third concern. They were readily available after the war, and many demobilized veterans retained theirs. The work done first by ONUMOZ and then by the Mozambicans themselves after 1994 was an extraordinary contribution to stabilization. ONUMOZ focused on creating the political conditions for a stable peace. It did not spend too much energy on chasing weapons that were at that time extremely cheap and omnipresent. Rather, it made the use of the weapons counter-productive. It also offered generous packages to those who were willing to move to civilian life. It followed an incentives strategy rather than a repressive one. It was clear to the UN officers that in the postconflict situation in Mozambique it was very difficult to even know how many weapons were in circulation. Renamo wanted to claim its power and relevance and did not have an incentive to reveal how few weapons it had. Even Frelimo forces did not always have all the equipment they were claiming. Light weapons were widely available to the fighters and the general population.

Managing the DDR process was also a challenge for the UN system. DPKO was in charge of demobilizing and disarming, but the United Nations Development Program had the reintegration mandate. The UN's Office for the Coordination of Humanitarian Affairs, which was just beginning to coordinate the humanitarian dimension, was underfunded and had no proper tools. As a result, though many from both Renamo and Frelimo decided not to enlist in the new military, largely for economic reasons, the implementation of the DDR process was at times poorly organized, with an occasional lack of food and proper lodging that led to unrest and violent expressions of dissatisfaction. In the end the tensions were diffused by appropriate responses, but the risk of local breakdowns was real.

HAS MILITARY INTEGRATION MADE LARGE-SCALE VIOLENCE LESS LIKELY?

Military integration may make large-scale violence in Mozambique less likely for the following reasons, taken from the causal chains suggested in the introduction. First, as a *mutual commitment*, military integration reassured both parties and created mechanisms of control that were future oriented. It also sent a message to other groups that the armed factions were willing to run political risks for a settlement, both by weakening themselves and by opening themselves to political attack from radicals in their own organizations. The debate was especially intense within Frelimo, and the choice of Armando Guebuza as its chief negotiator in Rome was the political masterpiece of President Chissano, who wanted to make sure that the radical fringe of the party (represented by Guebuza) was fully involved in the negotiation with Renamo.

Although military integration was indeed a costly commitment for both sides, the costs were calculated against the higher ones of continuing hostilities. Moreover, having recruits from both sides in the security system lessened the likelihood of its being turned against either side. Also, arming and training together changed the calculation for the belligerents regarding the benefits of renewing hostilities. The new military became a source of incentives in itself; if hostilities are launched again, very tangible things—salaries, rank, respect—are going to be lost. Renamo accepted dissolution into the new army, betting that security would be provided by the political agreement rather than by military control of a faction.

Second, the new military has sufficient political legitimacy and effectiveness to provide security, both for elites and for the population at large. Military integration proved that each side believes this power is unlikely to be used against it. This is the result of a shared political agreement that reconstituted the polity, designed new institutions, and allowed for a new social contract to be established throughout the country. The merged military has in many ways been the first product of this unprecedented (for Mozambique) collective enterprise. Though still in flux and not as effective as many had hoped for, government institutions have addressed underlying disputes politically, which makes renewed civil war less likely.

The increased security that was made possible by the political agreement and the integration of the military allowed a very significant economic development, which in turn will make renewed civil war less likely if the benefits are shared fairly. Although there has been significant economic progress, one of the emerging risks in Mozambique is that—much as in Kenya—the concentration of power in certain regions and tribal enclaves by Frelimo will fuel widespread dissatisfaction among underrepresented groups. Economic inequality, coupled with inadequate representation at the political level, may renew the push for more violent representation by the disaffected population.

Third, the new military did not originally have sufficient coercive capacity to provide security, but, because of the political pact that made military integration possible, it was able to give employment to many former fighters on both sides; this made it difficult for elites on either side to mobilize forces for renewed violence. Mozambicans, especially the combatants, were exhausted by sixteen years of civil war. The new military did not enforce peace by its coercive threat, but the enlistment of many former and potential fighters certainly made the recurrence of violent conflict less likely.

Fourth, Mozambique's military merger was an expression of political agreement that provided the security needed to reassure elites that they could indeed work together and successfully produce economic and social development. Indeed, the ability of the military to exist as an integrated institution in a divided country made it a symbol of national unity, inspiring other groups to emulate it. The existence of the integrated, merged military is important; its actual capabilities are less relevant.

Fifth, the process of negotiating military integration increased elite capacity and mutual trust. It was a crucial step in the process of learning to govern together—not through formal power-sharing agreements but rather through the establishment of a unified polity and the launching of institutionalized political dynamism. By supporting the peace agenda, the military also supported in a crucial way the emergence of a unified, independent, and peaceful Mozambique that had simply never existed before. This process was an affirmation of what the political leadership had been negotiating in Rome, and indeed it made renewed civil war less likely. The actual existence of the military is, in the case of Mozambique, somewhat irrelevant. But the agreement had an important military component, and the speed and quality of the merger had huge effects in establishing the viability and durability of the peace process.

ABBREVIATIONS

CCFADM	Joint Commission for the Formation of the Mozambican Defence Force
DPA	United Nations Department of Political Affairs
DPKO	United Nations Department of Peacekeeping Operations
FADM	Forcas Armadas da Defesa de Moçambique—new army

Frelimo	Frente de Libertação de Moçambique—government party
ONUMOZ	United Nations Organization for Mozambique
Renamo	Resistência Nacional Moçambicana—rebel party
SADC	Southern African Development Community

ACKNOWLEDGMENT

This chapter draws on many sources, including direct interviews with Father Matteo Maria Zuppi, who was one of the mediators of the Mozambique peace process; Aldo Ajello, who served as the special representative of the UN secretary-general to Mozambique and led ONUMOZ from 1992 to 1994; and Dmitry Titov, now the DPKO's assistant secretary-general for the rule of law and security institutions, who during the peace process was in charge of the DPKO's Mozambique dossier. It was also edited and commented upon by Leone Gianturco, who—as a member of the Community of Sant'Egidio—followed the peace process in Mozambique very closely. The authors thank them all for their outstanding contributions.

NOTES

1. A well-informed observer noted that it was at the time of the Mozambique peace process that demobilization, disarmament, and reintegration emerged as a fundamental function of complex operations, receiving sustained investment and institutional recognition within the United Nations.

2. The Agreement on Partial Ceasefire was signed by delegations of the Government of the Republic of Mozambique and Renamo, led respectively by Guebuza and Domingos. They met in Rome at the Community of Sant'Egidio headquarters in the presence of the mediators: Raffaeli, Goncalves, and Riccardi and Zuppi of the Community Sant'Egidio. Participants agreed to implement the agreements reached on November 9, 1990. For the details of the partial cease-fire agreement, see the Community of Sant'Egidio website, www.santegidio.org/archivio/pace/mozamb_19901201_EN.htm.

3. This is from www.gppac.net/documents/pbp/11/6_egidio.htm.

Chapter 11

BOSNIA-HERZEGOVINA: FROM THREE ARMIES TO ONE

Rohan Maxwell

The Republic of Bosnia and Herzegovina (BiH) declared its independence from a disintegrating Yugoslavia in April 1992, the third republic to do so and the one to suffer the most severe consequences. The population was almost entirely made up of three ethnic groups.[1] About half were Bosniaks (Bosnian Muslims); one-third were Bosnian Serbs (Orthodox Christian); and a little less than one-fifth were Bosnian Croats (Roman Catholic). In general, Bosniaks strongly supported independence from Serb-dominated Yugoslavia; Bosnian Serbs preferred to remain part of Yugoslavia, or at least to separate themselves from BiH; and Bosnian Croats vacillated between supporting independence and seeking their own separate part of BiH. A complex and vicious conflict consumed roughly 100,000 lives and displaced millions before the fighting ended in December 1995 with the signature of the General Framework Agreement for Peace (GFAP) and the establishment of a new State of BiH consisting of two largely autonomous entities. The Federation of BiH (FBiH) was almost entirely populated by Bosniaks and Bosnian Croats, who had begun the war as allies against Bosnian Serbs, then fought each other for a year, and then become uneasy allies again. The Serb Republic (Republika Srpska, or RS) was almost entirely populated by Bosnian Serbs. The new country was a complex creation dominated by mistrust among the entities and ethnic groups it contained.

Under GFAP, the state possessed no armed forces, but the entities retained large, conscript-based armies: the Army of the FBiH (Vojska Federacije BiH, or VF), with well-separated Bosniak and Bosnian Croat components (based respectively on the wartime Armija Republike BiH, or ARBiH; and the Hrvatsko Vijece Obrane, or HVO); and the Army of RS (Vojska Republike Sprske, or VRS). GFAP established two international organizations to oversee the peace. The Office of the High Representative (OHR), headed by the high representative (HR), would be responsible for the civilian aspects of implementation; and the Implementation Force (subsequently the Stabilization Force, or SFOR), led by the North Atlantic Treaty Organization (NATO), would be responsible for military aspects.

In 2001, BiH decided to seek membership in NATO's Partnership for Peace (PfP), and in response NATO set out specific political and military requirements, including democratic, state-level parliamentary oversight of armed forces (North Atlantic Treaty Organization 2001). This provided a positive but limited impetus for BiH authorities to begin to address such issues. However, the trigger for the externally driven, thirty-two-month process that replaced the entity armies with a single military force was a May 2003 decision by the HR that the lack of state-level command and control over the entity armies posed a threat to the civilian implementation of GFAP. Using his GFAP authority, he established the Defence Reform Commission (DRC) to develop recommendations and related legislation for the establishment of state-level command and control over the entity armies (OHR 2003).

The DRC, which was made up of members and observers from the international community and BiH, was chaired by a former US assistant secretary of defense.[2] He used intensive shuttle diplomacy to overcome the reluctance of BiH commissioners to engage in substantive work without clear guidance from the leaders (elected or otherwise) of their respective ethnic groups. Under this firm leadership, but using a consensus-based, inclusive approach, the DRC worked very quickly by BiH standards and submitted its recommendations and draft legislation in September 2003 (Bosnia and Herzegovina 2004). The inclusive approach paid dividends when key BiH political leaders supported the legislation (Haupt and Fitzgerald 2004).

The new BiH Law on Defence entered into force in December 2003, placing the VF and VRS under a minimal, state-level command-and-control structure: the tripartite BiH presidency, as the commander in chief; a small Ministry of Defence (MoD); and the Joint Staff. This hybrid structure became the Armed Forces of BiH (AFBiH). The practical authority of the new state-level organizations was limited to coordination and setting standards, and even this was to prove problematic in implementation, particularly because the ethnically balanced triumvirates established at all senior civilian and military levels ensured that wider ethnic agendas and disputes would be reflected in the defense establishment. However, in line with the DRC's recommendations, state-level parliamentary oversight was established by the creation of the Committee for Defence and Security in the BiH Parliamentary Assembly (Drewienkiewicz 2003; Haupt and Saracino 2005).

With the new minister of defense as a cochair, the DRC's mandate was extended to the end of 2004 so that the new legislation could be implemented. The legacy of the previous system made implementation very difficult, but nevertheless some useful technical progress was made during 2004. In addition, the VF and VRS were downsized, as agreed upon by the DRC in 2003, to a total of 12,000 full-time personnel and 60,000 reservists. However, NATO decided that it was not yet time to admit BiH to PfP, because of its lack of cooperation with the International Criminal Tribunal for the Former Yugoslavia. Despite this setback, BiH decided to pursue full membership in NATO, and NATO stated that this would require the transfer of all entity defense competencies to the state (Scheffer 2004). Shortly thereafter, the HR decided that despite the new arrangements the entity armies were still not under adequate state-level command and control. Accordingly, he renewed the mandate of the DRC for another year, to develop recommendations for establishing a single military force that would address both NATO's requirements and his own concerns (OHR 2004).

At the end of 2004, SFOR handed over its peace enforcement mandate to the European Union Force (EUFOR). NATO established a small headquarters with the primary role of supporting BiH's efforts in defense reform and NATO accession. NATO Headquarters Sarajevo (NHQSa) provided its political adviser (from the US Department of State) to cochair the DRC, and, through a combination of shuttle diplomacy and force of personality—backed by overarching NATO requirements—he forged agreement for a set of principles (Gregorian 2005a) and related legislation providing for a single, all-volunteer AFBiH of 10,000 full-time and 5,000 reserve members. This legislation entered into effect on January 1, 2006. The new AFBiH consisted of a Ministry of Defence with full authority in defense matters and a Joint Staff with full command of all military elements. In line with a DRC compromise regarding ethnic identification, the AFBiH adopted a modified version of the British regimental system in order to provide for a degree of ethnic identification in the AFBiH (Maxwell 2005; Gregorian 2005b).

In the proposed regimental system, infantry units would be grouped into three regiments—one Bosniak, one Bosnian Croat, and one Bosnian Serb—of three battalions each. The regimental structure would have no operational role; it would exist purely to provide a structure for ethnic military tradition, including such things as dress uniforms, ceremonies, awards, and museums. The battalions of the infantry regiments, the members of which would wear ethnically derived insignia (linked to those used within the former armies), would be mixed into multiethnic brigades made up of one battalion from each of the three infantry regiments plus multiethnic headquarters, signals, artillery, reconnaissance, and military police. All other elements of the armed forces except for the infantry would be multiethnic in composition. In line with the requirements for all BiH state-level institutions, the ethnic composition of the remainder of the defense establishment would reflect the country's, according to the most recent census (currently 1991), in line with the requirements for all BiH state-level institutions (Bosnia and Herzegovina 2006).

On this basis the BiH defense establishment entered 2006, ten years after the war, as a theoretically unified force. However, it was still necessary to dismantle the entity defense establishments and to replace them with a single, state-level defense system, beginning with the immediate tasks of developing the detailed force structure and selecting the members of the future civilian defense establishment. Planning for these tasks was carried out by a transition team established by the Law on Defence and staffed by personnel from the BiH defense establishment (selected with an eye to ethnicity as well as relevant qualifications), NHQSa advisers, and contractors funded by the United States. Thus, even though BiH was now in the lead for implementation, there was still very close coordination between BiH and its international partners, which was to ensure continuity in key relationships and expectations regarding implementation. The offices of NHQSa advisers, US contractors, and BiH members of the team were colocated in the MoD building, and the former DRC cochairs (the BiH minister of defense and the NHQSa political adviser) remained in close communication as their staffs worked together.[3] This long-term relationship proved crucial to keeping the implementation process moving and generally on track, even when the wider political environment deteriorated.

It took six months to agree on the new force structure, with most of the debate being based on political factors and parochialism (i.e., the VF vs. VRS ways of doing business). Unsurprisingly, the new force structure, which was adopted in July 2006 (Bosnia and Herzegovina 2006), had numerous flaws: little operational capability; too much infrastructure (with particular attention to placing infantry battalions where the respective ethnicity was in the majority[4]); a hollow manoeuvre force, in which empty positions (in theory to be filled by reservists[5]) gave the illusion of military presence and ethnic representation; and, as noted above, quotas to be filled by the three constituent peoples in line with the last prewar census in 1991. However, the census did not apply at the highest levels; a strict 1:1:1 ratio applied to senior military and civilian positions, so that a senior position in the AFBiH (above the brigade level) or the MoD (at the level of minister) was always balanced by a deputy position for each of the other two constituent peoples, with the top positions themselves being evenly distributed among the constituent peoples.

Despite its flaws, the new structure provided for a force that was multiethnic in all aspects save in the infantry battalions. However, there were differing views about the extent to which infantry regiments perpetuated the former Bosnian Serb, Bosniak, and Bosnian Croat forces, and how they related to noninfantry elements of the AFBiH. Notably, some wanted *all* members of the AFBiH to belong to ethnically defined regiments, regardless of the branch of service. This would have resulted in three distinct ethnic mini-armies in the AFBiH, rather than limiting such identification to the infantry only. It became clear that if the regimental system was not implemented with great care, it would be used to continue ethnic division in the AFBiH—as could manipulation of assignments to nominally multiethnic units.

Although the force structure had been agreed upon within half a year, selecting which members of the former entity defense institutions would join the new organization took two years, in part because elections took place in the first year. (Despite this delay NATO invited BiH to join PfP at the end of 2006.) After a lengthy process of political compromise played out within the transition team and the senior military and civilian leadership, the minister of defense established selection commissions to determine who should be retained and who should be released in order to reduce military personnel from approximately 12,000 to 10,000 and civilians from 1,800 to 1,000. Although complex and contentious, this process was completed by the end of 2007, albeit (and inevitably) not to the satisfaction of all concerned.

ORIGINS

The peace agreement allowed two separate militaries to continue to exist: the VF (Bosniaks and Bosnian Croats) and the VRS (Bosnian Serbs). These forces were not merged until well after the settlement, and this occurred because of international pressure and after extensive negotiation.

The Historic Role of the Military

The VF and VRS were initially offshoots of the Yugoslav People's Army (Jugoslavenska Narodna Armija, or JNA), which was deliberately used as an integrative (albeit Serb-dominated) tool by Tito, the commander of the Partisan guerrilla force that emerged on top following World War II and strongman leader of Yugoslavia until his death in 1980. Tito was not himself a Serb—he was half Croat, half Slovene—but the bulk of his communist partisans were, and Serbia was the most populous and dominant of the republics that made up postwar Yugoslavia. Tito did his utmost to submerge ethnic divisions in a sea of common Yugoslav identity, in large part through the nurturing to near-mythical status of a particular version of the Yugoslav experience during the war and through the iron grip that he maintained for most of his reign. The JNA played a key role in this process.

Conscript service ensured that most males experienced a common rite of passage, serving with soldiers from all over Yugoslavia and in most cases serving outside their own part of the country, and transferring to the reserve force upon completion of their conscript service. The JNA, as the touted shield and guardian of the Yugoslav state (against both NATO and the Warsaw Pact), occupied a powerful, respected position in society and dominated all defense-related activities, with the minister of defense being a military officer. Military expenditures were not subject to effective oversight and were not transparent; the JNA essentially ran its own show.

The post-1995 VF and VRS followed the JNA model. The "rite of passage" aspect remained important—to the leadership and to politicians, if not to the increasingly disenchanted, unpaid, and maltreated conscripts—and each ethnic group of soldiers saw itself, and was generally seen, as protectors of its constituent people. The senior leadership of both armies remained heavily influenced by their JNA background. However, though the VRS remained organized largely along JNA lines, with JNA weapons and doctrine, the VF acquired a different veneer from an extensive US bilateral training program, and it continued to suffer from divisions between Bosniaks and Bosnian Croats.

The Emergence of Military Integration as an Issue

The merger took place long after the negotiations that ended the conflict. The treaty that had ended the Bosniak/Bosnian Croat subconflict joined but did not merge the ARBiH and HVO into the VF, and GFAP subsequently preserved the VF and VRS. Despite international pressure to create a degree of state-level coordination, the existence of the VF and VRS was not challenged until the HR established the DRC in 2003 and the entity armies were brought under limited state-level command and control. At that stage the existence of the entity armies was not in question; even NATO had conceded that the country could join PfP with two armies. However, by the end of 2004, BiH's desire to join NATO and HR's reaction to more incidents that demonstrated inadequate state-level control led the international community to decide that the entity armies had to be merged and to BiH's acquiescence to this hitherto unacceptable requirement. Although the "pull" of NATO was undeniably a factor, it was the HR's decision that forced the entities to the table.

Supporters, Opponents, and Neutrals

The international community in general supported the replacement of the two entity armies with a single AFBiH. Given the propensity for conflicts in the region to spread, the replacement of two conflicting forces by one oriented toward NATO accession was seen as a positive step. International support proved critical to the merger process, from the conceptual stage through to the still ongoing implementation of the merger.

Within the various BiH constituencies, Bosniaks supported the merger because, as the largest of the constituent peoples, they sought (and seek) to strengthen the authority of the state, and thus they would have preferred a more complete merger, with no provision for ethnic representation or ethnic checks and balances.[6] Bosnian

Serbs did not greet the merger with enthusiasm, but they worked to ensure that it took place with minimum damage to their perceived interests. The period from 2003 to 2005 was one in which the international community in BiH, particularly OHR, was disposed to take firm action, and the Bosnian Serbs were often on the defensive. In addition, within the Bosnian Serb leadership there was a sense that military misbehavior was bringing opprobrium on the RS for no net gain, and so Bosnian Serb politicians risked little by participating in the merger.

The principal concern of Bosnian Croats was ensuring a level of representation in the new structure that was acceptable to them; they did not oppose the merger per se. As the least numerous of the constituent peoples, Bosnian Croats were concerned not just with representation but also with visible symbols, and the regimental system was developed in response to concerns first voiced by the Bosnian Croat FBiH minister of defense and quickly echoed by the RS. On balance, therefore, wholehearted support for the merger was largely limited to Bosniaks (and the international community), but neither the Bosnian Croats nor the Bosnian Serbs offered determined opposition.

Compromises in the Process

The three key compromises were the regimental system, the reserve system, and the quota system. The first compromise, developed to satisfy Bosnian Serb and Bosnian Croat concerns regarding ethnic identity in the new army (and, less explicitly, concerns regarding the future availability of at least some units from their respective ethnic groups), allowed for the retention of ethnically distinct infantry battalions in the merged force but in mixed brigades, while establishing that the rest of the AFBiH would be mixed. Thus the merger contained elements of integrating individuals into mixed units and, at the brigade level, of integrating ethnically distinct units into mixed formations. The second compromise, which addressed the Bosniak desire to retain conscription and large, inactive reserve forces, provided for a small but usable reserve force. (Both these compromises were proposed and brokered by NATO.)

The third compromise, addressing Bosnian Serb and Bosnian Croat concerns about a Bosniak-dominated force, allocated a fixed percentage of the positions in the armed forces to each of the three constituent peoples. This was agreed to by BiH interlocutors despite NATO's concerns about the effect it would have on personnel management (all positions are "tagged" by ethnicity, which results in personnel management difficulties) and operational effectiveness (there is no provision to permit an empty position to be filled by a person of another ethnicity, so it is possible that many units—particularly in the infantry—may become hollow shells).

There were other, less obvious compromises, notably about the force structure and the locations of units. The force structure did not meet any rational capability requirements. It was designed to preserve as many of the elements of the VRS and VF as possible, and thus it was neither cost-effective nor operationally effective. (However, this approach did mean that, for better or for worse, the new force was neither weaker nor stronger than the combined precursor forces.) This compromise was not opposed by NATO, which took a longer view—in essence, "First bring them together, then we'll worry about making it better." In due course NATO and BiH agreed that within its PfP commitments, BiH would undertake a thorough defense review (initiated in 2009 and still in progress at the time of writing) in an attempt to address force structure requirements analytically. Similarly, NATO did not oppose the operationally dubious dispersal of AFBiH elements throughout BiH; this, too, was a problem best left for later, and one that is also to be addressed through the defense review.

CREATION

The military merger was driven initially by the international community and would probably not have happened a few years later. However, the process involved both internationals and Bosnians, with the local role increasing over time. Elaborate rules for quotas were developed, and there was limited screening for human rights violators. Training remains an issue, because of differences among the various components and limited funding.

Shaping the New Military

The new AFBiH was designed by the international community and BiH through a formal process under the DRC. The DRC worked primarily on the basis of the requirements of NATO and the Organization for Security and Cooperation in Europe's Code of Conduct on Politico-Military Aspects of Security (as a member of the latter organization, BiH was bound by that code), and under the HR's overarching direction to bring the entity armies under a state-level command-and-control structure. Although the DRC had a strong, externally imposed mandate and a high level of international community representation, BiH's representatives were fully involved. The BiH minister of defense cochaired the DRC for most of 2004 and all of 2005 and had sole responsibility (albeit with strong international community support) for implementation and for subsequent efforts within PfP.

There was (and is) a tendency on the part of some international actors to offer solutions that replicate those of their home countries. Multilateral actors such as NATO were able to control this by the multinational nature of their advisory teams; however, bilateral actors were likely to make suggestions based on their national backgrounds, with varying degrees of consideration for the specific circumstances of BiH. Fortunately, the DRC provided a mechanism through which all proposals could be analyzed, assessed, and coordinated. In addition, the long-term goal of joining NATO meant that NATO's requirements—real or perceived—were paramount, and in this sense NATO, by providing a positive focus for planning, filled the role that in some other cases has been filled in a negative sense by the existence of an enemy common to all parties.

Personnel Selection

The DRC did not specify how the members of the merged AFBiH or the enlarged BiH MoD would be selected; it provided guidance, but the details were left for the minister of defense. Those details were shaped by two factors: the quota system described above, and the personnel selection process.

The personnel selection process itself satisfied no one but was an achievable compromise, in large part because much downsizing had taken place before the merger so the number of people who would lose out this time around was relatively small. The system of rank-based selection commissions, many held in rotation at various locations around BiH, was logical enough, but the volume of work, the lack of time, and the high level of mistrust combined to make virtually everyone unhappy; the only saving grace was that *everyone* was unhappy. The quota system meant that the selection process was not a zero-sum game pitting the constituent peoples against each other, and selection commissions were carefully balanced on ethnic lines, so there was little justification for claims of bias on ethnic grounds. (The remaining grounds for complaint, including nepotism and corruption, undoubtedly had some validity.) At the request of the minister, most selection commissions were attended by observers from NATO or EUFOR, but despite a genuine effort this proved only a limited confidence-building and transparency measure.

NHQSa and EUFOR, exercising their GFAP authority, screened and continue to screen candidates for all senior military and civilian defense positions (excluding, since 2006, those being considered for minister of defense and deputy ministers of defense). However, in recent years the screening process has come to be based on a higher standard of proof than under IFOR/SFOR. NATO commanders of those forces had used very wide discretion in deciding whether a person was suitable to hold a defense position, the broad criterion being whether that person represented a threat to the peace process. Given the ease with which allegations of war crimes, corruption, and so on could be made, and the paucity of investigative and judicial resources to resolve such allegations, wide discretion was necessary if appointments were to be confirmed or denied within a reasonable time.[7] Since 2005, commanders of NHQSa and EUFOR, exercising a joint authority, have tended to seek a higher degree of certainty before denying an appointment, in part because of growing unease with the exercise of such authority over the personnel of a nominally sovereign country. Nevertheless, the commanders of the international forces remain alert for cases in which human rights violators are found in the ranks of the AFBiH and in the MoD.

Rank Allocation and Training

Like all military personnel, officers and NCOs were drawn from the former VRS and VF and went through the same personnel selection process used for all ranks.

Training as such was not part of the merger process, and capacity-building efforts focused on the MoD, Joint Staff, and subordinate commands. These efforts were conducted multilaterally (by NATO) and bilaterally (primarily by the United States). Broad tactical-level training had ceased with the termination of US support for the establishment of the VF. SFOR had conducted limited training with the VF and VRS, but this was largely confidence building rather than a consistent training program. EUFOR continued a similar effort with the AFBiH, and in 2010 established a specific element for capacity building and training. Bilateral training support activities continue, including courses conducted both outside and inside BiH.

In 2003, a twelve-nation consortium led by the United Kingdom, in partnership with BiH, established the Peace Support Operations Training Center just outside Sarajevo, with a multinational staff and the mandate to provide training in peace support operations to junior and mid-level officers from BiH and the region. The center continues to operate, with its core course being essentially a junior staff course with peace support overtones. However, control and responsibility have now been transferred to BiH, and funding is a continuous concern.

BiH's decision in 2004 to contribute an explosive ordnance disposal platoon to the multinational coalition in Iraq led to a US bilateral training program designed to impart the necessary technical skills to AFBiH members for the first few years of six-month deployments. This effort was successful, and in due course BiH assumed full responsibility for training, including, for the final six-month period, the training of an additional infantry platoon for static security duties.

In 2009, BiH began to contribute individual staff officers to the NATO-led force in Afghanistan. In 2010, the decision was made to expand this commitment with an additional infantry platoon for static security duties, and the first such platoon deployed in late 2010, with some training under Danish auspices. This infantry deployment was augmented subsequently with a platoon-sized deployment of military police after training under the auspices of the United States, and with small numbers of instructors in various fields.

During the period leading up to the final merger, the development of common training was based on finding areas in which common standards would be helpful, particularly when it became apparent that the entity armies were going to come under some sort of state-level framework. This included SFOR- and OHR-led efforts to define common standards for recruit training and agree on other training requirements (e.g., for UN military observers or troops deploying to Iraq). Since the merger, the key determinant of training standards has been NATO's requirements and expectations within PfP.

The training system in the new AFBiH is still being developed, but the challenges are primarily technical rather than political. The United States–based system of the VF had never been fully implemented, and the VRS had remained more or less along JNA lines. The requirement now is for the AFBiH to develop and implement a system that meets its specific requirements, rather than trying to import a system from elsewhere.

Philosophical differences aside, the key problem was and remains BiH's lack of internal resources and its commensurate reliance on bilateral and multilateral support. Although the effectiveness of individual bilateral and multilateral efforts varies, the overarching problem is that each such effort is likely to use a different standard, which leads to an untenable mixture of practices. This is equally the case at all levels, whether for command and staff training or for technical training.

Unless BiH develops its own capability in every area—which would be improbable, and not necessarily desirable—the logical solution is to use one external provider for each area of training; for example, BiH could send its helicopter pilots to be trained in Country X, and its mid-level officers to be trained in Country Y. However, such decisions carry political and ethnic baggage, and so the current, unfocused approach is likely to continue for some time. Meanwhile, with its own system still under development, BiH suffers from a lack of clear training goals and priorities. Even if such goals were forthcoming, it would be difficult to find all the necessary financial and personnel resources with which to implement them.

OUTCOME

The new military is formally under civilian political control. Ethnic tensions remain, especially at the senior level. The military deploys around the country and has some symbolic value as an integrated institution. Its operational capabilities are limited, but because it is almost surrounded by NATO members and prospective members and is not designed to be used domestically, this does not seem to be a major problem. Economically, it is small enough to be sustainable if it were reorganized efficiently. Because it has foreign troops stationed in the country, it has not really been tested.

Political Control

The AFBiH exists within a legally sound, state-level command-and-control framework, with full provision for civilian control and no legally authorized avenue for entity or ethnic interference. The structure incorporates ethnic checks and balances at each level, specifically by providing for a triumvirate in the MoD, the Joint Staff, and commands above the brigade level. The BiH Parliamentary Assembly's Defence and Security Committee, also headed by a triumvirate, provides parliamentary oversight, and as a last resort NHQSa and EUFOR retain the authority under GFAP to intervene if necessary. To date there has been no lasting challenge to the state-level command-and-control structure, but ethnic incidents originating from both inside and outside the defense establishment demonstrate the fragility of the merged force.

In 2005, while the entity armies remained separate but under a state-level MoD and Joint Staff, a group of new conscripts in RS refused to say "Bosnia and Herzegovina" while reciting the recruitment oath to BiH (as required by the 2003 Law on Defence, replacing oaths to entities). Instead, in what was certainly a scripted incident, they substituted "Republika Srpska," in the presence of numerous senior defense officials. The international community, particularly NHQSa and EUFOR, responded strongly, and in due course a BiH investigation led to disciplinary action against a small number of VRS officers. More pointedly, the chief of the VRS general staff was removed from his position by NHQSa and EUFOR, using their GFAP authority. No similarly overt events since then have originated within the defense system; the closest call to date was a canceled protest by Bosniak soldiers in early 2010 against their impending redundancy under age/rank limitations in the 2005 defense legislation.

Nevertheless, there is constant ethnic tension in the upper levels of the defense establishment. This tension is not manifest as a refusal to obey orders or to carry out instructions; instead, it appears in the form of arguments presented by staff officers and civil servants that are based on ethnic rather than technical concerns. The institutionalized parallel chains of command that existed in the VF (Bosniak and Bosnian Croats) do not exist in the AFBiH, but unofficial parallel chains of influence remain, now expanded to three rather than two constituent peoples.

Overt external interference in the chain of command is rare but dangerous, because it leaves no space for quiet, internal negotiation and instead places highly visible stress on the system. For example, in 2009 the RS prime minister called for Bosnian Serb members of the AFBiH to leave a contingent that was on its way to participate in a PfP exercise in Georgia, so as not to slight Russia (which was generally viewed as favoring the RS). Fortunately, this call was not heeded, and the public response of the Bosnian Serb chief of the Joint Staff was correct in all respects.

To cite another incident, in the spring of 2010 the Bosniak minister of defense, under pressure from Bosniak veterans' groups from the 1992–95 conflict, issued orders that resulted in uniformed Bosniak AFBiH personnel providing military honors at the funeral of a former Bosniak general officer of the ARBiH who had been convicted of war crimes (but was free pending appeal). This generated a significant negative response, particularly from Bosnian Serbs, despite the minister's claim that he had actually managed to reduce the scope of the veterans' demands. The public exchange of recriminations between the Bosniak minister and his Bosnian Serb deputy, and between the Bosnian Serb chief of the joint staff and his Bosniak deputy (the Bosnian Croat deputies at both levels stayed out of the dispute), did not reflect well on the merged force. It is clear that although the unified structure is in place, it is far from being irreversibly established.

Military Capabilities

The AFBiH has significant momentum and justification for existence (overseas deployments, NATO accession) and, all other things being equal, in time ethnic issues may play a decreasing role as the force continues to focus on its long-term professional goals. However, if the wider environment deteriorates, it would be unreasonable and unwise to expect the AFBiH to remain intact in the face of widespread ethnic violence.

The AFBiH is distributed throughout BiH, much of it in multiethnic units.[8] Training activities are largely confined to garrisons and training areas and thus are not visible to the public. (A United States–sponsored exercise in Banja Luka, capital of the RS, in September 2009 was a notable exception.) However, the presence of AFBiH facilities throughout the country, with the BiH flag on display (and no entity flags) has a symbolic value. Further value has been garnered from military assistance in civil emergencies such as floods, forest fires, and aero-medical evacuations. The AFBiH is also capable of conducting symbolic deployments in BiH, although this has not been deemed necessary to date. However, elements of the AFBiH have conducted field training in BiH alongside EUFOR units, with some resulting visibility.

The AFBiH is not designed to provide stand-alone territorial defense against regional neighbors. Beyond the fact that the various constituent peoples would expect external assistance from different sources (and eventually from NATO, if BiH were to join the Alliance), with the exception of Serbia the regional neighbors are either in NATO, are on the verge of joining, or have declared their desire to join. (Serbia has not expressed a desire to join NATO, but its current path is to join the EU and increase cooperation with NATO, which is not consistent with attacking its neighbors.) With this in mind, the AFBiH's operational ambitions are modest and are focused on being deployed as part of NATO-led or other coalition operations. There is no official expectation that it would be necessary to fight closer to home, and this is reflected in the security and defense policies. As well, the BiH has foreign peace enforcement troops on its territory and can expect them to deal with any threats. In fact the presence of these troops has permitted BiH the luxury of undertaking an extensive restructuring of its armed forces without any need to maintain a day-to-day operational capability.

Given the recent historical context, the missions of the AFBiH, as specified in the Law on Defence, do not include the use of force against groups in BiH society. Planning for the use of troops within BiH is focused on disaster response and limited support for civil authorities in the event of a terrorist attack. Legal constraints aside, the AFBiH is simply too new and fragile to be used against groups within the country. To the extent that such domestic threats arise, they are handled by the police; for example, ethnic incidents surrounding the annual commemoration of the Srebrenica genocide or a violent rioting by disgruntled veterans in the FBiH. The police forces in BiH are far larger than the AFBiH, and their training and equipment are better suited to such this type of response, as long as incidents remain relatively small. If larger incidents lead to sustained violence, all bets will be off.

Sustainability

Assuming that there is no drastic reduction in GDP, the current 10,000-person structure is affordable, but it is not cost-effective.[9] It could be if the age/rank pyramid and force structure were rationalized, and this is—or should be—a goal of the ongoing defense review. An agreement to reduce to below the current number of infantry battalions and brigades is unlikely, and with that as a baseline an overall force of 10,000 is not unreasonable. BiH is roughly between the two NATO members in the region. Slovenia, with roughly half the population of BiH (2 million vs. 3.8 million), has an active duty strength approximately 75 percent of that of BiH. Croatia, with a somewhat larger population (4.5 million) than BiH, is moving to an active-duty force that will be about 50 percent larger than that of BiH. It should also be noted that GFAP contains arms control provisions that are still in force, and thus serve as a check on any excessive expansion; however, in practical terms all the affected countries (BiH, Croatia, and Serbia) are well below the stipulated ceilings and will in all probability remain so.

As noted above, the force is economically sustainable if logic prevails. Its political sustainability is linked to the wider viability of BiH. At present the AFBiH is seldom the direct target of ethnic dissension, and if such dissension remains below a level that would break up the country, the AFBiH will survive. But if BiH were to dissolve, the AFBIH would dissolve as well.

Alternative Outcomes

Three scenarios could have produced a drastically different outcome. First, a complete military victory by either side in the 1992–95 conflict would have either eliminated the merger issue (if the Bosnian Serbs had won) or resulted in a somewhat less complex merger (if the Bosniaks and Bosnian Croats had won).

Second, the architects of GFAP could have insisted on a demilitarized BiH, reasoning that the large NATO-led Implementation Force would deal with any security requirements. However, at the time the intention was to withdraw that force quickly, and in the absence of a long-term guarantee demilitarization was a nonstarter (and would likely have been a nonstarter even with such a guarantee).

Third, if the international community had not been able to maintain strong pressure (positive and negative) on BiH during the 2003–5 period of the DRC, or if the entities had felt strong enough to defy the international community (as the RS did by 2007), the merger would not have been possible. It took place because of a fortuitous combination of circumstances, deftly exploited by a small group within the international community as part of a wider state-building effort, during a period when ethnic political leaders were able to concede entity defense competencies without suffering significant political damage.

HAS MILITARY INTEGRATION MADE THE RESUMPTION OF LARGE-SCALE VIOLENCE LESS LIKELY?

The introduction hypothesizes five causal paths through which military integration can reduce the likelihood of large-scale violence. The first path would be to bring both sides into an arrangement that requires commitment to a system that is less easily subverted, thus imposing a political and military cost on each side that it is reluctant to write off by returning to violence, and setting an example of taking risks for peace that others may follow. The second path would be to provide a degree of security that in turn fosters the establishment of effective institutions and the resolution of underlying disputes. The third path would be to employ significant numbers of former fighters. The fourth path would be to demonstrate that integration is possible and thus foster integration in other institutions and in society. And the fifth path would be to generate trust through negotiations over military integration that can then permit negotiation on other issues. None of these descriptions applies wholly to the BiH case, but some have a degree of salience:

- *Commitment and cost.* Creating the AFBiH did require all sides to commit to the new system, and the ethnic checks and balances in that system make it unlikely that elements of the AFBiH could be misused without warning. However, the remaining elements of this causal path do not apply to BiH. The political risk of integration was less than might have been anticipated. Despite some rhetoric at the time of the integration, it is unlikely that many votes were lost at the next election because of military integration, if only because the armed forces had already become too small to be of any real importance to voters. The risk to the militaries was also low because in practical military terms there was little cost to the deal; all sides had roughly the same numbers of "their" troops as they did before integration, even though many of those troops were in mixed units; and all sides retained access to large numbers of former soldiers who could be rearmed readily if a perceived need arose. Finally, progress on other aspects of governance such as police, justice, and constitutional reforms remains limited; military integration has not increased willingness to make compromises in other areas.

- *Security*. Military integration in BiH had little relation to increasing security in the country, either directly through military force or indirectly through employment for former combatants. The presence of an external military force and steady reductions in the indigenous armed forces meant that security was generally associated with NATO and subsequently with EU troops, and with the various police forces in BiH. Meanwhile, the addition of a few thousand more demobilized soldiers to the hundreds of thousands who had already been demobilized did not create any significant additional security risk from disgruntled ex-soldiers.

• *A symbol of integration.* Military integration is touted by both BiH and international organizations as the most successful reform process in the country. The MoD and AFBiH are the largest integrated institution in postwar BiH and are perceived as leading the way towards BiH's future NATO membership. They are, therefore, symbols that integration is both possible and desirable. Some hope that this spirit will spread to other institutions, but despite the increasing requirements of the NATO and EU accession processes, this is occurring at a glacial pace at best. Nevertheless, there are signs of increased professionalism and trust among military personnel, particularly those who have served in BiH's overseas contingents. Although there is no evidence that this type of interethnic trust has been transmitted to the civilian population to any degree, and the military itself is not free from tension, relations within the military appear to be reasonably good, particularly as new recruits flow in to replace the wartime generation.[10]

• *Trust as a by-product of the negotiation process.* The military integration negotiating process relied heavily on external pressure, and indigenous negotiators were aware that integration would be conducted while foreign troops remained in BiH as guarantors of the peace. Thus, it proved possible to agree on military integration without a significant increase in the level of trust among the ethnic groups, and there has been no crossover of trust to other issues. In fact, a subsequent attempt under international community pressure to attain a degree of integration in policing met with crippling opposition and ended in 2008 with a minimal and as-yet-unimplemented compromise.

On the whole, military integration is not a significant positive or negative factor in the context of renewed violence in BiH. The risk of renewed violence is independent of the armed forces and has been for many years. Initially, this was because both the VF and VRS were subject to the control of foreign military forces. Subsequently, the steady downsizing of both armies made them even less likely to be the source of any trouble, and military integration has placed the AFBiH under layers of ethnic checks and balances. If the ever-present tensions in BiH continue to increase to the point of disintegration, the AFBiH may become a source of personnel, weapons, and equipment, but it will not be the cause of that disintegration. The high proportion of people in the country with military experience, and the large quantities of small arms and ammunition that are available to those people (under Tito, BiH was the arsenal of Yugoslavia), are far more relevant to any discussion of renewed violence.

CONCLUSION

Given the unique circumstances in BiH, particularly the strong role that was available to the international community and used during the key 2003–6 period, it would be difficult to replicate the BiH military integration process elsewhere. Nevertheless, the BiH case demonstrates that external supporters and enablers can help to overcome internal tensions and can facilitate dialogue and consensus, thereby helping to achieve ambitious and sustainable results. Carrying out this role effectively requires a comprehensive vision that is common to all external actors and is communicated effectively and consistently to internal actors in order to generate a mutually understood and accepted goal. In addition, external actors must be prepared to mount a sustained, long-term effort, combining political engagement with civilian and military technical support through the conceptual, development, and implementation stages of the integration process. Finally, external actors must be agile, able to respond quickly and flexibly to changing circumstances, and able to develop and exploit a sense of urgency at critical junctures.

ABBREVIATIONS

AFBiH	Armed Forces of BiH—new, integrated army
ARBiH	wartime Bosniak army
BiH	Bosnia and Herzegovina, composed of FBiH and RS
DRC	Defence Reform Commission—group that planned military integration, created by High Representative with international and local membership
EUFOR	European Union Force—successor to SFOR
FBiH	Federation of BiH—Bosnian/Croatian entity of Bosnia-Herzegovina
GFAP	General Framework Agreement for Peace—peace agreement
HR	High Representative—chief executive of the Office of the High Representative, established by NATO to supervise the peace
HVO	wartime Bosnian Croatian army
IFOR	implementing force—NATO peacekeeping force immediately after the war, succeeded by SFOR
JNA	Yugoslav People's Army
MoD	Ministry of Defence
NATO	North Atlantic Treaty Organization
NHQSa	NATO Headquarters Sarajevo
OHR	Office of the High Representative—established by NATO to supervise the peace
PfP	Partnership for Peace—NATO program to prepare governments and militaries to apply for membership
RS	Republika Srpska, Serb Republic—Serb entity of Bosnia-Herzegovina
SFOR	Stabilization Force—NATO peacekeeping force, successor to IFOR, succeeded by EUFOR

VF   wartime FBiH army, coalition of Bosniaks and Bosnian Croatians
VS   wartime Bosnian Serb army

NOTES

1. The term "ethnic" is used here in the regional shorthand whereby religious affiliation is equated with ethnic distinction, even though the purportedly distinct groups are genetically extremely similar. Furthermore, many do not practice the religion through which they are defined. However, while prewar BiH contained many who chose not to define themselves in ethno-religious terms, postwar BiH does not.

2. The members include the BiH State and Entities, OHR, the Organization for Security and Cooperation in Europe (OSCE), SFOR, and NATO. The observers include the United States, the Russian Federation, Turkey (representing the Organization of the Islamic Conference), and the European Union.

3. The transition team was dissolved at the end of 2007. NHQSa advisers and a few US contractors remain co-located in the MoD building.

4. Some infantry battalions also suffered from being split between two locations, the better to address political desires to show the (ethnic) flag in as many places as possible, and of course to spread the economic benefits of military facilities to the surrounding communities.

5. The Law on Defence provided for a reserve force half the size of the 10,000-person full-time force. However, to date there is no mechanism and no funding in place with which to produce those reservists.

6. The veterans' groups from the 1992–95 conflict wielded limited influence. Anecdotally, RS veterans were vociferously opposed to the merger, Bosnian Croat veterans less so but still opposed on the whole, and Bosniak veterans apparently supportive. However, all veterans' groups generally expressed concern about the fate of demobilized personnel.

7. War crimes investigations in BiH that could lead to indictments, by both by the International Criminal Tribunal for the former Yugoslavia and by BiH authorities, are ongoing. The number of cases is very large and processing times are very long; thus, the absence of an indictment is not necessarily an indication of innocence.

8. In most cases soldiers have chosen not to move their families to areas in which they are an ethnic minority, choosing instead to travel home on weekends. The perceived level of ethnic tension is low enough to allow soldiers to work and socialize but not low enough that they are willing to move their spouses and children to these areas.

9. The affordability of the 5,000 reservists is still open to question.

10. This is a far cry from a joint VRS–VF training exercise in April 2004, when the officer commanding a VRS company stated that he had not met or spoken with anyone in the VF since the end of the war, more than eight years earlier.

Chapter 12

BRINGING THE GOOD, THE BAD, AND THE UGLY INTO THE PEACE FOLD: THE REPUBLIC OF SIERRA LEONE'S ARMED FORCES AFTER THE LOMÉ PEACE AGREEMENT

Mimmi Söderberg Kovacs

Almost a decade after the end of the civil war that ravaged the country for most of the 1990s, Sierra Leone has come a long way. The same small country in West Africa that for years was known to many outsiders for its brutal conflict, numerous coups, and flagrant human rights abuses, along with the almost complete collapse of its state institutions, has witnessed a remarkable process of transformation. State authority has been reestablished, and the security situation throughout the country has improved considerably. In 2007, the second postwar elections were followed by a peaceful turnover of government, and slow but steady progress has been made in a range of vital development areas. In a country where poor governance and abuse of the armed forces were once the norm rather than the exception, the almost complete overhaul of the security-sector institutions of the state has been very much at the core of the peace-building efforts. Crucially, the various armed groups—the rebel forces, renegade army soldiers, and militias that tore the country apart during years of armed conflict—are now joined together in a single national army.

The purpose of this chapter is to take a closer look at the integration of the armed forces in Sierra Leone after the civil war and the establishment of the new Republic of Sierra Leone Armed Forces (RSLAF). More specifically, the aim is to discuss why a military merger was decided on in the Lomé peace agreement, how this merger was subsequently carried out, and some of the achievements and shortcomings of the reform efforts almost ten years later. Ultimately, the aim is to cast some light on the relationship between the process of military integration and the outcome of the peace process. The argument is made here that although there is much evidence that the transformation of the armed forces in Sierra Leone represents a success story in the making, there are also reasons for concern. The true test is likely to come in a few years, when the last representatives of the International Military Advisory and Training Team (IMATT) are withdrawn. Finally, it is suggested that although the relatively successful transformation of the armed forces in

Sierra Leone represents a critical stepping-stone in the process from war to peace, the relationship between the military merger and the durability of peace is far from straightforward.

This chapter is divided into four main sections. First, it provides a brief background on the civil war in Sierra Leone, with a focus on the roles of the warring actors and the disintegration of the national army. This is followed by a discussion of the Lomé peace negotiations, with particular attention given to the provisions in the peace settlement concerning the integration of the various armed factions into a single military force. Second, it discusses the process of military integration, from the early reforms carried out in the immediate postsettlement period to the training and reconstruction of the RSLAF following the ending of the armed conflict. The third section discusses the outcome of the military integration, highlighting some of the major achievements and shortcomings of the still-ongoing reform process. The fourth and last section offers concluding thoughts regarding the causal pathways linking the military integration process and the durability of peace in Sierra Leone, along with some policy implications.

ORIGINS

In early 1991, very few people understood that they were witnessing the beginning of a civil war when they received the news that a small group of insurgents had crossed the border from Liberia and attacked villages in the eastern and southeastern provinces of Sierra Leone. Due to the initially isolated effects of these attacks, the one-party regime of the All People's Congress (APC), which was largely dominated by Temne speakers from the north, did not take the threat seriously (interview with a representative of the APC regime, Freetown, October 15, 2004). It was soon announced over the radio that a group called the Revolutionary United Front of Sierra Leone (RUF) had led the invasion for the purpose of overthrowing the corrupt regime and restoring multiparty democracy to Sierra Leone (Richards 1998, 7; for an in-depth account of the origins of the RUF, see Abdullah 1998). Despite their small numbers, the rebel forces quickly advanced into the country.

The Historic Role of the Military

The government failed to contain the rebellion in its early years, in part because of the poor state of the Sierra Leonean Army (SLA), which at this time consisted of approximately 3,500 personnel. In order to prevent it from posing a threat to the political elite, it had deliberately been denied the necessary means and resources to play much more than a ceremonial role (Albrecht and Jackson 2009, 43). In addition, it was heavily corrupt, and recruitment and promotions were based almost exclusively on patronage and ethnic affiliation (Ebo 2006, 483; Ginifer 2006, 793). The pressure on the army to counter the invasion, in spite of poor condition of service and lack of equipment, soon led to mounting internal dissent, and in 1992 a group of lower-ranking frontline soldiers staged a coup to overthrow the government. A military junta, the National Provisional Ruling Council (NPRC), was established, which initially gained widespread popularity due to its populist rhetoric about ending corruption and the abuse of state power, as well as its promise to bring peace to the country (Abraham 2004; Zack-Williams and Riley

1993). A massive recruitment campaign was launched, and the rebels were soon pushed back to the border regions (Gberie 2005, 64; Keen 2005, 95).[1]

However, the expansion of the forces strained the army's already limited resources. The recruitment process had also attracted a large number of people with little training or experience. Faced with declining conditions of service and inadequate training, the army soon experienced a collapse in discipline and morale (Albrecht and Jackson 2009, 44). In addition, the rebel forces were again making military advances after having taken control of Kono, the country's premier diamond district, which enabled them to buy weapons and supplies and to recruit new soldiers. Allegations soon emerged that army soldiers were cooperating with the rebels in the illegal diamond trade and in the looting of the civilian population, giving birth to the term *sobels*—soldiers by day and rebels by night (Keen 2005, 107–31). From 1994, violence increased across the country. Responding to the deteriorating security situation, civil militias that were formed out of traditional hunting societies became the most significant resistance force against both the RUF and turncoat soldiers. Most prominent among these was the southeastern-based militia known as the Kamajors (Muana 1997).

By early 1995 the rebels controlled many of the rural areas outside the capital, and rumors spread that the capital was about to be attacked (Adebajo 2002, 84). In response, the military government called in a private South African security firm, Executive Outcomes, to help the civil militias and the remains of the army to counter the rebels. Although the progovernment alliance was soon able to push back the RUF and initiate negotiations with the rebel forces, the government's popular support had by now begun to dwindle, and it was forced by growing domestic and international pressure to announce elections and the return to civilian rule. In early 1996, elections were held, and Ahmed Tejan Kabbah—leader of the Sierra Leone People's Party (SLPP), which was primarily supported by the Mendes in the southeastern regions—was elected president (Kandeh 2004). The civilian government resumed negotiations with the rebels, and on November 30, 1996, a peace settlement was signed in Abidjan (Bangura 1997).

The RUF leadership, however, soon stalled on its commitment, and after the withdrawal of the Executive Outcomes forces, the rebels renewed their war efforts. Meanwhile, Kabbah had increasingly come to rely on the civil militia rather than the national army, whose loyalty he questioned, for the protection of himself and his government, and plans were announced to significantly reduce the size of the army (Gberie 2005, 86, 104–6; Keen 2005, 197–99). The violent response from the army to these plans came on May 25, 1997, when a new military coup took place in Freetown. Major Johnny Paul Koroma was declared the leader of the new Armed Forces Revolutionary Council (AFRC), which invited the RUF to share power with them in the capital (Gberie 2005, 99–106).

The AFRC/RUF junta was, however, met with strong resistance both domestically and internationally. The Economic Community of West African States (ECOWAS) committed itself to reinstating Kabbah and his cabinet, and Nigerian troops under the banner of the Economic Community of West African States Military Observer Group (ECOMOG) were dispatched to Freetown. These troops effectively came to act as Sierra Leone's national army. In early 1998, in cooperation with the civilian militias, which were now formally united as the Civil Defence Forces (CDF), they were able to drive the junta out of Freetown and restore Kabbah to office. In January 1999, however, the capital came under attack again, primarily by the AFRC but supported by the RUF. Although ECOMOG, together with the CDF, soon managed to drive the insurgents out of the capital a second time, the event clearly signaled the continued military strength of the armed opposition.

Following the January 1999 attack, President Kabbah came under strong regional and international pressure to initiate peace negotiations to end the war (Bangura 2000, 564; Gberie 2005, 157; Francis 2000, 364; interview with a former government representative in the Kabbah administration, Freetown, October 19, 2004). As a result, negotiations were initiated with the RUF during the spring of 1999 in Lomé, the capital of Togo (Rashid 2000). The forty-five days of peace talks were presided over by the United Nations, some foreign diplomats, and a handful of civil society representatives (Hirsch 2001, 81). The AFRC was not, however, present at the table; nor was it officially named in the agreement. Most observers suspected that there had been a split between the former allies at this time, and it was later revealed that Koroma was being held hostage by the RUF in Kailahun during the time of the negotiations (Adebajo 2002, 97; Keen 2005, 252; interview with civil society representative present at the Lomé negotiations, Freetown, April 22, 2010). Unlike the previous negotiations in Abidjan, the rebels held the upper hand at the negotiation table in Lomé and made extensive demands in exchange for peace. In addition, it was clear that the Nigerian forces were preparing to leave the country, which increased the urgency to find a political solution to the conflict (Keen 2005, 248–50).

The Emergence of Military Integration as an Issue

The Lomé agreement was signed on July 7, 1999 (Sierra Leone 1999). It was a far-reaching power-sharing agreement. It provided four cabinet posts for the RUF in the government and four deputy-ministerial positions for the duration of the term of office of the government. It also lifted the death sentence on Foday Sankoh, the leader of the RUF, and gave him the chairmanship of the Commission for the Management of Strategic Resources, National Reconstruction, and Development. The Lomé accords also gave all RUF combatants and collaborators "absolute and free pardon" for any actions committed during the armed conflict, a provision that was strongly criticized both domestically and internationally.[2]

Article XVII of the agreement called for the reconstruction of the Sierra Leone armed forces, "with a view to creating a truly national armed forces." It explicitly stated that those ex-combatants of the RUF, CDF, and SLA who wished to be integrated into the new restructured national army would be able to do so provided that they met "established criteria," without explaining what these criteria were. In addition, recruitment into the new force was to reflect the "geo-political structure of Sierra Leone within the established strength."

Supporters, Opponents, and Neutrals

The merger of the armed forces into a single army was a negotiation demand by the RUF, which had serious security concerns for its troops. But it was also beneficial to the SLPP government, which wished to see the CDF included in the new armed forces. Hence the idea of a military merger and the formulation of Article XVII did not spark any particular debate or controversy at the negotiation table. In fact, no alternative to such a merger was ever seriously discussed, though President Kabbah was known to privately endorse disbanding the armed forces altogether, as he feared another coup (interview with civil society representative present at the Lomé negotiations, Freetown, April 22, 2010). This option had first been discussed in the circles around Kabbah during his time in exile, and for a short period after his return to Freetown in 1998 the army was officially disbanded, but it was reinstated again shortly thereafter due to security considerations (Albrecht and Jackson 2009, 22–23; cf. Nelson-Williams 2010, 123). Although the idea of a military merger was never opposed at the negotiation table, many representatives of civil society voiced their opposition when the Lomé agreement was announced and the content of Article XVII became publicly known (Hanson-Alp 2010, 190; interview with a civil society representative present at the Lomé negotiations, Freetown, April 22, 2010).

CREATION

Military integration began according to the terms of the Lomé accords, but it broke down when the war started again. British military intervention defeated the RUF, aided by an ad hoc integration of AFRC personnel. Integration resumed after the war, but only a few people enlisted. Some former rebels now occupy high positions, and there seems to be little tension within the ranks of the new military.

Shaping the New Military

The first attempts at reforming security-sector institutions had been initiated long before the signing of the Lomé agreement.[3] The signing of the accord, however, provided a timely opportunity for a more fundamental and coordinated approach under British supervision. At this point the national army consisted of about 6,300 troops, of which about 2,000 had been recruited since Kabbah's return to office in 1998. The army also consisted of a large group of former junta soldiers, with little loyalty to the government and with an appalling human rights record (International Crisis Group 2001b, 7).

Detailed plans for a complete reorganization and reconstruction of the Ministry of Defence (MoD) based on the British model of joint civil and military management were drawn up. It was suggested that international military and civilian personnel be posted in advisory roles to support and assist in implementing the reforms, in what was subsequently to become the International Military Advisory and Training Team (IMATT). The reform proposals were submitted to the Sierra Leonean government and endorsed by President Kabbah in March 2000 (Albrecht and Jackson 2009, 46–49). In addition, plans for the establishment of an IMATT-assisted Military Reintegration Programme (MRP) were outlined in April 2000, in response to a formal request from the government of Sierra Leone and the National Commission for Disarmament, Demobilization, and Reintegration (NCDDR) to implement the provisions of the Lomé peace accord regarding the integrations of ex-combatants from the various former warring factions into the new armed forces (Albrecht and Jackson 2009, 63–64).

The political realities on the ground were, however, soon to radically change the course and pace of these reform efforts. During the implementation of the Lomé peace agreement, the RUF tried to maximize the benefits of its newly acquired position of government power, while stalling on the disarmament and demobilization process. A UN peacekeeping force, the United Nations Mission in Sierra Leone (UNAMSIL), was deployed to oversee the implementation, replacing the ECOMOG troops as these were gradually withdrawn. However, the UN forces came under attack as they attempted to assist the disarmament and demobilization process and deploy in RUF-controlled territories. This culminated in the ambush and abduction of five hundred newly arrived Zambian troops near the northern town of Makeni in May 2000 and a renewed effort by the RUF to retake the capital of Freetown by force (Gberie 2005, 161–66).

This became a critical turning point in the peace process. UNAMSIL and the existing remnant forces of the national army—having already partly disarmed and demobilized under the terms of the Lomé peace agreement—were in acute need of direct military assistance in resisting the RUF. Seven hundred British paratroopers and members of the Royal Marines were dispatched to Sierra Leone, originally for the purpose of evacuating civilians, but soon engaged in direct combat against the rebel forces (*Africa Research Bulletin* 2000b, 139–48).

President Kabbah also made a critical decision to support the formation of an "unholy alliance" of progovernment forces: the existing SLA forces, the AFRC (primarily its main splinter group, known as the West Side Boys), and the CDF, joined together under government command and supported by the British (Albrecht and Jackson 2009, 51–53; Dorman 2009, 93; *African Research Bulletin* 2000a, 139–48). Johnny Paul Koroma in particular seized the opportunity to get back at his former allies and regain some political credibility. In return, many ex-AFRC soldiers received a political promise to be reinstated into the SLA (Nilsson 2008, 150). The result was a hasty and informal military merger that had not been envisaged in the peace agreement, with thousands of former AFRC soldiers rejoining the national army, which nearly doubled in size. For security reasons, it was decided that the SLA was not to reenter the disarmament and demobilization process after it had successfully pushed the rebels back from the capital; instead, British-led short-term training teams were deployed to assist and train the SLA in intensive training courses, while IMATT was gradually deployed during the summer and fall of 2000 (Dorman 2009, 98–100).

By the spring of 2001, the RUF had been marginalized and essentially faced a military defeat.[4] Following a renewed cease-fire agreement signed by the parties in Abuja in May 2001, in which the RUF finally dropped its demand that the SLA should also disarm, the disarmament and demobilization process was finally reinitiated, and in January 2002 the war was officially declared over. At the same time, the new Ministry of Defence was inaugurated, and President Kabbah announced that the army would be unified with the tiny Sierra Leone Air Force and the Sierra Leone Navy to form the new armed forces, officially renamed the Republic of Sierra Leone Armed Forces (RSLAF).[5]

Personnel Selection

All ex-combatants who participated in the renewed disarmament and demobilization process were briefed on the existence of the MRP and given the option to seek entry into the new armed forces (Albrecht and Jackson 2009, 65, box 9). However, only about 3,000 former combatants, out of the total 72,000 registered in the disarmament and demobilization program, eventually decided to enter the MRP.[6] War weariness was one stated reason for the lack of interest in pursuing this option; another may have been the high expectations that many ex-combatants generally had for the civilian reintegration program, although these proved to be quite unrealistic (interviews, Freetown, April 20–30, 2010). Out of the 3,000, only about 2,350 were eventually posted into the new army, having completed the training program (the number is put at 2,600 by Malan 2003, 99). About two-thirds of these volunteers came from the RUF, and the rest came from the CDF (Albrecht and Jackson 2009, 66, table 2). One hundred and fifty of the volunteers were officers (Nelson-Williams 2010, 128). At this time, the existing army—consisting of former SLA soldiers and former AFRC junta soldiers—numbered "about 12,500" (Malan 2003, 96). With the implementation of the MRP, the RSLAF expanded to about 14,500 military personnel, including its maritime and aviation wings (Malan 2003, 99; Albrecht and Jackson 2009, 57).

After having signed up to join the MRP, the potential recruits were put in temporary holding camps to undergo basic military exercises and screening processes. The police and army intelligence agencies performed background checks to rule out any with formal criminal records. All potential recruits also underwent a full medical examination, and physical, educational, and military tests were carried out (Albrecht and Jackson 2009, 65, box 9). In addition, the local paramount chiefs from the recruits' home areas were consulted (interview with a representative of the Human Rights Commission, Freetown, April 22, 2010). However, the public and civil society raised concerns about the screening process, as it appeared to pay little attention to human rights abuses committed during the war, the potential recruits' psychological health, or their willingness to submit to civilian oversight and control (Albrecht and Jackson 2009, 66; Keen 2005, 284). It has been suggested that, in reality, "virtually no one" was turned away on human rights grounds (International Crisis Group 2001a, 12). British soldiers involved in the screening process acknowledged that it primarily focused on whether the individual had actually been discharged previously from the army or had a criminal record (Keen 2005, 284). The reasoning was most likely that it was better to keep people in the army where they could be monitored and controlled than to have them roaming the streets causing trouble, an argument still being voiced by IMATT personnel a decade later (International Crisis Group 2001a, 12; interview with an IMATT adviser, Freetown, April 22, 2010).

Training

Next, everyone who wanted to enter the RSLAF had to attend a formal selection tribunal. These tribunals were normally chaired by a UNAMSIL colonel and included RUF and CDF liaison officers employed by the NCDDR. IMATT officers acted as secretaries and sometimes chaired the sessions. The selected recruits were subsequently put in platoons to undergo a nine-week program of basic infantry training before joining existing units (Albrecht and Jackson 2009, 65, box 9). There were deliberate efforts to break up the ex-combatant groups into different units and subunits and to avoid the formation of units made up of members of any particular armed factions (Albrecht and Jackson 2009, 67; Malan 2003, 99).[7] After six months, subject to individual performance and recommendation, the recruits' temporary ranks were confirmed (Albrecht and Jackson 2009, 65, box 9). It has been suggested that promotion did not seem to correlate with past allegiance to any of the former warring factions (Nelson-Williams 2010, 128). The last ex-combatants graduated from the MRP on May 17, 2002 (Malan 2003, 99).

During the MRP there were no reports of violence among ex-combatants from the former warring factions (Malan 2003, 99). To the extent that the new army experienced any internal conflict, it appears to have stemmed from tensions emerging between new recruits and the older generation of ex-SLA soldiers who had remained in the army throughout the war. This was probably partly due to old civil war rivalry and partly to jealousy on the part of some of the ex-SLA soldiers, as many of the new recruits were better trained and had better career prospects (International Crisis Group 2002, 10). Ten years down the road, no disgruntlement stemming from past civil war allegiances seems to linger in the RSLAF, in spite of the fact that many former rebels are holding high-ranking positions (interview with a group of RSLAF officers, Freetown, April 28, 2010, and with an IMATT adviser, Freetown, April 22, 2010). The military has been decommissioning some officers and recruiting new ones to achieve a gradual renewal of the forces. However, the core of the RSLAF today still consists primarily of former combatants and former army soldiers (interview with an IMATT adviser, Freetown, April 22, 2010).

OUTCOME

The new military seems capable of handling likely internal and external threats. Some military/civilian tensions remain, and the army seems too large to be sustainable in the long run.

Military Capabilities

The Constitution of Sierra Leone identifies the primary tasks of the army. In descending order of importance, these are to safeguard the territorial integrity of the country, to create and maintain a safe environment for the people, and to assist with development of the country (Malan 2003, 93). Most observers agree that the RSLAF in 2010 had at least the basic capacity to defend the country against the most likely external and internal security threats (interview with IMATT advisers, Freetown, April 22, 2010, and with a high-ranking RSLAF officer, Freetown, April 28, 2010). The operational capacity of the armed forces is widely considered to have improved under IMATT supervision. The army is much better trained and equipped to perform its duties than it was at the start, although a number of problems remain: shortages of heavy military equipment, communications, transportation, and accommodation (Ebo 2006, 487–91; Ginifer 2006, 799–800; Horn, Olonisakin, and Peake 2006, 119–20; International Crisis Group 2007, 12–13; interview with an IMATT adviser, January 13, 2011).

The security situation in the greater Mano River region remains a source of potential threat to the stability in Sierra Leone. A number of incidents in 2003 raised concerns about the capacity of the new force, as Liberian rebel forces were able to cross over the Sierra Leonean border (International Crisis Group 2003, 6). Following the end of the civil war in Liberia in late 2003, however, the situation improved considerably.

The election turbulence in both Guinea and Côte d'Ivoire does not appear to have had any effect on the security situation in Sierra Leone, although it clearly points to the volatility of the region. Some unresolved border disputes between Guinea and Sierra Leone date back to the civil war, but so far the RSLAF appears to have managed them professionally (interview with an international security adviser, Freetown, April 26, 2010). However, it must be noted that notwithstanding the improved capacity of the armed forces, Sierra Leone lacks the general military capacity to counter a large-scale military intervention by another country, due to its lack of modern weapons technology and an effective air force (interview with an IMATT adviser, Freetown, April 22, 2010, and with a civilian representative with the MoD, Freetown, April 29, 2010).

The most prevalent security threats to Sierra Leone, however, remain internal in nature. Most observers agree that one of the most pressing is the country's large number of marginalized and disillusioned youths, many of whom are former combatants or child soldiers. Another source of concern is the government's failure to provide basic services, such as electricity and water, or to sufficiently address problems related to decaying infrastructure, high unemployment, and widespread corruption (interviews, Freetown, April 20–30, 2010). In addition, although the 2007 elections and the subsequent peaceful change of government power were largely interpreted as signs of democratic progress, the election results also testify to the return to traditional politics characterized by a strong north/south divide with an ethnic dimension (Kandeh 2008; Wyrod 2008). Some see worrying signs that the APC government in power is deliberately taking steps to conflate its party with the state, in an attempt to consolidate its electoral gains (interview with an international security adviser, Freetown, April 26, 2010). Most observers agree that the polarization of the two dominant parties—the APC and the SLPP—has grown stronger since the APC's victory in 2007 (Adolfo 2010; interviews in Freetown, Bo, Pujehun, Kenema, Kono, and Makeni in January 2011).

At the same time, another large-scale civil war seems unlikely. In 2003, some ex-soldiers and civilians attacked an RSLAF armory in Wellington outside Freetown, in what was allegedly an attempt to overthrow the government and reinstate Johnny Paul Koroma (International Crisis Group 2003, 6–7). Since then, the security situation has improved considerably throughout the country, and the security institutions of the state are in a much better position to counter low-level incursions and unrest before they can escalate (interview with international security adviser, Freetown, April 26, 2010). One critical part of reforming the RSLAF has been coming up with new roles and tasks for the army in times of peace. In 2004 the Military Aid to Civil Power (MACP) policy was introduced, making it possible for the army to legally provide support to the Sierra Leonean Police (SLP) when this was deemed necessary and appropriate due to security concerns (Ebo 2006, 486–87). Most observers testify to the professional behavior of the army in providing this support (interviews in Freetown, Bo, Pujehun, Kenema, Kono, and Makeni in January 2011). RSLAF also aspires to contribute to and participate in international peacekeeping missions.

A company for Peace Support Operations for ECOWAS, the AU, and the UN has been set up. A Sierra Leonean reconnaissance company was also deployed to Darfur as part of UNAMID in 2009, and the experience of this operation so far is claimed to be positive. There are also plans to contribute troops to the ECOWAS Standby Force (Ebo 2006, 486; interview with a civilian representative with the MoD, Freetown, April 29, 2010).

Political Control

In addition to external and internal security threats, one of the most critical threats to the Sierra Leonean state has traditionally come from the military itself. A great concern in the early years of the RSLAF was the lack of loyalty in the army toward the government, and the continued risk of another coup (Keen 2005, 283–87). In the 2002 elections the security forces voted separately, and the results showed overwhelming support for the ex-AFRC leader Johnny Paul Koroma's People's Liberation Party (PLP) (International Crisis Group 2002, 7). Although the relationship between the political elite and the army has improved over the last few years, and the politicization of the RSLAF is considered remarkably low in light of the historical experience of the country in this respect, those at the executive level of the government are still suspicious of the armed forces (Albrecht and Jackson 2009, 149; interview with IMATT adviser 2, Freetown, January 13, 2011). This suspicion may not be completely unsubstantiated.[8]

At present, three different coup scenarios are at least theoretically possible. First, the initiative could come from lower-ranking soldiers if they do not receive their expected salaries and rice allowances, and if they see few improvements in the general conditions of their service. Second, it could come from middle-ranking soldiers who are not promoted according to expectations because of a bottleneck in the ranking system. Third, it could come from the top and be driven by issues of personal power and prestige (interview with IMATT adviser 2, Freetown, April 22, 2010). However, it is commonly believed that the risk of a military coup in Sierra Leone is relatively low for the moment (interviews, Freetown, April 20–30, 2010). A new open and competitive recruitment procedure has been introduced, and a new policy is in place for making sure new soldiers are recruited from all four major provinces in Sierra Leone (interview with a group of RSLAF officers, Freetown, April 28, 2010). There has also been a review of payment, pensions, allowances, leave, and resettlement packages, and some improvements have been made to the living conditions of soldiers, although much remains to be done (Ebo 2006, 487; Gbla 2006, 8; Ginifer 2006, 799–800; Horn, Olonisakin, and Peake 2006, 119–20; interviews with a high-ranking RSLAF officer, Freetown, April 28, 2010, and a civilian representative with the MoD, Freetown, April 29, 2010). A perception survey carried out among RSLAF staff and published in early 2007 noted significant changes in the attitudes of RSLAF soldiers and in their perceptions of their own role and identity (Albrecht and Jackson 2009, 149).

As part of the reconstruction process, there have also been efforts to improve the public perception of the armed forces (Malan 2003, 99). In the early years of the RSLAF, civilian trust in the armed forces was low, and many believed that the army was still violent and enjoyed impunity (International Crisis Group 2004, 16–17). Ten years down the road, public surveys showed that perception of the security forces more generally had improved and that they enjoyed a better reputation than before or during the war. RSLAF is no longer considered a security threat, and soldiers enjoy more trust among the population (Albrecht and Jackson 2009, 194–95; Smith-Höhn 2010). In addition, though some clashes and hostilities did occur between civilians and army soldiers during the early years of the RSLAF, few incidents have occurred since (Horn, Olonisakin, and Peake 2006, 120). The legal framework for effectively addressing such issues within the army has also improved in recent years, although the weakness of the judiciary in Sierra Leone remains a key obstacle to improving the rule of law (interview with a legal adviser with the RSLAF, Freetown, January 4, 2011).

Although the reformation of the RSLAF, including the integration of former combatants into a single army, has largely been successful in addressing the security needs of the Sierra Leonean state, a few concerns remain. The RSLAF is not yet fully under civilian and democratic control, as envisaged by the Lomé peace agreement. Although significant progress has been made in that direction, not least in terms of official decision-making structures and procedures, much remains to be done before these institutional changes become self-reinforcing in the Sierra Leonean political culture. Certain areas of responsibility, such as budget and procurement, are decisively in the hands of civilian personnel, though other areas, such as policy, still remain partly under military control. This is probably due both to deep suspicion and resentment on the part of the military and to a lack of capacity and willingness on the part of civil servants to enforce their control (interview with an IMATT adviser, Freetown, April 22, 2010).

The problem of how to retain and pass on institutional memory has also been identified as a major obstacle to civil servants' taking on more control, given that many MoD officials who were trained in the early years have subsequently left the ministry. There is also a lingering perception in the military that civilian oversight is primarily a political instrument to prevent another coup, and not an integral part of a democratic state (Albrecht and Jackson 2009, 147–48). In addition, there are few if any effective and functioning democratic oversight mechanisms (Ebo 2006, 494). Even when such institutions are in place, such as the parliament and its committees, their capacity and willingness to enforce their roles are generally very limited (Gbla 2006, 88; Ginifer 2006, 802; interview with an IMATT adviser, Freetown, April 22, 2010). Civil society and the media also frequently lack the information, capacity, and political independence necessary to function as the democratic watchdogs they are intended to be (Horn, Olonisakin, and Peake 2006, 120; interviews with an IMATT adviser, Freetown, April 22, 2010, and with a Sierra Leonean university professor, Freetown, April 29, 2010).

Sustainability

The ambitious nature of the transformation of the RSLAF, as well as the current size of the armed forces, has raised questions about the long-term sustainability of the security-sector reform process in Sierra Leone (see Horn, Olonisakin, and Peake 2006, 120). The size of the army in particular is a matter of concern. Most international observers, including IMATT representatives, argue that the size of the army in 2010, a little short of 8,500 men, is too large for the country, in light of both long-term affordability and current threats to the state (interviews with IMATT advisers, Freetown, April 22, 2010). It has been suggested that in terms of what the government could afford to sustain without significant external funding, the optimal force size is likely to be closer to two or three thousand troops, which corresponds to the size of the army before the civil war. In terms of operational requirements, however, the right size might be somewhat larger, perhaps as many as five thousand, considering the still-volatile security situation in the greater Mano River region and the promising potential for further Sierra Leonean participation in international peacekeeping operations (interview with an IMATT adviser, Freetown, April 22, 2010).

Downsizing the army is, however, a sensitive political issue. It was not until late 2002 that the Defence Council gave its approval for the RSLAF to begin a process of downsizing from its estimated size of approximately 14,500 men (Nelson-Williams 2010, 129 suggests about 15,500) to 10,500 within four years. The first reductions in the force were based on voluntary discharges with the help of substantial donor support to offer attractive retirement packages (Albrecht and Jackson 2009, 155). In January 2004, 784 people were retired; in 2005, another 1,000; and in 2006, an additional 1,092 (Nelson-Williams 2010, 128). The new APC government set the target at 8,500 after coming to office in 2007, and a new round of downsizing was initiated (interview with a group of RSLAF officers, Freetown, April 28, 2010). Although the MoD seems to understand the need for further downsizing of the army in the years to come, any such proposal is likely to be met with great resistance from military leaders, who generally disagree that the force is too large (interview with a civilian representative of the MoD, Freetown, April 29, 2010, and with a group of RSLAF officers, April 28, 2010). A decision to downsize also carries with it certain security risks, as most of those asked to leave are likely to face bleak employment prospects. The third round of downsizing in particular was met with loud protests and official complaints about its management and procedures.[9]

The third and last concern is the question of local versus international ownership of the reconstruction of the RSLAF and the MoD. International personnel have occupied high-level executive and advisory posts for more than ten years. Some Sierra Leonean officers have complained that they have been sidelined or insufficiently consulted during this process, while IMATT has claimed that the lack of capacity and willingness displayed by RSLAF officers has made it necessary to bypass them (Ginifer 2006, 801). Clearly, the substantial and long-term commitment to the security-sector reform process in Sierra Leone on the part of the international community, not least the UK government, is something of a two-edged sword in this respect (interview with an international security adviser, Freetown, April 26, 2010). However, international personnel have gradually moved from command and executive posts to advisory and support roles. In 2007, the last remaining executive powers were handed over from international control (Albrecht and Jackson 2009, 145). In early 2010, forty-five international IMATT personnel remained in Sierra Leone. The plan is to maintain the current mission and mandate until after the 2012 elections, at which point a review will be carried out (correspondence with an IMATT adviser, May 24, 2010).

HAS MILITARY INTEGRATION MADE LARGE-SCALE VIOLENCE LESS LIKELY?

The armed factions that made up the core of the new RSLAF at the time of its inception—the former SLA, the AFRC, the CDF, and the RUF—were all in various stages of disarray after having survived a decade-long armed conflict known for its brutality and human rights abuses. Ten years down the road, they were working together side by side for a common purpose in an army that has been completely reconstructed and reformed under close international supervision. Although the transformation process has seen some serious shortcomings, which may undermine the long-term benefits of these changes, there can be no doubts about the relative success of the military integration process in Sierra Leone.

Has this success made the resumption of civil war in Sierra Leone less likely? The most accurate answer to this question is probably both a qualified yes and a qualified no. Yes, because the army has always constituted one of the gravest threats against peace and security in Sierra Leone and was one of the key perpetrators of the indiscriminate abuse of the civilian population that took place during the civil war. The creation of a much-better-trained and more professional army under civilian and democratic control is therefore a critical step in ensuring the sustainability of peace. Integrating ex-combatants from all former warring parties into a new single army may also have decreased the risk of renewed warfare, at least in the immediate postwar period, with the important caveats that relatively few former RUF soldiers and CDF militia members eventually joined the RSLAF, and that these did not include the most notorious or highest-ranking among them.

However, on closer examination the causal relationship between the military integration process and peace becomes more complex. The military integration process succeeded for the very same reason that war did not resume: The RUF suffered a de facto military defeat between the signing of the peace agreement in 1999 and the signing of the second and last cease-fire agreement in May 2001. Hence, although the peace agreement provided the legal and political framework for both the military merger and the ending of the civil war, it would be a mistake to perceive Sierra Leone's case as a typical negotiated war ending. It was, rather, a military victory in disguise. In this respect, the case bears some similarity to that of Rwanda, in which a power-sharing agreement failed and was followed by a military victory and a subsequent process of military integration. However, the contexts are radically different. For one thing, unlike in Rwanda, the peace settlement did not completely break down; although the RUF returned to war, the provisions of the Lomé agreement still provided the framework for the military merger once the rebels had been defeated. In addition, the entire peace-building process in Sierra Leone, including the reconstruction of the army, took place with significant and extensive international guidance and control. Nevertheless, we have to take postwar military power relations into account when evaluating the causal pathways between the military integration and renewed warfare suggested by this research project.

Let us first consider the commitment path. This causal mechanism may carry some weight in the case of Sierra Leone, albeit with important reservations. Military integration was a key demand by the RUF during the Lomé peace process, probably at least in part because of genuine security fears. Although a military merger was a costly commitment by Sankoh in many respects—subsequent developments did reveal both serious tensions between the RUF's political leadership in Lomé and the military officers in the field, and attacks on RUF combatants during the disarmament and demobilization process—it is possible that the RUF's leadership at the time of the negotiations never intended to carry out its commitment to the accord and simply argued in favor of military integration for strategic purposes.

More important, however, the decision to opt for a military integration process can only partly be considered a costly commitment on the part of President Kabbah, who at the time did not trust his own forces and had spent the war relying on the protection of either foreign troops or the civil militias. His primary security fear was not, therefore, a military merger, but rather maintaining the status quo. The fact that Kabbah did not insist on dissolving the army altogether and agreed to the creation of a new army with coercive capacity, in light of his previous war experience and the threat that the armed forces posed against his office, made the military integration process somewhat costly to him. But unlike the case of the RUF, it is very doubtful that the government felt more secure because of the military merger. In the long term, however, and certainly once the military merger was effectively under way, the military integration process may have helped to increase the confidence of the government in pursuing further reform efforts, which in turn made renewed warfare less likely.

The third mechanism regarding employment is also relevant. The lax screening process for entry into the new army, particularly regarding applicants' human rights records, was deliberately instituted by donors keen on getting as many ex-combatants as possible off the streets, where it was thought they would be easy recruitment targets for potential spoilers. However, due to the high expectations that most ex-combatants had for the civilian reintegration package at the time of the disarmament and demobilization process, relatively few of them decided to enter the RSLAF. That said, it must be remembered that the official MRP targeted only former combatants from the RUF and the CDF. Many former AFRC soldiers had already been reintegrated into the SLA following the May 2000 events, when immediate security concerns forced the government to rally all available progovernment forces against the RUF. For many junta soldiers, one of the main goals of the armed struggle had been reinstatement or employment into the army, given that many of these combatants faced bleak prospects for civilian employment (Nilsson 2008, 155).

The fourth causal mechanism speaks to the symbolic relevance of the military integration process, and seems to apply in the case of Sierra Leone. The RSLAF is a symbol of national unity in a country in which mismanagement, violence, and abuse of and by the various armed groups—including the national army—has been at the very core of the dynamics of the civil war. However, and of particular importance, it is not merely the existence of the new and integrated army that appears to matter in this respect, but equally its improved capacity. It is because the RSLAF is better trained, more professional, and under civilian and democratic control that the relationship between the political elite and the army has gradually improved. Wartime identities appear to matter little among the military personnel, and the new army has gained some trust from civilians, who are generally more supportive of the military reintegration process now than they were at the time of the peace accord.

It is much more difficult to find evidence in favor of the fifth causal path, which identifies the negotiation process leading up to a military merger as a critical confidence-building measure. Most observers agree that there was very little trust among the parties during the Lomé negotiations, something that did not change with the signing of the agreement. The turbulent and occasionally violent postaccord period further eroded existing relationships and, if anything, saw an increase in the risk of renewed civil war. It was only thanks to the military intervention by British forces following the breakdown of the disarmament and demobilization process in May 2000 that such a scenario was prevented.

Several policy implications can be drawn from this study of Sierra Leone. First, it clearly points to the critical role that the international donor community can play in the process of military integration. There can be no doubt as to the importance of IMATT in all aspects of the reform process, from the training provided in the MRP to the strategic placement of advisers high up in the hierarchy of the completely reorganized MoD. The commitment of the British government to the Sierra Leonean peace process, particularly in the area of security-sector reform, has been both remarkably extensive and unusually long-term. However, Sierra Leone is something of a special case in this respect, with the British engagement stemming from strong historical ties between the two countries, a proactive and vocal Sierra Leonean diaspora in the United Kingdom, and a series of personal commitments by both influential British politicians and people on the ground (Jackson and Albrecht 2010). It is much less clear that domestic support in Sierra Leone played any significant role. The military integration process decided upon in the Lomé agreement was actively opposed by many civil society organizations, due to the appalling human rights records of all the armed groups (including the army) during the civil war. The RSLAF has subsequently gained trust among the population, but there is little evidence that the relative success of the integration process was based on such domestic support.

Second, there are few indications that the lack of more thorough screenings for human rights violations has had any obvious negative effect on the behavior of the new armed forces. Given the history of most of the ex-combatants involved, remarkably few reports have been made of violence or intimidation between soldiers and civilians. Third, little suggests that the lack of quotas for the various ex-combatant groups made any difference in the process of military integration. This is probably because the armed conflict was not primarily about identity or ethnicity. Fourth, violence and tension were largely absent during the training period of the MRP, and former conflict divisions did not seem important to the individual ex-combatants who joined the new army. Fifth, although the approach chosen in the case of Sierra Leone—to bring as many ex-combatants as possible into the new army in order to keep them off the streets and in the barracks where they could be monitored and controlled—did initially lead to an increase in the total number of troops, the RSLAF has subsequently been significantly downsized. Although the size of the army continues to be a politically sensitive question, the long-term commitment of IMATT and an unusual consensus across the Sierra Leonean political party divide with respect to this issue have so far ensured the viability of the process.

ABBREVIATIONS

AFRC	Armed Forces Revolutionary Council
APC	All People's Congress
CDF	Civil Defence Forces
ECOMOG	Economic Community of West African States Military Observer Group
ECOWAS	Economic Community of West African States
IMATT	International Military Advisory and Training Team
MoD	Ministry of Defence
MRP	Military Reintegration Programme
NCDDR	National Commission for Disarmament, Demobilization and Reintegration
NPRC	National Provisional Ruling Council
RSLAF	Republic of Sierra Leone Armed Forces
RUF	Revolutionary United Front of Sierra Leone
SLA	Sierra Leonean Army
SLP	Sierra Leonean Police
SLPP	Sierra Leone People's Party
UNAMSIL	United Nations Mission in Sierra Leone

NOTES

1. The exact size of the SLA during the war is difficult to determine. Le Grys (2010, 42) suggests that the army numbered about 16,000 at the height of the war. Nelson-Williams (2010, 133) puts the number at "over 15,500."

2. Sierra Leoneans and human rights activists openly expressed their opposition to the amnesty clauses, and Kabbah was strongly criticized from within his own cabinet. The UN, for its part, added a separate disclaimer to the amnesty article in the agreement (Francis 2000, 365–66).

3. A series of minor reform programs sponsored by the British had been set up following the Abidjan peace accord in 1996, but they were abandoned with the May 1997 coup. More serious reform initiatives followed the restoration of Kabbah to office in 1998, only to be thwarted again by the January 1999 attack on Freetown. After the attack, new reforms were introduced to bring the armed forces under civilian control (Albrecht and Jackson 2009).

4. The group also suffered economically from Charles Taylor's forced disengagement from the RUF following diplomatic and economic pressure on the Liberian president. By this time there had also been a leadership change in the rebel organization, and the new leaders were more willing to accept that the RUF was now a spent military force and embrace the option to transform into a political party (International Crisis Group 2001a; Söderberg Kovacs 2007).

5. In 2001, there were also plans for the creation of a Territorial Defence Force, a reserve security force of some 7,500 men to provide a backup for the military and the police. However, these plans were eventually abandoned by the Sierra Leonean government (International Crisis Group 2002, 11–12; Albrecht and Jackson 2009, 61–62).

6. The total number of disarmed RUF soldiers was 24,352; for the CDF it was 37,377 (Thusi and Meek 2003, 33).

7. Ten years later there were still few if any units that are readily identified as "belonging" to any one former warring faction (interview with an IMATT adviser, Freetown, April 22, 2010).

8. A press release issued from the Office of the President on August 24, 2010, announced that the army chief of the Defence Staff was retiring. However, rumors in the corridors of the MoD speak of a planned coup attempt as the real reason that the chief, along with some other key personnel, was removed. *Awareness Times Newspaper*, August 25, 2010; author's interviews, Freetown, January 2011.

9. On January 1, 2009, disgruntled soldiers calling themselves the Detective Reconnaissance Emergency Action Mission Team (the DREAM Team) wrote an open letter to the president, accusing the RSLAF leadership of mismanagement, corruption, and tribalism; *Cotton Tree News*, February 6, 2009. These grievances were later confirmed in an author interview with a representative of the Human Rights Commission, Freetown, April 22, 2010.

Chapter 13

MILITARY INTEGRATION IN BURUNDI, 2000–2006

Cyrus Samii

This chapter describes military integration in Burundi associated with the 2000–2006 peace process that ended the civil war that began in 1993. Burundi is a small, impoverished, landlocked country of approximately 8 million people (as of 2010) in central Africa. It has been racked by political violence since it gained independence in 1962. Like Rwanda to the north, Burundi has a society marked by a caste-like stratification that has historically privileged a Tutsi minority relative to majority Hutu and a very small third group, the Twa.[1] Burundi is also like Rwanda in that its people have struggled to escape a conflict pitting custodians of this "ranked ethnic system" (Horowitz 1985; Lemarchand 1970) against those ostensibly seeking to remove barriers to Hutu mobility.[2] This is the context within which military integration has taken place in Burundi.

ORIGINS

This section first examines the historic role of the military in Burundi. Then it turns to the emergence of military integration as an issue. In this process, it considers the roles of supporters, opponents, and neutrals—along with compromises in the process.

The Historic Role of the Military

Burundi's national army, known as the Forces armées burundaises (FAB) from 1962 to 2004, has featured centrally in the country's bloody political drama since just after independence. In the first four years after independence in 1962, Burundian politics suffered a series of assassinations, an abortive coup by Hutu officers, repressions, and reprisal massacres. The events culminated in a purge of high-ranking Hutu officers and a 1966 military coup led by the defense minister, Captain Michel Micombero, who ended the ruling monarchy.

Thus began a period of de facto military rule and intensified concentration of economic opportunities and power, particularly within the army, into the hands of a Tutsi clique from the southern Bururi province. The clique oversaw a dramatic intensification of Hutu exclusion and a degree of exclusion of nonsouthern Tutsi.

A 1972 insurrection coordinated by Hutu expatriates and Hutu army members escalated to involve massacres of Tutsis, mostly in the southern part of the country. This triggered a barbarous crackdown by the army, which sought to prevent future uprisings by "decapitating" Hutu society. The army targeted for execution Hutus in positions of authority or exhibiting leadership potential (e.g., those who wore glasses). The estimated number killed that year—mostly Hutu, it is thought—was 150,000 to 200,000; and tens, if not hundreds, of thousands of Hutus were driven into Rwanda and Tanzania (United Nations 1996a).

Competition among clan-based factions of southern Tutsi military officers shaped the next twenty years of politics. Micombero was overthrown in a 1976 coup by his deputy chief of staff, Colonel Jean Baptiste Bagaza, who was also a Tutsi from Bururi Province. Bagaza presided over a decade of fast-paced development, largely to the benefit of the Bururi elite.

The Emergence of Military Integration as an Issue

During this period movements emerged to contest for Hutu rights. These included the Burundian Workers Party (UBU), organized among Burundian émigrés in Rwanda; the Parti pour la libération du peuple hutu (PALIPEHUTU) and its armed wing, the Forces nationales de libération (FNL), headquartered in Tanzania; and the Front pour la libération nationale (FROLINA) and its armed wing, the Forces armées du peuple (FAP), also headquartered in Tanzania. Within Burundi, the army contained threats from these groups handily.

In 1987, Bagaza was overthrown in a coup led by Major Pierre Buyoya, also from Bururi. A rural uprising in the north in 1988 gave rise to another cycle of civilian massacres and another vicious army crackdown, with an estimated 5,000 to 20,000 killed (Loft 1988; Chrétien, Guichaoua, and Le Jeune 1989). This event, international pressure, and increasing agitation by Hutu intellectuals convinced then-president Buyoya to initiate reconciliation and democratization. Buyoya oversaw the promulgation of a National Unity Charter in 1991 and a new Constitution in 1992, setting the stage for 1993 elections. Places in the national officer academy, the Institut supérieur des cadres militaires (ISCAM), were also opened to Hutus.

But this gesture masked a more general resistance toward army reform. The 1992 charter declared that "the truth is that there is no discrimination within the army" (Lemarchand 1996, 139). A beneficiary of this slight opening, a Hutu from Bururi named Jean-Bosco Ndayikengurukiye, was part of an early integrated ISCAM class. He would eventually defect to become a leader in the rebellion in 1993.

Members of the southern Tutsi officer class were not easily pried from power. In peaceful and fair elections in 1993, Melchior Ndadaye defeated the incumbent Buyoya by a large majority in the presidential race. Ndadaye was a Hutu civilian, a former UBU leader, and the founder of the newly formed political party Front pour la démocratie au Burundi (FRODEBU). FRODEBU also took a large majority in the national assembly. Its platform called for removing barriers to Hutu mobility, including within institutions such as the army. The Ndadaye administration called for the rapid promotion of Hutu officers, ostensibly to align the officer corps with the civilian government.

After only three months in power, Ndadaye was assassinated in a bungled coup attempt on October 21, 1993. The events remain shrouded in mystery, although Burundians commonly attribute them to associates of former president Bagaza.[3] The assassination triggered what a United Nations commission described as genocidal reprisals by Hutu mobs against Tutsi men across the country, followed by army massacres of Hutus (United Nations 1996a).

In the ensuing ferment the government became increasingly beholden to members of the southern Tutsi "old guard." At the same time, members of FRODEBU left government to establish the rebel movement, the Conseil national pour la défense de la démocratie (CNDD), and its military wing, the Forces pour la défense de la démocratie (FDD).[4]

The CNDD and FDD declared their intention to be the "restoration of democracy." Explicit in their strategy for achieving this aim, and voiced publicly as early as 1994, was the defeat and dismantling of the *armée mono-ethnique*, so called because the officer corps was the near-exclusive domain of southern Tutsis.[5] Hutu men had long been admitted, and during war conscripted, into the rank and file.

But conservative members of the Tutsi elite tended to associate Hutu officership with threats like the abortive 1965 coup and 1972 insurrection. Those Tutsi elites rejected the characterization of the army as an instrument of Tutsi dominance. Rather, they saw Tutsi control over the army as a necessary protection against "genocidal" or "revolutionary" tendencies among the Hutu masses.[6]

The FNL and FROLINA had called for a reshaping of the army as part of more general Hutu liberation. Ndadaye's assassination concentrated minds intensely on the issue. The phrase *armée mono-ethnique* reflected a sense among those sympathizing with the rebellion that, for decades, the army and police had been instruments of southern Tutsi domination and the main obstacle to freedom, particularly among Hutus.

The army (FAB), FDD, FNL, and FROLINA fought a civil war that lasted until 2004. Small-scale residual violence caused by a splinter FNL faction continued until 2008. Wartime violence touched all regions. Conventional estimates put the death toll by 2004 at 300,000 and the number of displaced at 500,000.[7] A peace process had begun initially in 1996. Agreements signed by the warring parties in Arusha in 2000 and Pretoria in 2003 ushered in genuine peace. The FDD forces were largely successful on the battlefield, although the FAB forces were not defeated outright. Rebel successes are reflected in the agreements, whose provisions constitute a near-revolution in the country's distribution of power, including the creation of a new military integrating FAB and rebel forces. This outcome was consolidated when the CNDD-FDD (the party formed from the politico-military movement) won large majorities in the national assembly and local councils in the 2005 elections.

Supporters, Opponents, and Neutrals

As discussed above, the rebels set as an explicit goal that the army be reformed. The issue was not easily resolved in negotiations, which opened in 1996. Tanzanian president Julius Nyerere mediated a formal process among members of major political parties that had not gone underground. The Catholic Comunita di Sant'Egidio facilitated covert meetings between the CNDD-FDD leadership and government representatives. In these talks, the rebels issued their demand that the army and police be dismantled and a wholly new security force created. Members of the old forces would be no more or less eligible to join the new force than others.

Obviously, the ruling faction rejected this idea. The FAB elite scoffed at the idea that anyone other than them, with their formal training, could manage security affairs. At the same time, the message became clear that some military reform would be necessary. However, ruling faction members initially took this to be a "problem of merely integrating a few rebel elements into the armed forces" (Nindorera 2007, 11).

The "integration" agenda moved forward once talks picked up again in 1998 in Arusha. By that time the ruling faction and army were being pressured from various directions, including by international sanctions, a lack of battlefield progress, and growing fatigue in the population. The rebel movements had begun to splinter by this time. The largest rebel factions—that is, the core FDD and FNL factions—remained defiant and spurned the formal negotiations.

When talks resumed in Arusha, facilitators had delegates form committees on (1) the nature of the conflict; (2) democracy and governance; (3) security; and (4) reconstruction and development.[8] It was in the third committee that discussions turned to reforming the security sector. Nyerere's death in October 1999 prompted Nelson Mandela to take up the facilitation role, reinvigorating the talks.

Agreement was elusive initially. But continued international pressure, including condemnation for the *regroupement* (concentration camp) strategies used in the counterinsurgency (United Nations, Office for the Coordination of Humanitarian Affairs 2000), and continued degradation of the domestic situation spurred on the negotiations. As of April 2000, the International Crisis Group reported that no agreement had been reached on military reform, although the Tanzanian and South African facilitation team had put forward proposals to integrate members from the "armed political groups" (i.e., rebel groups) into a reconstituted security force (International Crisis Group 2000a, 7). The delegates to the third committee reached out to the FDD and FNL factions outside the talks, though with no success.

At this point it is worth recounting how the interests of the various factions were arrayed relative to the proposal to integrate the armed forces. At one end were the large FDD and FNL factions that remained outside the negotiations. The FDD faction's armed forces numbered about 25,000, and they had the political support of the largest segment of Burundians.[9] As of 2000 they had not signaled a willingness to accept anything short of a "dismantling" of the army. This defiance apparently reflected confidence in battlefield prospects.

The armed FNL faction, whose forces were much smaller—around 3,000—seemed to bank on a type of "outbidding" strategy. Hard-line opposition to the integration proposal was a means by which the armed FNL faction could maintain its ideological purity as the true champion of "Hutu liberation." With bases secure in rugged enclaves around Burundi's capital of Bujumbura, FNL hard-liners could rely on spoiling and terrorist tactics to coerce a deal out of whatever powers would come to be. They could play a game of one-upmanship, taking as given concessions won by the CNDD-FDD and then bidding for favor among Hutus by demanding even more.

At the other end of the spectrum were the Bururi-based Tutsi elite. These included the UPRONA faction led by Charles Mukasi, which also remained outside the peace process. Although their numbers were few, they had considerable wealth and connections. Their opposition to integration rested on beliefs already outlined above: Controlling the security sector was essential to guard their privilege; less cynically, it was necessary to protect against genocidal or retributive tendencies among Hutu masses. These hard-liners found allies among the military officer class, many of whom felt they were uniquely qualified to run the armed forces. With their education in European military academies, they scoffed at sharing ranks with rebels from "the bush."

But this professionalism cut both ways; those who were not deeply tied into Bururi privilege would likely accept orders to integrate were they issued by political authorities with the endorsement of the international community.[10] It was this sense of professionalism that South African facilitators eventually leveraged in moving the process forward.

Between these two extremes were those whose willingness to compromise was largely caused by their need to find some way to secure their political futures without having at their disposal any military means. These included the heads of small rebel splinter factions (including the CNDD's founder, Léonard Nyangoma, who was ousted in 1998, and the heads of FNL splinters) and most members of the embattled political parties FRODEBU and UPRONA.

With heavy cajoling by the South African mediation team and the personal intervention of US president Bill Clinton, the parties in Arusha reached a formal compromise in August 2000. Relative to other cease-fire agreements and peace accords, the Arusha Accords were extraordinary in the elaborateness and enlightened manner in which they were crafted to address Burundi's political ills. This happened for several reasons. First, the Tanzanian and South African mediators worked in good faith and drew on a wealth of transition experience. This international assistance should not be underappreciated. The South Africans were well positioned to help, having recently integrated rebel African National Congress Umkhonto we Sizwe (MK) forces into a new postapartheid military. The South African mediators could use this similar experience to gain the trust of the Burundian protagonists.

Second, though many accused the participating Burundian elite of venality, there were some strong intellects among them.[11] With the main rebel factions outside the process, the mediators and Burundian elite crafted a document illustrating what "could be." It was essentially a new Constitution, although without the political backing necessary for implementation. Nonetheless, the thoroughness of the proposals could be used as a strategy to win over the masses as well as a critical share of the outstanding FDD and FNL membership.

Compromises in the Process

The accords provided extensive guidance on military reform. They established a rule of ethnic balance that posts would be allocated equally to Hutus and Tutsis; the overall composition of the security forces was to be balanced in this way "in view of the need to achieve ethnic balance and to prevent acts of genocide and coups d'état" (Protocol II, Chapter 1, Article 11). The army and police were not to engage in politics. The new police forces should be "community oriented," which meant that each police contingent should be composed mostly of people from the community it patrolled.[12] War criminals and coup plotters should be excluded from the reformed security forces.

In addition, the accords opened the door for members of the "armed political parties" (i.e., the rebel factions participating in the Arusha process) to be directly integrated into a reconstituted national security force. To ensure that qualification requirements were met, the accords called for accelerated training for those from the armed political parties. Some portion of both FAB and the armed political parties would be demobilized. A joint technical committee would specify numbers to be admitted from FAB and armed political parties, as well as overseeing the harmonization of ranks and establishing timelines. The committee would include members of FAB and the armed political parties, as well as representatives from the African Union and UN.

The signing of the Arusha Accords did not mark the end of the war or immediately draw the outstanding FDD and armed FNL factions into the peace process. For three years the Arusha Accords remained mostly a paper exercise. The rotation of transitional heads of state was implemented, with FRODEBU leader Domitien Ndayizeye succeeding Buyoya as president of the transitional government in April 2003. But the proposed elections, peacekeeping operation, and transitional justice processes were put off as war continued.

Nonetheless, the accords were a large step forward relative to the denialism that had marked the National Unity Charter a decade earlier. The accords were unequivocal about integrating rebel factions into a new military. In contemporary conflicts in which rebel forces have fought government armies to a draw on the battlefield, such has not always been the case. For example, Côte d'Ivoire's 2003 Linas-Marcoussis Agreement and Nepal's 2006 cease-fire emerged in roughly similar strategic settings, but each was terribly vague about rebel integration into a new army. In both cases this vagueness developed into a major sticking point that held up the political transitions.

The Arusha Accords' 50 percent rule, whereby post assignments and the overall composition of the military would be ethnically balanced, addressed the discrimination problem in a remarkably frank manner. One can contrast this to the postwar situation in neighboring Rwanda, where the government under Paul Kagame strictly forbade reference to ethnic identity.

Of course, FNL sympathizers in Burundi are quick to point out that a 50 percent rule is unfair because it does not reflect the country's ethnic balance. From my own conversations with Burundians from different parts of the country and walks of life, I believe that empathy for this perspective is widely held among those who identify as Hutu. However, I have found that most of those empathizing with this view also appreciate that the costs of pushing for a larger share are simply not worth it.

CREATION

Using the Arusha Accords as guidelines and with assistance from South Africa, the FDD was eventually brought into the integration process. A joint committee directed the process. Quotas were established for various ranks, and in fact FDD officers proved to be roughly equivalent to those from the FAB. Many foreign governments contributed to training.

Shaping the New Military

With the Arusha Accords in the background, the creation of an integrated military occurred through power sharing among the CNDD-FDD, the transitional government, and the high officer corps of FAB. The Arusha Accords continued to structure the negotiations. South African mediators and experts worked with Tanzanian and Ugandan representatives to encourage and offer ideas.[13] On the battlefield the FDD controlled large swaths of the countryside, but FAB also showed that it would not be vanquished anytime soon.

Politically, the CNDD-FDD began to sense, by late 2002, that while it retained the support of large segments of the population, it risked losing such backing because of spreading war fatigue.[14] The CNDD-FDD nonetheless remained reluctant to accept the Arusha Accords formulas—particularly a 50/50 ethnic balance in the military. CNDD-FDD leadership initially viewed with great suspicion South African mediators who pushed for this formula, accusing them of pro-Tutsi bias. An important task for the mediation team was to gain the trust of the CNDD-FDD and persuade it of the practical necessity of accepting the formula (Southall 2006, 119–22). Increasingly, the FNL took on the role of spoiler.

In December 2002, the government and the CNDD-FDD agreed to a cease-fire, in which they reaffirmed the Arusha Accords' provisions on military restructuring. They called for a two-phase process of first integrating the armed forces and then rationalizing the size and structure of the resulting force according to a new security doctrine. Remarkably, the cease-fire also called for the FAB and the CNDD-FDD to establish "joint military units" to defeat the FNL.

The cease-fire broke down within a month, however. Some have claimed that this breakdown was entirely avoidable, given that it was caused mostly by a spiral of mistrust that could have been prevented by a peacekeeping force (Boshoff 2003). Indeed, as they resumed their mediation in the summer of 2003, the South Africans aggressively laid plans for an African Union intercession force that could be deployed to stabilize a cease-fire.

The transitional government, the FAB high officer corps, and the CNDD-FDD agreed on a detailed plan for establishing an integrated military through negotiations in Pretoria in late 2003. Intense South African mediation in autumn 2003 led the transitional government and CNDD-FDD leadership to strike a new cease-fire agreement in October. This was quickly followed by negotiations over power sharing. As in Arusha, technical experts from South Africa provided substantial input into the process.

Personnel Selection

Among the agreements signed in Pretoria was the November 2003 Forces Technical Agreement (FTA).[15] The FTA laid out in considerable detail the terms under which rebel forces and FAB would be integrated into a new military. The FTA upheld guidelines from the Arusha Accords and filled in details. Specifically, it mandated the creation of a supervisory body, the Joint Ceasefire Commission (JCC), and a third-party African Union peacekeeping mission, the African Union Mission in Burundi.

The JCC was a subsidiary of the Implementation Monitoring Committee, which included representatives of the transitional government, the CNDD-FDD, and the other smaller armed factions that were parties to the Arusha Accords. The JCC brought together the military leaders from these parties. The JCC operated under a United Nations mandate with facilitators from the UN and the African Union.

The FTA stated that the FAB rank structure would be maintained and that the JCC would rule on the rank qualifications of those integrated from the rebel forces.

Members of the rebel factions would be subject to a cantonment and verification process supervised by the JCC. The FTA spelled out specific education requirements for commissioned officers, noncommissioned officers, and the rank and file.[16]

The FTA called for an integrated top officer echelon consisting of 60 percent FAB officers and 40 percent CNDD-FDD officers, and a 65/35 FAB-to-CNDD-FDD breakdown for the top officer echelon of the integrated police. Throughout the ranks, a principle of "ethnic equilibrium (50/50)" would be observed.[17] A small share of lower posts, approximately 10 percent, would be allocated to officers from the other armed parties. The FTA also provided for the establishment of joint military units; approximately 7,000 CNDD-FDD soldiers would join FAB for approximately two months of training to fight the FNL.

Within the JCC a two-phase integration process was foreseen. The first was the "integration" phase, which would run from 2004 to approximately 2006. During this phase, some 26,000 members of the rebel armies and 40,000 members of FAB would be integrated to form a new national army and police force. A total of 14,000 soldiers would be demobilized immediately, of whom 5,000 would come from FAB and 9,000 from the rebel forces.[18] The second was the "rationalization" phase, to begin in 2006 and last for an indefinite period. The national army would be trimmed down to a target of 25,000 as quickly as possible, and the community-oriented police force would be trimmed to about 20,000.

Rank Allocation

As the JCC went about its work in early 2004, an initial sticking point was in the harmonization of ranks for rebel soldiers with the rank structure of FAB. As discussed above, FAB elite officers often scoffed at the idea that they should share ranks with "officers" from rebel forces, few of whom had formal training. The distribution of ranks within the rebel forces suggested a degree of "rank inflation," in that early rank promotions were used within the rebel forces as a way to boost morale, given that other incentives, such as pay raises, were not available.[19] As Jackson (2006, 22) notes, "Given the Tutsi dominance of the old army's command, this was a wedge issue to the peace process as a whole."

Ultimately, FAB officers and CNDD-FDD officers worked among themselves outside the JCC to fix a rank-harmonization formula, with the small groups then being "forced to fall in behind the new arrangements" Jackson (2006, 22). Thus, the CNDD-FDD and FAB pushed potential spoilers and fundamentally irrelevant parties to the side in order to solve this difficult problem. For the most part, rebels were downgraded based on a fixed schedule to ranks in the FAB structure that were more appropriate to their respective levels of training and experience (BBC Monitoring 2005).

Surveys of members of the new armed forces suggest that the conditions for integration were rather favorable because of the characteristics of the ex-rebel and ex-FAB soldiers themselves. The 2007 "Wartime and Postconflict Experiences in Burundi" survey included interviews with a random sample of 496 members of the new army, of whom 134 were from the rebel forces, and 618 members of the new police force, of whom 231 were from the rebel forces.[20] In general, the former rebels had higher satisfaction with the peace accords than their ex-FAB counterparts, and they were much more hopeful about prospects for peace. But there was a near-consensus among ex-rebels and ex-FAB that "war fatigue" made ending the war necessary and desirable.

Ex-rebels tended to be two years younger than ex-FAB (a median of twenty-eight vs. thirty), and ex-rebels were 24 percent less likely to have completed primary school. However, there was no education gap at higher levels (secondary, university), indicating that any difference in educational qualifications was concentrated among the rank and file. The ex-FAB tended to have more soldiering experience, but the gap was pronounced only among armed forces members with more than twelve years of experience; this gap is mostly the result of the timing of the war's onset.[21]

Although 96 percent of ex-FAB indicated that they had seen written codes of conduct, 79 percent of ex-rebels indicated the same. Most of the ex-rebel respondents were from the CNDD-FDD. The high rate of exposure to written codes of conduct reflected a high degree of formality within the CNDD-FDD. Indeed, with its congressional assemblies and highly organized military, the CNDD-FDD had evolved into a provisional state by about 2001.

Overall, the human capital gap between the ex-rebels and ex-FAB was not so great. The education gap among the rank and file might appear to have been the most problematic of the differences, but even here the gap was not so large. The difference in years of service at the top echelons was not very consequential given near-continuous battlefield exposure and eventual training from European partners.

Training

Training programs were an important element of the implementation process, with many foreign partners contributing to the effort (United Nations 2004; United Nations, Office for the Coordination of Humanitarian Affairs 2005; Nindorera 2007, 12; Powell 2007, 26–27). To achieve rank harmonization, the Belgian government provided bridging training to rebel officers at its Royal Military Academy and Royal Defence Institute; the Netherlands, France, China, Sudan, South Africa, Rwanda, and Egypt ran other training programs either in Burundi or at their own military academies.

The Belgians and the United Nations Operation in Burundi (ONUB, by its French acronym) conducted "professionalization" programs to inculcate national purpose that transcended ethnic and political divisions. The US Agency for International Development sponsored the Burundi Leadership Training Program, which brought elite political and military figures from different factions together in confidence-building seminars.

UN peacekeepers were present at the training of the joint CNDD-FDD/FAB units, and Belgian and Dutch officers were temporarily reassigned to the Burundian Defense Ministry to help "keep an impartial eye" on things. My interviews with both FAB and rebel officers indicate that both sides initially approached the integration process with skepticism and, at times, resentment. But as far as I understand from those engaged in the process, there was no violence and, in general, things proceeded smoothly.[22]

Like Tanzanian and South African assistance in the peace process, these international contributions to the training process were generally helpful. Above, I compared the thoroughness of the military integration agreement in Burundi with the troublesome vagueness of the agreements in Côte d'Ivoire and Nepal. Another difference is noteworthy. External assistance to the latter two countries has been complicated by the looming presence of hegemonic outside countries—France in the case of

Côte d'Ivoire and India in the case of Nepal. No such condition complicated matters for Burundi, although Belgium's historical colonial ties to the country probably provided special motivation for its involvement. Perhaps the most important factor in driving international support was the sense that while Burundi's crisis was primarily domestic, a stable Burundi could make negotiating transitions in neighboring Democratic Republic of the Congo and Rwanda much less complicated.[23]

OUTCOME

As described above, the CNDD-FDD's grand political strategy required military integration as a condition for peace, and by all accounts the vast majority of Burundians supported this aim. The CNDD-FDD's battlefield success preempted a replay of 1972. Military integration also provided spoils that were used to build a consensus. In these ways, integration was crucial to the ending of the war. In a technical sense the reintegration process was successful at the time of writing in early 2011. Once an agreement was reached in Pretoria, the process advanced according to a rational, deliberative process. At times, however, attempts by political parties to either use the military for political gain or accuse it of being so used have threatened this success.

Military Capabilities

The new armed forces have demonstrated effectiveness on the battlefield, but at the same time they have been involved in documented instances of abuse. The combined ex-FAB and ex-CNDD-FDD forces fought determinedly against the FNL to bring it to the negotiating table. Abuses were documented in these campaigns (Human Rights Watch 2007; All Africa 2007). These abuses included the execution and imprisonment of civilians suspected of supporting the FNL. Most notorious was the massacre of thirty-one civilians in a 2006 incident in Muyinga province. On a positive note, however, the military responded to pressure from the United Nations and international nongovernmental organizations, which launched an investigation and ultimately convicted the key perpetrators in 2008 (Human Rights Watch 2008).

The new Burundian army deployed two battalions in 2007 to the African Union Mission in Somalia (AMISOM). Their contribution has been substantial. After three years in the mission, the Burundian troops were exposed to a fair amount of active engagement, suffering twenty-nine killed and seventeen wounded; among the dead was the Burundian major general Juvenal Niyonguruza, AMISOM's deputy commander, who was killed in a suicide attack in September 2009. Back home, since 2005 there have been no reported incidents of violence within the military indicative of systematic problems, although there have been reports of some isolated, idiosyncratic events.[24]

Political Control

More worrying are the ways in which the army, police, and intelligence services have been implicated in struggles between the major political parties. Their involvement suggests an unscrupulous elite. During the 2005 elections, the head of the United Nations mission, ONUB, accused Tutsi members of the army of committing small-scale attacks to sow a sense of insecurity and depress voter turnout (Associated Press 2005).

The intelligence services have been implicated in the violent repression of critics and opponents of the ruling CNDD-FDD (Human Rights Watch 2006, 2009b, 2010a). Affiliates of the CNDD-FDD within the intelligence services, army, and presidential offices also launched a series of arrests and investigations associated with an alleged coup plot in 2006. Among those implicated were a few members of the armed forces not affiliated with the CNDD-FDD (including one ex-FAB officer), FNL splinter group leader Alain Mugabarbona, and former (Hutu) president Domitien Ndayizeye. All those arrested were eventually let go, and as a noteworthy sign of his unwillingness to play according to the CNDD-FDD's script, Defense Minister Lieutenant General Germain Niyoyankana (ex-FAB) claimed that the members of the army had been falsely accused (Reuters 2006).

At the time of the 2010 elections, opposition leaders and human rights advocacy groups accused CNDD-FDD-aligned intelligence forces of intimidation and brutality against civil society activists and opposition party members (Human Rights Watch 2010a, 2010b). The elections themselves increased domestic tensions. Opposition parties accused the CNDD-FDD of fraud in early local elections, and they used that accusation to justify a near-total boycott in the subsequent parliamentary and presidential polls. International observer missions rejected these accusations of electoral fraud, but the result of the process was nonetheless a worrying slide toward a single-party state. Aside from the intelligence forces, however, the army seemed to stand apart from these events.

Overt ethnic tensions have occasionally surfaced. For example, in 2007, members of the UPRONA party accused the ruling CNDD-FDD of not respecting the 50/50 ethnic balance principle in the assignment of police posts, although nothing came of these accusations (Agence France-Press 2007). And while the Arusha agreement puts forward the principle of political nonengagement for security force members, slots have in fact been allocated on the basis of past political affiliation (namely, with FAB, the CNDD-FDD, or FNL). When speaking with soldiers, one gets the sense that such affiliations remain highly relevant.

The defiance of the defense minister relative to the 2006 coup-plotting accusations made by those aligned with the ruling CNDD-FDD brings into sharp relief the issue of control over the military. At the dawn of integration, ex-FAB officers constituted the bulk of the officership, although former CNDD-FDD members were placed in key positions and have been elevated over the years.[25] The new military operates under the scrutiny of foreign officers temporarily reassigned from the Netherlands and Belgium to Burundi's Defense Ministry. The authority of these foreign officers is boosted by the substantial aid that their countries provide to Burundi. This balance of ex-FAB presence and CNDD-FDD presence, and of domestic presence and international presence, reduces the risk of any one political group's gaining what Huntington (1957) calls "subjective" control of the military institutions.

Major deployments such as the AMISOM mission are decided upon, in effect, by a consensus among these actors. Legitimate authority over the military as an institution is embodied by this coalition of ex-FAB, CNDD-FDD members, and bilateral partners. But as suggested by the cases of abuse listed above, factions inside the army, and to an even greater extent inside the police and intelligence services, remain tied to political bosses and have proven willing to do their bosses' dirty work.

Sustainability

"Right sizing" has been a secondary priority, and this raises questions about sustainability. Phase I, integration, was completed by the fall of 2005. Phase II, downsizing, was still in effect at the time of writing in mid-2010. A South Africa–brokered cease-fire with the remaining FNL faction in 2008 resulted in 3,500 of its members being integrated into the new army and police forces in 2009. Adding this number to estimates of the size of the national forces in 2007 (Nindorera 2007, 12), I conclude that the FDN's size in 2010 was approximately 30,000, with about one-third originating from the rebel groups. The police forces numbered about 22,000, again with about one-third originating from the rebel groups. The number of those who were demobilized totaled approximately 30,000, with about two-thirds having originated in the rebel forces.

The current size of the military is thus above the target of 25,000 set out in the FTA. A military of this size, or even of the size prescribed in the FTA, is unsustainable without major financial assistance from donor governments. These donors have regularly expressed their concern about Burundi's army, police, and intelligence services being too large and too costly (e.g., BBC Monitoring 2008). Military expenditures hovered around 4 to 5 percent of GDP from 2006 to 2008, which is more than double the average share spent by other members in the East African Economic Community: Kenya, Tanzania, Rwanda, and Uganda (SIPRI 2010).

HAS MILITARY INTEGRATION MADE THE RESUMPTION OF LARGE-SCALE VIOLENCE LESS LIKELY?

Might the apparent success of Burundi's military integration process have made the resumption of large-scale violence less likely? This book proposes a few ways in which it might. In my estimation many of these causal pathways apply to Burundi's situation, mostly because of the 50/50 ethnic quota. If this is true, such a quota can be a crucial instrument for building peace.

Costly Commitment

As a commitment device, the quota provided a credible means by which the CNDD-FDD leadership could signal that its goal was not the annihilation of Tutsis, despite accusations to that effect by Tutsi radicals. This has certainly eased interethnic tensions.[26] The quota, combined with the army's apparent capacity to maintain order, reassures Tutsi elites of their security. This makes them more willing to keep their assets and remain in the country and to work with the new elite. This is crucial to Burundi's economic development.

However, two factors muddy the picture. First, it remains to be seen whether this wealth and elite cooperation will reduce mass grievances by creating opportunities for those who were shut out previously. In addition, the events surrounding the 2010 elections suggest that the CNDD-FDD has kept under its direct control some of the most dangerous elements in the intelligence services (Human Rights Watch 2010a, 2010b).

An integrated military may contribute not only to *managing* inter-elite and interethnic conflict but also to *transcending* it as a symbol of national unity. In Burundi, the military was integrated all the way down to the level of individuals. A deeply integrated army may serve as a national symbol of interethnic cooperation. For Burundi, public opinion data for assessing this theory rigorously is unavailable, although the theory strikes me as plausible. Evidence indicates that interethnic contact in the integrated military has reduced prejudice among soldiers (Samii 2013).

However, this "transcending" effect occurs very incrementally. Only if there are no shocks to the system for a good twenty years or so do I think that these small changes in attitude will accumulate to appreciably affect the potential for large-scale violence. In the short to medium terms, this pathway is far less significant in reducing the likelihood of conflict than the 50/50 quota, which signals the intentions of the CNDD-FDD elite and reassures the Tutsi elite.

Trust Inspired by Negotiations

Another causal pathway proposed in this volume is that success in striking a deal on military integration may inspire trust among elites that facilitates their tackling other problems. In Burundi's case, this pathway appears to be even less relevant than the previous one. As discussed above, negotiations over military integration were *the* sticking point, and so integration was not to serve as a bridge to anything else. Rather, all other agreements were a bridge to tackling this prime issue.

Is the 50/50 ethnic balance within the military likely to stick? So long as the memories of war remain fresh, it would seem so. Evidence comes from the negotiations that brought the last armed FNL factions into the peace process in the period 2006–9. At that time the FNL leadership once more demanded that the armed forces be entirely restructured to reflect the country's true ethnic balance. Although much of the Burundian public may have been sympathetic, the FNL did not inspire any public demonstrations of support for this proposition. It is likely that war-weariness and a desire to move on caused the vast majority of Burundians to oppose such restructuring, at least for the moment. The FNL factions had to accept integration on essentially the terms provided in the Arusha Accords and FTA.

However, war-weariness may fade. It is conceivable that gains in Hutu mobility will not be broad enough throughout society. At the same time, Tutsis might feel their prosperity threatened by Hutu gains if overall economic growth sags. Thus tension over ethnic balance in the military could reemerge as part of a broader resource struggle fought again across ethnic lines. A national-level reconciliation process would mitigate this risk, complementing the incremental "transcendence" process occurring within the military. Hopefully, the 50/50 quota will provide a basis for cooperation until inter-ethnic divisions become considerably less pronounced.

ABBREVIATIONS

AMISOM	African Union Mission in Somalia
CNDD	Conseil national pour la défense de la démocratie—Hutu rebel movement

FAB	Forces armées burundaises—government army during the war
FAP	Forces armées du peuple—military wing of FROLINA, Hutu rebels
FDD	Forces pour la défense de la démocratie—military wing of CNDD, primary Hutu rebel military
FDN	National Defence Force (Burundi)
FNL	Forces nationales de libération—military wing of PALIPEHUTU, Hutu rebels
FRODEBU	Front pour la démocratie au Burundi—government political party
FROLINA	Front pour la libération nationale—Hutu party organized in Tanzania
FTA	Forces Technical Agreement—military integration agreement
ISCAM	Institut supérieur des cadres militaires—national military academy
JCC	Joint Ceasefire Commission
ONUB	United Nations Operation in Burundi
PALIPEHUTU	Parti pour la libération du peuple hutu—Hutu party organized in Tanzania
UBU	Burundian Workers Party—Hutu party organized in Rwanda
UPRONA	hard-line Tutsi party led by Charles Mukasi

ACKNOWLEDGMENT

This chapter draws on findings from the multipurpose survey "Wartime and Post-Conflict Experiences in Burundi (2006–2010)," sponsored by the Folke Bernadotte Academy (Sweden) and the US Institute of Peace. My coinvestigators on that project were Michael Gilligan, Eric Mvukiyehe, and Gwendolyn Taylor. The project was implemented in partnership with ITEKA—Ligue Burundaise des Droits de l'Homme (Burundian Human Rights League). I am also grateful to Henri Boshoff and Nicole Samii for comments and advice.

NOTES

1. Conventionally, Tutsi are said to constitute 14 percent of Burundian society; Hutu, 85 percent; and Twa, 1 percent. These figures are from a 1956 colonial-era census of dubious methodological quality. Even if they were correct for the time, the current distribution is likely to differ, not least due to imbalances in mortality rates in the various crises since independence. Analysis of survey data collected by my research team in 2007 suggests that the distribution

may slightly overstate the Hutu proportion, although the margins of error are quite large. However, to the extent that electoral results from 1993 and 2005 largely reflect ethnic preferences, the 14/85/1 distribution may not be so far off.

2. This brief background is based on two types of sources. The first is field notes from extensive interviews and focus groups conducted by my research team in Burundi from 2006 to 2009. The second is secondary accounts, including those of Lemarchand (1970, 1996), Reyntjens (1993), United Nations (1996a), and Ngaruko and Nkurunziza (2000).

3. This is based on the author's own experience in Burundi from 2006 through 2009. Some Bagaza associates within the army had been arrested but then released as part of what the government declared a coup attempt in 1989 (Lemarchand 1996, 140).

4. Ndayikengurukiye led the FDD.

5. Author's interviews with Léonard Nyangoma, founding leader of the CNDD, and Colonel Jean-Bosco Ndayikengurukiye, founding leader of the FDD, Bujumbura, 2006.

6. In 2001, the leader of a hard-line faction of the long-dominant UPRONA party, Charles Mukasi, declared in a letter to the United Nations Security Council, "Au Burundi, il n'y a pas de loi interdisant aux Hutu l'accès à l'armée. La réalité est que l'armée burundaise n'est pas monoethnique. Il y a des circonstances historiques qui ont fait que, depuis 1972, les Hutu soient minoritaires au sein de ces corps: (i) la répression contre le génocide de 1972 fut aveugle, maladroite et coupable; (ii) l'exclusion ethniste et régionaliste sous les régimes Bagaza, Buyoya fut pratiquée; et (iii) des campagnes du Frodebu et du Palipehutu sous le régime Ntibantunganya, destinées à empêcher le Hutu de se faire enrôler dans l'armée pour nourrir la propagande raciste et génocidaire du Frodebu furent menées."

7. These are the figures commonly reported in the press and are well within the credible bounds that I am able to establish from survey data collected by my research team in 2007. The survey included more than a thousand adult civilians and over two thousand former civil war combatants.

8. A fifth committee was eventually established to discuss implementation of the terms of the agreement.

9. Force size estimates are based on figures determined in the postwar demobilization and reintegration process (World Bank 2004).

10. From my interviews with members of the Burundian officer corps, I believe that as of 2000 many understood that international norms had changed, and that their freedom to act in 2000 was not the same as it had been when, for example, the Algerian military was essentially free to cancel the 1991 elections in that country.

11. A posh new suburb in northern Bujumbura is called the "Arusha quarter," because the sparkling mansions there were allegedly built with the stipends received by participants in the Arusha talks.

12. In the past, police contingents were made up of officers drawn from other parts of the country—often from Bururi. This was seen as a reason for the brutality with which the police sometimes enforced the government's will locally.

13. After Arusha, the South African mediation team was led by Deputy President Jacob Zuma, with President Thabo Mbeki stepping in at key moments.

14. This understanding comes from our research team's interviews with CNDD-FDD officers, including Colonel Prime Niyongabo, Brigadier General Silas Ntigurirwa, and Colonel Manasse Nzobonimpa, as well as many lower-ranking officers.

15. Boshoff, who served as military technical adviser to the South African mediation team, called this agreement the step that allowed the parties to "overcome what has been perhaps *the* major stumbling block in the Burundi peace process" (Boshoff and Gasana 2003, 1).

16. All members of the forces were required to be volunteers, physically fit, and Burundian nationals. As for education, officers were required to have an advanced degree or experience as an officer, noncommissioned officers were to have a high school degree or experience as a noncommissioned officer, and the rank and file were to have completed primary school or have had experience as a soldier. Experience was to be judged during the verification process, which would involve demonstration of military skills before representatives of the JCC. Informally, FAB also used an age cutoff of forty-five years, although some exceptions were allowed.

17. Note that the ethnic balance rule did not imply a 50/50 split along rebel/FAB lines, because some (about 17 percent, according to our survey data) of the FAB rank and file were Hutu.

18. The outcomes of this reintegration process are studied by Gilligan, Mvukiyehe, and Samii (2013).

19. Interviews with Ndayikengurukiye and Niyoyankana, Bujumbura, 2006.

20. See the acknowledgment note in this chapter.

21. Though the FNL and FROLINA had been fighting for decades, the mass rebellion led by the CNDD-FDD started in 1994, meaning that most rebels had no more than ten years' experience as of 2004.

22. Interview with Colonel Henri Boshoff (retired), who was ONUB's lead military technical adviser.

23. Indeed, certain units from the CNDD-FDD had been active as mercenary forces for the Joseph Kabila government of the DRC some years before the Pretoria peace process. Interview with Jean-Bosco Ndayikengurukiye, who led CNDD-FDD units that fought on behalf of Kabila, Bujumbura, 2006.

24. E.g., in 2006, a dispute between a private and a sergeant resulted in some shooting (Net Press 2006), and a dispute over pay among members of the AMISOM contingents resulted in thirty-three soldiers' being arrested for mutiny in 2010 (BBC 2010).

25. E.g., the CNDD-FDD nominated Senate-appointed ex-FDD commander, Major General Godfroid Niyombare, to succeed Major General Samuel Gahiro (ex-FAB) as the army chief of staff in April 2009.

26. Agreement by the CNDD-FDD with the quota triggered an attempt by the FNL to "outbid" the CNDD-FDD by continuing with demands for more radical reform, but thus far such attempts have not attracted substantial adherents.

Part IV

ALTERNATIVE PERSPECTIVES

Chapter 14

THE INDUSTRIAL ORGANIZATION OF MERGED ARMIES

David D. Laitin

The chapters of this volume focus on mergers between state armies and their former enemies—insurgent militias. These mergers are a key, although insufficiently studied, element of civil war settlements. The motivation of international peacekeepers in promoting these mergers is not difficult to discern. A country with a divided military is a recipe for conflict and dysfunction, as the prior forces will inevitably find themselves arming one against the other over issues of policy and resource allocation. The follow-up questions are more difficult to answer. What makes the United Nations or NATO think that merger is a feasible element of a peace agreement? What are the downsides of merging formerly antagonistic forces? How can we evaluate the prospects of military forces joined by outsiders in a way that goes beyond the textured narratives offered in this volume?

With the goal of shedding some light on the possibilities and perils of these merger efforts, this chapter first surveys findings on mergers and acquisitions in the industrial organization (IO) literature that underline the extraordinary challenges they present, and their concomitant failures. Although the disanalogies between industrial acquisitions and military mergers may seem huge—and thus undermine the premise of this chapter—these very differences help sort out the generic problems as well as the particular opportunities of post–civil war merged armies. After the survey of the relevant IO literature, this chapter then relies on the case literature in this volume (ignoring issues of case selection). The materials presented in this volume reveal that the merged armies face some of the same hurdles as do merged firms. Yet there are differences such that the peace settlements under review achieve some measure of unexpected success, and serve as lessons for those who would engineer future mergers of armed forces.

THE INDUSTRIAL ORGANIZATION PERSPECTIVE

The basic justification for mergers in industrial organizations is that a firm seeks to increase its market power in a certain product class and thereby increase the possibility of charging monopoly prices for its products. But in a related class of mergers, called *horizontal acquisitions*, firms seek to create value through economies of scale or the elimination of what would become, after the merger, redundant assets (Capron 1999). If downsizing through merger cuts excessive use of resources, those resources can be allocated more productively (Jensen 1986).

The more theoretical reason for these horizontal mergers comes from transactions theory economics, with its focus on market failure (Williamson 1975). From the point of view of this literature, acquisitions enhance performance by allowing businesses to obtain preferential access to resources that cannot be purchased in a competitive market. This market failure argument plays a central role in explaining why firms pursue acquisitions. Indeed, several imperfections may exist in markets for intangible resources, including workforce immobility, information asymmetries, and associated moral hazards. Mergers help overcome these imperfections by creating a new hierarchy such that each firm in the merger is protected from the moral hazards of imperfect markets.

But the theoretical call for hierarchy ignores psychological factors that impede efficiency. Consider the effects (as described by Capron 1999, 992) of a merger and the subsequent asset divestiture on the target's workforce: "Altogether, the likely target asset divestiture, and the colonization of the target by the acquirer, nurture a feeling of diminished relative status of target managers and create high target executive turnover after an acquisition. A similar behavioral rationale also applies to redeployment of resources, because the acquirer sometimes prefers to invest its idle resources in the target and eliminate the target's resources. In the process of selling off the target's resources, acquirers may make choices that destroy the competencies of the target firm and make the target less likely to provide valuable resources to the acquirer." In everyday English, this means that in the process of rationalizing the merged form, the skills of the target managers are undervalued by the acquiring managers, even if it was these managerial skills that induced the acquiring firm to see the acquired firm as a valuable target.

Empirical results (Farrell and Shapiro 1990) indicate that the process of rationalizing the merged business does not lead to systematic cost savings, and can even hurt profits. Capron's essay also suggests that the acquirer is more effective in rationalizing its own assets than in rationalizing those of the target. The data also show that the target is three to five times more likely to be downsized than the acquirer. Capron suggests that these realities, along with the magnitude of the target's divestiture, are likely to lead to the target's disruption and to offset the benefits expected from a rationalization process.

Case study research reveals broader problems in the postmerger firm. One study shows that as the firm reduces its organizational slack, asset divestiture risks damaging capabilities and reduces the new firm's propensity to innovate and develop new markets (Hamel and Prahalad 1994; Kanter 1989). Other studies show that downsizing can also violate employee trust (Shleifer and Summers 1988), inhibit risk taking (Staw, Sandelands, and Dutton 1981), and break the network of informal relationships used by innovators, which leads to reduced innovation. These problems can be ameliorated (as suggested by Kreps 1990; and by Akerlof and Kranton 2000) if the new firm establishes a corporate culture creating common knowledge and thereby facilitating efficient coordination.

If it is in general harmful, however, why are mergers and acquisitions such big business? A cynical explanation for industrial mergers is that they create enormous fees for lawyers and management who must work out the details of the integration of separate firms.[1] A report on the merger of two mega-airlines revealed the enormous challenges to management—down to determining the number of lime slices provided in the drink service—of making one out of two (Mouawad 2011). Although shareholders may suffer from these deals, management and its lawyers swim in cash.

When the IO literature is juxtaposed with the case studies of merged armies in this volume, at least six patterns become clear.[2] Three of them are consistent with the expectations generated by the IO literature. First, governments have the most incentive to negotiate with rebel militias when those militias are better fighting forces (controlling for size and resources) than state armies; but as with mergers of firms, once military mergers are implemented, the skills of the rebel fighters are undervalued while the ranks of the state officers are overvalued. Though this might not be the case for senior officers, at lower levels in the officer corps rebels lose points for lack of formal education but do not gain points for education in the school of hard knocks. Generally, a merger yields a less efficient fighting force. Second, gains are rarely made from reductions in personnel due to the merger, as demobilization is costly in the short run. Third, as is recognized in the literature on corporate culture, ideology has a key role in organizational success, and armies with common knowledge of how it is appropriate to act under unforeseen conditions are able to overcome many of the misunderstandings and internal conflicts that are rife when two organizations are joined as one. Such a culture could also help reduce fears of disruption by fifth columnists—those soldiers who are initially disloyal to the newly installed government—through their reeducation.

But three patterns make military consolidation different from inter-firm consolidation. First, the efficient production of state security ceases to be the goal of the state army once the civil war is politically resolved. The role of the army changes more or less as in *Radetzky March*, Joseph Roth's acclaimed novel satirizing the Austro-Hungarian army in the early twentieth century: It becomes a symbol rather than a tool of the state. Therefore, issues of efficiency are less consequential for merged armies in most postconflict situations than for competitive firms after a merger. Second, because the failure of a post–civil war merger creates a "public bad"—high negative externalities that affect other states—those other states (which international relations scholars once called competitors in the international system) are generous in funding mergers, unlike competitive firms. Third, unlike merged firms, which tend to have bloated payrolls that decrease efficiency, merged armies in post–civil war situations have an outlet: They can perform as troop-contributing countries (TCCs) in internationally sanctioned peacekeeping operations (PKOs). This outlet serves three goals: TCCs earn considerable funds for each troop serving in a PKO, which help support a bloated military; countries and their leaders who provide troops to PKOs earn prestige; and serving in a PKO keeps restive officers from intervening in local politics, thereby lowering the probability of a military coup undermining the postwar political equilibrium.[3]

CASE MATERIALS AS ILLUSTRATIONS OF THE SIX PATTERNS

This section first considers the undervaluation of militias and the overvaluation of state soldiers. Then it looks at the bloating of postmerger armies, the role of corporate culture, the irrelevance of efficiency and the value of symbols, the interest of "competitors" in merger success, and the triple value of the outsourcing payoff.

ALTERNATIVE PERSPECTIVES

The Undervaluation of Militias and the Overvaluation of State Soldiers

In Burundi, as explained by Cyrus Samii in chapter 13, the integrated army, as would have been predicted in theory, overvalued formal rather than fighting skills.[4] The FDD (the military wing of the Hutu insurgency, with about 25,000 soldiers) was largely successful on the battlefield (despite the 30,000 soldiers in the European-trained state army, the FAB). In the negotiations that began in 1996, the rebels requested that the state army be disbanded, and a new army created that would disqualify FAB soldiers. But in international negotiations, with strong advocacy from the South African mediators, the FAB became the force into which the rebels would merge, and its officers "scoffed at sharing ranks with rebels from 'the bush.'" In the negotiations, criteria for rank were based on formal education, not success in the field of operations, and despite their ranks in the rebellion, "rebels were downgraded based on a fixed schedule" that focused on formal training. The rebels had at least the same human capital as the members of the state army, and were surely better fighters, but on average the criteria for rank favored the former FAB.

In Sudan, a big issue in the integration of forces was which of the Anyanya would get commissions (with 6,000 being the agreed-upon figure), and this turned into a tribal conflict, with the Nuer, Dinka, and Shilluk (who had formed the core of the rebel army) feeling that they were getting the short end compared with the Equatorians (who were Southerners with many officers in the state army). These feelings of devaluation of guerrilla expertise were exacerbated by a rather ad hoc interview process for inclusion in the state army. On the other side, regular soldiers were angry that rebels, who had little military background, were getting ranks and status that the regular soldiers did not think they deserved. Meanwhile, many former Anyanya militiamen went into the bush, complaining of lack of pay, training, and promotions. The worst outcome was for those called the "leftovers," who were given demeaning positions in the civil service; some got no positions at all. In sum, the absorbed Anyanya rebels did not get positions in the new army that were commensurate with their fighting success.

Chapter 8 of this volume, on South Africa, citing Frankel 2000, reports on the near-irrelevance to the integration process of the military skills acquired by the Umkhonto we Sizwe (MK, or Spear of the Nation) soldiers. Their guerrilla skills were not easily transferrable to the professional, high-technology context of the South African Defense Force (SADF). The MK, though it had little cash, responded to this rapidly, and sent thousands of its personnel abroad for conventional training so they would be competitive in the integration process. Those who got officer status from the placement boards had little influence initially, because they were again sent abroad for further training, or because they were not in the elite SADF networks, or because of language differences. However, some 17 percent of the officers in the new army were from the irregular forces (the MK and military arm of the Pan-Africanist Congress, at the time of the merger called nonstatutory forces, or NSF). And so, even though their skills were not fully recognized, NSF combatants were not, like the Anyanya, sent into civil service doghouses.

Similar stories were elaborated (in personal communications) by case authors. Paul Jackson writes concerning Nepal that the Maoists created an "effective insurgent army" while the Nepalese state forces specialized as "ceremonial agencies. In particular, the value of marching up and down, cleaning boots and having swords appears to be more valuable [to them] than actually fighting." As for the Philippines, Rosalie Arcala Hall noted to me that while no definitive study has demonstrated higher relative skill levels within the MNLF than within the state army, local commanders often claim that the integrated soldiers from the MNLF are more effective than those from the state army in conducting operations in Mindanao.

To be sure, guerrilla armies are not always efficient military entities. In many cases rebel militias work merely to survive and are far less skilled than state forces, even if the latter are unable to defeat the insurgents completely. (The Northeast Indian autonomy-seeking language groups, analyzed by Lacina 2011, are a prime example.) In other cases, such as that of the SADF in South Africa, state troops were highly skilled but were defeated due to the untenable political regime they were defending. Nonetheless, when a state army is juxtaposed with guerrillas in a merger operation, the skills of the guerrillas tend to be undervalued.

The Bloating of Postmerger Armies

Postmerger demobilization is difficult in both firms and armies. After the peace agreement in Sierra Leone, even with international support for severance packages for demobilized soldiers, and with no external enemies knocking at its doors, the army is oversized—it has 8,500 men, versus a need for about 2,000.

In Zimbabwe, newly elected president Robert Mugabe needed to demobilize 30,000 former fighters. The program Soldiers Employed in Economic Development and the subsequent Demobilisation Directorate led to resentment on the part of those chosen to be demobilized, as it was perceived as a black mark on their service record. As a result of the resentment of those slated to be demobilized, some 65,000 army personnel remained on the regular payroll by the end of 1981, a year after liberation—more than twice the number necessary for security needs.

The exception to this bloating principle is Mozambique. At war's end, most of the soldiers were exhausted and hated military life. They much preferred demobilization (especially with a generous two-year severance package that in some cases included a land grant). As a result (and because both sides inflated the figures of their combat forces), neither side could fulfill its quota of soldiers to be merged; consequently, the quotas soon became defunct. And because demobilization was attractive—see below on international interest in the success of mergers—military expenditures as a proportion of GDP plummeted from 3 percent in 1988 to .85 percent in 2008.

The Role of Corporate Culture

Chapter 6, on Rwanda, reports on a stunning integrative success, a top-down incorporation of thousands of ex-state soldiers into the victorious rebel army imbued with an integrating ideology that guided the indoctrination program of *ingando*.

In Lebanon there was a similar role for ideology. Fouad Shihab had been Lebanon's first commander in chief and was credited with superseding confessional divides in the officer corps. General Emile Lahoud, leader of the post–civil war army, tried to replicate this success by developing a credo of military professionalism that he called Shihabisme. With this clear message of professional and nonconfessional organizational culture, Lahoud was able to integrate a demoralized state army with a wide array of contending militia forces (though not Hezbollah) into an army of mixed confessional brigades. A requirement to adopt absolute internal neutrality for professional advancement, with the understanding that Israel was the common enemy, created a shared understanding across confessional groups of the need for a disciplined and united armed force.

An ideology of inclusion, though not a very compelling one, was that of *brassage* (brewing) in the new Congolese army, as described in chapter 9. This ideology, under which no unit represented a former faction exclusively, was the guiding principle of integrated brigades. To be sure, those militias that most strongly played the ethnic card (e.g., the Nkunda-led faction) refused to take part in the integrated military. Perhaps because they were rarely paid, the heterogeneous units could not act collectively and therefore were not an effective challenge to the government. Their course for survival was to establish local strongholds and steal from the population. Thus the political status quo—in which the state apparatus, the military elite, and the local militias are united in theft—has been sustained. This equilibrium hardly needs an ideology of *brassage* to be self-enforcing, yet it remains the case that lack of major fighting within units, whether ethnic or factional, may partly be explained by this cultural trope.

The lesson taught by President Paul Kagame of Rwanda and General Lahoud of Lebanon is a valuable one: Formal rules for collective action are rarely sufficient for organizational success. A corporate ideology, one that provides expectations of what fellow soldiers will do under conditions that have not been foreseen, reduces misunderstandings and permits coordination. Here the IO literature and our case histories of post–civil war armies are in concert.

The Irrelevance of Efficiency and the Value of Symbols

Lebanon is the perfect example of this principle. The state army held to neutrality as long as possible during the horrific civil war, and did so by not seeking to restore order. After the war, with its (rather empty) professionalism as its only card, it gracefully accepted a complex confessional quota system for both the officer corps and enlisted men. The result in Florence Gaub's judgment, as discussed in chapter 5, is that the "Lebanese Army embodies the end of the civil war rather than actual military force."

In Mozambique, as examined by Andrea Bartoli and Martha Mutisi in chapter 10, the effectiveness of the army was never an issue in the peace negotiations; what was "relevant is that the military institution is integrated, inclusive and subject to legitimate political control." The army became "a symbol of national unity, inspiring other groups to emulate it." This was made possible by the nature of the civil war, in which Frelimo controlled the cities while Renamo controlled the countryside. Ethnic cleavages did not divide the sides. Thus there were no regions to which soldiers in the new army refused to be deployed. As a result, the army was at peace with itself even if it was not organized to assure peace against an external threat.

In Bosnia-Herzegovina, three constituent nationalities fought in the war in various combinations and subsequently formed two separate entity armies (one of them itself divided in two) that were maintained for ten years after the war. Subsequently, each was awarded a quota of one-third of the infantry positions, and a census-based proportion in the overall force including the infantry, in the newly merged state army. Because of the quota system, the former entity armies (though hardly credited with tactical efficiency) were much more effective killing machines than is the current army. But the new military was purposefully inefficient, at least in the short term, so the costs of the merger in lowered efficiency had no practical consequences. Indeed, as Rohan Maxwell summarizes in chapter 11, what the new force can actually do is "remain in existence without its members killing one another in large numbers," despite many sparks that could have set off fires. The symbolic value of not breaking apart is all it has achieved. The merger, he concludes, "resulted in a force that was neither cost-effective nor operationally effective." But this hardly matters; because no outside force has any ambitions in Bosnia-Herzegovina (Serbia wants to be in the European Union and would not reopen hostilities), the new army at its point of inception faced hardly any security issues for which operational efficiency would be useful, though professionalism is now being encouraged not for national security purposes but rather to meet NATO's standards, in the hope of future membership.

The story of the Democratic Republic of the Congo (DRC) merger is a frightening one. The postagreement army, the Forces armées de la République démocratique du Congo (FARDC), has no fighting ability, as was revealed in its failure in a confrontation with the Tutsi-led army in North Kivu in 2007. Here 20,000 FARDC troops were resoundingly defeated by 4,000 militiamen. The FARDC has also been unable to prevent regular incursions of armed groups into DRC territory. Rather than being a symbol of the nation, Judith Verweijen notes in chapter 9, in its very weakness the FARDC serves the interest of the transitional government and a potpourri of entrepreneurs (criminal and quasi-legal), all profiting from the lack of an efficient state army. Verweijen calls the FARDC a "leaking giant" that uses violence mostly to prey on the population (its real source of income) rather than to assure stability. Its inefficiency, she concludes, served the interests of power.

In Sierra Leone, army weakness was also the political goal, as Mimmi Söderberg Kovacs points out in chapter 12. President Ahmad Kabbah did not demand an efficient state army, as he did not trust the state army to protect him.

In Sudan, as described by Matthew LeRiche in chapter 3, the army was neither efficient nor a symbol of national unity. Such symbolism was almost made possible when the Absorbed Forces based in Khartoum saved President Gaafar Nimeiri's life in a coup attempt shortly after the Addis Ababa accords that had created a semi-integrated army. But the president never gave credit to the former Anyanya for this national service. Moreover, the Absorbed Forces were by treaty deployed in their own region; and as a result, the new force was a symbol of continued division. Years later, when Nimeiri sought to bring the Southern forces under a unified (Northern) command, he created the conditions for a renewed civil war.

The Filipino merger of one faction of the Moro movement with the state army had little effect on ending the civil war, as explained by Rosalie Arcala Hall in chapter 7. Rebel army soldiers sent their relatives into the state army (which was regarded as an employment provider) rather than abandoning their own units. The only consolation was that the integration had symbolic value because it introduced diversity into the army. Even if they have not yet introduced halal meats into combat rations, the Armed Forces of the Philippines built mosques in the training camps, reduced exercises for the Muslims during Ramadan, and developed sensitivity programs in training. The incorporation of Moro-supported combatants helped develop a model of Muslims and Christians working within the same units, even if only in uneasy cooperation.

In Zimbabwe, as described by Paul Jackson in chapter 4, intra-army hostilities and the rise of a competitive party challenging Mugabe's rule induced the president to worry about his security. Indeed, during the local elections of 1980, a renewal of armed hostilities between remnants of the ZIPRA and ZANLA forces nearly took place. The hostilities continued in 1981, when three hundred were killed, with one-third of the integrated battalions disintegrating in faction fighting. Mugabe first tried to turn the state army into a de facto ZANLA force. Outraged by the politicization of the army in favor of ZANLA's interests, many former ZIPRA guerrillas deserted and returned to their homeland. Without a personally loyal army, Mugabe relied on funds and military training from North Korea to establish new units outside the integration process—the infamous Fifth Brigade and the Zimbabwe People's Militia. Over time, through its operations in the DRC, the Zimbabwe Defense Forces has become a leader in diamond smuggling, turning its officers into mining entrepreneurs. They cling to their military roles and turn security agencies into paramilitary organs of state repression. This provides support to Mugabe's kleptocratic autocracy. The army is not as inefficient as those discussed in the other cases reported in this volume. Indeed, it has performed well in countering insurgency in eastern Zimbabwe. But Mugabe need not worry about the coup implications of its efficiency, as due to smuggling, officers have no interest in overthrowing Mugabe, nor in demanding regular salaries.

The big exception to the pattern of operationally useless merged armies is Rwanda. After the RPF victory in the civil war, the new government faced a real security threat in 1996 from Rwandan Hutu forces exiled in the eastern DRC. The victorious RPF, as Stephen Burgess explains in chapter 6, sought Hutu participation in a new professional army, and therefore integrated members of the defeated state army in waves. Responding to the threat from within the DRC, the newly integrated army took the offensive, and marched 2,000 kilometers on to Kinshasa, taking part in the overthrow of President Mobutu Sese Seko in 1997. Later on, amid a daring airborne raid in Kitona, west of Kinshasa, and after nearly succeeding in overthrowing the regime of President Laurent-Désiré Kabila, the RPA retreated and remained in limited control of nearly half the DRC, though unable to eliminate the Hutu-led extremist Forces démocratiques de libération du Rwanda (FDLR). As discussed, the top-down structure and the ideology of *ingando* were not used merely for symbolic reasons. These efforts helped overcome the typical merger outcome of a symbolic army incapable of performing as an efficient and impartial military organization, as was the case of the UN-sponsored merger of the state army and the RPF in 1993, after the Arusha Accords.

The Interest of "Competitors" in Merger Success

Contrary to the vision of anarchy and zero-sum power relations between states that is portrayed in some accounts of international conflict, peaceful states do not look on with competitive joy as fellow states collapse into civil war violence. Civil wars breed terrorists, viruses, and refugees—all of which are public "bads" that are costly for richer and more peaceful states. Thus the rich states (and their organizations) invest in the success of merged armies, quite unlike competitive firms hoping for merger failure.

Consider post-Yugoslav Bosnia-Herzegovina. The progressive merger of the entity armies in BiH into a single state army was initiated by the high representative designated in the General Framework Agreement for Peace, and received support from the Organization for Security and Cooperation in Europe, NATO, the European Union, the Organization of the Islamic Conference, and several embassies. In 2003 the high representative established the Defense Reform Commission to begin what turned out to be a thirty-two-month, two-step process: first establishing a degree of state control over the entity armies, and second merging those armies. The High Representative relied on his coercive resources but was aided by elite desire in Bosnia-Herzegovina to join NATO, which insisted that the entity armies be replaced by a single army under the control of the central state structure. In sum the merging was exogenously proposed and enforced.

In Zimbabwe, the British were intimately involved in securing civil peace. The British Military Advisory Training Team oversaw the integration of the two competing forces. In 1980, under the Commonwealth Monitoring Force, the two rebel forces received combined training.

In Burundi, foreign forces act as surveillance agents to protect the agreements. At the time of Cyrus Samii's writing about this case, fifteen years after the successful peace negotiations, foreign officers from Netherlands and Belgium were still working in Burundi's defense ministry. Similarly, in Sierra Leone international personnel remained in high positions in the Republic of Sierra Leone Armed Forces and the Ministry of Defense in 2011, nine years after the merger. Even in South Africa, where external funding was hardly needed, British military personnel served as arbiters. Britain was not a competitor with South Africa for regional dominance; it shared the new government's wish to ensure no outbreak of a renewed racially based civil war.

In Mozambique, the extended civil war after independence was not at all seen by Portugal or other European powers as an opportunity to restore colonial rule. To be sure, the white-dominated regimes in Rhodesia and South Africa had an interest in extending the civil war to demonstrate to the world that African rule in their countries would threaten security for all. However, the great powers sought peace. Peace initiatives came from Rome under the joint auspices of Italy and the Catholic Community of Sant'Egidio. Overseeing the creation of the new army, the Joint Commission for the Formation of the Armed Forces for the Defense of Mozambique (Comissão Conjunta para a Formação das Forças Armadas da Defesa de Moçambique) in the Ministry of Defense included representatives of the government, of Renamo, and a four-country set of foreign powers, the United States, the United Kingdom, France, and Portugal.

In the Congo, Verweijen reports, a new force was created with salaries and training funded by international powers. However, the Congo case demonstrates that there are limits to how much rich states are willing to pay through the UN for the diminution of public bads. The amounts given by a variety of Western states turned out to be very little. Verweijen concludes that it is partially because of this (but largely because of embezzlement and inefficiencies) that training facilities often failed to meet basic health standards, creating their own humanitarian crises.

The Triple Value of the Outsourcing Payoff

Outsourcing bloated armies by employing them in PKOs serves at least three interests for post–civil war governments.[5] First, the UN and other international organizations cover salaries of the troops, allowing for relief from high recurrent expenditures, especially for armies that have no operational tasks at home. Second, outsourcing keeps armies away from state capitals, where bored officers easily develop visions of presidential grandeur that could undermine the postwar peace. Third, states that become troop-contributing countries (TCCs) get high international rewards for their services. In the halls of the UN, when a country seeks the status of a temporary Security Council seat, it advertises its role as a TCC to demonstrate its internationalism (Fearon and Laitin 2004, 25). And TCCs, as in the case of Uganda with its service in Somalia for the African Union, can leverage their good deeds in procuring financial and military aid from the wealthy countries that do not want to put their own soldiers' boots on the ground in dangerous zones of low strategic value. Merged firms can only downsize to become efficient; merged armies can send their personnel to international organizations.

The Bosnian army contributed to the multinational coalition in Iraq (with an Explosive Ordnance Disposal platoon) and later in Afghanistan. Meanwhile, in seeking to meet NATO's strict criteria for qualifying for deployment, which would bring resources and status to the state, Bosnian leaders have strong incentives to focus on external missions as peacekeepers rather than internal enemies. The oversized Burundi army does not *do* anything militarily at home, but it does serve in Somalia, and the government uses that service to cover recurrent defense expenditures. The new Sierra Leonean army serves as a TCC in Darfur and is integrating units into the ECOMOG Standby Force. The Zimbabwean army has proven capable in PKOs in Angola, the DRC, and Mozambique. South Africa's postapartheid army contributed to UN PKOs with success in the DRC, Burundi, and Sudan. Rwanda, despite its own national security needs, has supplied troops to the UN mission in Sudan, with up to 3,500 in Darfur; building on that, the country has developed a close relationship with the United States–led African Command. In Mozambique, the new army has no security concerns but contributes to the Southern African Development Community, a regional economic organization with such concerns.[6]

Private employment for demobilized soldiers who keep their eyes off presidential palaces also helps save countries from renewed internal combat. In Lebanon the Forces Libanaises, the principal Christian militia, was the hardest to incorporate into a confessionally neutral army, which its leaders derisively called the "Army of the Muslims." But these troops caused little trouble, in part because many found business opportunities in France. And although this is not noted in chapter 8, on South Africa, demobilized officers not only got generous severance packages but also have found even more lucrative employment in the security firm Executive Outcomes, or one of its many successors, as mercenaries protecting (or overthrowing) other African regimes. These opportunities have kept restive soldiers from dreaming obsessively about the military restoration of white rule in South Africa.

CONCLUSION

This volume provides raw material for a model of the industrial organization of post–civil war merged armies. Like the IO literature on mergers and acquisitions in the world of competitive firms, these case studies point to an asymmetrical valuation of skills—one in which the skills of the target employees are undervalued even though the firm became a target due to the high quality of those skills. This feature of mergers, in which the leaders of the highly bureaucratized acquiring firm (or state army) want the dynamism of a poorly organized start-up (or guerrilla militia) but are unable to recognize raw talent that has not come up through the bureaucratic ranks, appears to me to be a general psychological bias, though I have not seen it identified as such in the IO literature.

Furthermore, as in the world of firms, demobilization of army personnel is a risky business, and the morale costs of layoffs are often greater for the firm or army than the costs of a bloated bureaucracy. A third similarity between merged firms and merged armies is that in some cases. common knowledge of the purposes and goals of the new organization, articulated through ideology, can help entities that were previously in conflict overcome recurrent problems in coordination.

But the IO of merged armies in the current international system makes them more likely to succeed than shareholders of acquiring firms for three reasons. First, unlike firms, post–civil war armies typically have no need to become efficient as fighting forces, so that barriers to success (not killing each other; serving as models of integration) are quite low. Second, the external environment is far more permissive with merged armies. Powerful states (unlike powerful competitive firms) have an interest in the success of the merger because it may prevent public bads. In fact, in most of the cases discussed herein, powerful states helped fund the merger to promote success. Third, unlike firms that must downsize or face shareholder revolt, states with merged armies can send battalions overseas as troop-contributing countries, and receive handsome salary relief for those troops. This has a triple potential effect: It permits the retention of larger force size, thereby not threatening morale; it promises international prestige to contributors; and it keeps potential coup makers away from state capitals with useful and engaging (and occasionally enriching) activities.

In sum, a study of the IO of post–civil war army mergers reveals in part why moderate successes have been reported in this volume, despite an organizational challenge that many would have seen as too great for fragile states to meet. However, the principal question driving the empirical studies in this volume is whether the balanced incorporation of rebel militias into a state army—measured by a merged army at peace with itself—would facilitate political accommodation and ultimately development. This chapter suggests a quite different interpretation for an army at peace with itself, and it is in line with evidence presented in the empirical chapters of this volume. This evidence suggests that the observed successes reflected an institution that had no real tasks, but could produce tax revenue for the state by being outsourced to peacekeeping missions. The implications for political accommodation and development of my "industrial organization" interpretation are consequently less sanguine than approaches that inferred broader societal and political returns for the success of a post–civil war merged army.

ABBREVIATIONS

Anyanya	Sudanese rebels
DRC	Democratic Republic of the Congo
ECOMOG	Economic Community West African States Monitoring Group
FAB	Burundian government army
FARDC	government army of the DRC
FDD	Burundian rebels
FDLR	Rwandan rebel group based in the DRC
IO	international organization
MK	military arm of the African National Congress, South Africa
PKO	peacekeeping operation
Renamo	rebel group in Mozambique
RPA	Rwandan Patriotic Front—military arm of RPF, now Rwandan government army
RPF	Rwandan Patriotic Front—successful rebels led by Paul Kagame
SADF	South African military during apartheid
TCC	troop-contributing countries
ZANLA	one of two rebel armies in Rhodesia led by Robert Mugabe
ZIPRA	one of two Rhodesian rebel armies, which lost out to its rival, ZANLA

NOTES

1. For a description of how mergers and acquisitions enriched a generation of upstart lawyers, see Gladwell (2008, chap. 5).

2. There are exceptions to the patterns identified here, and, as will become evident, the DRC is the egregious outlier.

3. The UN provides generous packages per troop to the TCCs for participation in its PKOs. Even TCCs authorized by regional organizations receive financial rewards that help sustain a bloated army. With financial support from the EU and the United States (through its Global Peace Operations Initiative), the African Union gives TCCs $500 a month for every soldier serving in Somalia (serving in the African Union Mission in Somalia). In West Africa, although Nigeria shouldered most of the Economic Community of West African States Monitoring Group (ECOMOG) burden even for the 42 percent of the soldiers who were not Nigerian, in 1993 the United States contributed $31 million to defray the costs of Ugandan and Tanzanian troops in ECOMOG (Hull and Svensson 2008, 42; Shaw et al. 1966, n41).

4. Nonreferenced descriptions and quotations in this section refer directly to the relevant chapters of this volume.

5. "Outsourcing" because states are sending their producers of order (i.e., brigades) overseas to realize higher returns on their investments in training.

6. Lest we overemphasize the particular usefulness of becoming a TCC for states that have bloated armies due to mergers, reliance on Hartzell's data set on mergers (updated from her 2009 publication), and merging it with statistics on the United Nations Department of Peacekeeping Operations website (and thereby excluding non-UN operations), we see that countries whose civil wars ended in decisive victories and with no subsequent mergers were bigger contributors to UNPKO—Pakistan, Sri Lanka, Indonesia, Jordan, Bangladesh, and Ethiopia are the leading examples. Only Nepal, among those states with a postwar merged army, contributed at the levels of these leaders. The correlation between army merger and the sum of soldiers committed to a UNPKO from 2010 to 2012 (calculated for each October) is negative and significant ($p = .09$). Though the case materials in this volume point to the outsourcing of bloated merged militaries, statistical examination suggests that this is more potential than real, at least compared with postwar armies that have not been merged.

Chapter 15

MILITARY DIS-INTEGRATION: CANARY IN THE COAL MINE?

Ronald R. Krebs

The rising tide of civil wars in the second half of the twentieth century, and especially the post–Cold War surge, gave rise to a postconflict peace-building industry. One of that industry's guiding faiths has been that the future stability of these devastated polities is determined in part by whether a new national army can rise from the ashes. Indeed, there is no question that, in the wake of civil war, negotiations over the future of the various armed forces, state and rebel, are often central to peace negotiations, and with good reason. Yet we should not take it on faith that the way in which the conflicting parties resolve this problem is crucial to the duration and quality of the postwar peace, nor that the ideal is a *national* army—that is, a professional and communally representative force that serves aims both strategic (so that all can feel secure and lay aside their arms) and symbolic (so that all can identify with a larger national project).

The findings of the eleven case studies in this volume do not conclusively prove that those beliefs are irreparably flawed, but they do give us little reason to subscribe to those beliefs—although admittedly many of the authors would not agree with this conclusion. In fact, the cases suggest that, while military integration often (though not always) features prominently in negotiations to end civil war, the failure of military integration projects is more a reflection of underlying distrust or incompatible win-sets than it is a fundamental cause of peace's breakdown. It is tempting to see failed integration as the key issue, and thus to develop and recommend a set of universal "best practices" that local actors might employ and that the international community might promote. It is tempting to conceive of military integration as a tractable technical problem. But to blame military integration for peace's downfall is a bit like saying the dead canary is at fault for the coal mine's having become uninhabitable. When we observe military integration fall short of the ideal or even collapse, it is normally an early warning sign of peace's demise, not its fundamental cause.

Those who seek to define military integration best practices have admirable intentions; if scholars could figure out what constitutes "good" military integration—that is, military integration that contributes to postconflict peace and stability—the international community could impose it on local actors, entice them to jump aboard, or seek to persuade them of its merits. But to think of military integration this way is to imagine it as a mere engineering problem and to see failures of integration as rooted in a lack of technical know-how. Such practical problems can be overcome—perhaps

TABLE 15.1 Country Military Integration Details

Country	Resumption of Fighting?	Communal Conflict?	MI = Cause? Author	MI = Cause? Krebs	Alternative Cause, from Chapter
Bosnia-Herzegovina	No	Much	No	No	—
Burundi	No	Some	Yes	No	War weariness; postwar distribution of power
Democratic Republic of Congo	Yes	Yes	Yes	No	Insecurity and greed
Lebanon	No	Yes	Yes	Undecided	—
Mozambique	No	No	Yes	No	Battlefield stalemate; war weariness
Philippines	Yes	Yes	No	No	—
Rwanda	No	No*	Yes*	No	Tutsi victory
Sierra Leone	No	Yes	No	No	—
South Africa	No	No	Yes	No (or small)	Unsustainability of apartheid regime
Sudan	Yes	Yes	Yes	No	Lack of socioeconomic development in South; introduction of Sharia
Zimbabwe	Yes	Yes	No	No	—

Note: Entries in this table are based on the information contained in this volume's chapters—unless otherwise indicated.
*Entries in these cells are based on personal communications from the chapter author or volume editor.

TABLE 15.2 Country Military Integration Outcomes

Country (Chapter)	Resumption of Fighting?	Communal Conflict?	MI = Cause? Author	MI = Cause? Krebs	Alternative Cause (from Chapter)
Bosnia-Herzegovina (11)	No	Much	No	No	—
Burundi (13)	No	Some	Yes	No	War weariness; postwar distribution of power
Democratic Rep. of Congo (9)	Yes	Yes	Yes	No	Insecurity and greed
Lebanon (5)	No	Yes	Yes	Undecided	Battlefield stalemate; war weariness
Mozambique (10)	No	No	Yes	No	—
Philippines (7)	Yes	Yes	No	No	—
Rwanda (6)	No	No*	Yes*	No	Tutsi victory
Sierra Leone (12)	No	Yes	No	No	—
South Africa (8)	No	No	Yes	No (or small)	Unsustainability of apartheid regime
Sudan (3)	Yes	Yes	Yes	No	Lack of socioeconomic development in South; introduction of Sharia
Zimbabwe (4)	Yes	Yes	No	No	—

Note: The entries in this table are based on the information contained in this volume's chapters—unless otherwise indicated.
*Entries in these cells are based on personal communications from the chapter author or volume editor.

not easily, but with good faith and hard work and adequate knowledge. Ignorance is the root of all evil, declares the alluring liberal, perhaps peculiarly American, creed (Hoffmann 1968; Packenham 1973). But the flaws of post–civil war militaries seem to be rooted more in politics than in ignorance, as this volume's case studies, stretching impressively across time and space, repeatedly show. The same interests, institutions, ideas, and deep historical structures that drive a given peace settlement, warts and all, appear also to shape the corresponding military integration. It is these forces, more than the particulars of integration, that seem to affect the duration of the peace. Building a post–civil war military is fundamentally different from, and a lot harder than, building a better mousetrap.

A glance at the data presented in this volume reinforces the point. I have constructed tables 15.1 and 15.2 from the case material, examining a series of factors sometimes associated with the success of military integration. As I do not have extensive knowledge of the cases, I simply report the authors' findings.[1] Table 15.1 presents what I gleaned regarding the nature of post–civil war military integration. To what extent was military integration central to the warring parties' negotiations of the peace settlement?[2] Was military integration based on a detailed agreement between the warring parties or on one that laid out only broad principles? What role did international actors play in helping the parties integrate their forces? Did military integration take place more or less coincidently with the peace settlement, or was it implemented substantially later? Was the resulting military an entirely new institution, or did the process entail integrating former combatants into an existing institution? Was the resulting military a truly national force, drawing more or less proportionately from the various communal (e.g., ethnic or religious) groups and regions and treating these populations equally, or was it a communally or geographically discriminatory force? Was the resulting military smaller than at the war's conclusion? Did the integrated military recruit and/or promote using communal or geographic quotas? How effective was the new army at its designated military tasks?

Table 15.2 presents what I learned regarding the duration and quality of the postwar peace and its relationship to integration. Did fighting resume in reasonably short order? Even if fighting did not resume, did intense communal conflict persist? (The distinction is important; though Bosnia-Herzegovina has remained peaceful since the 1994 Dayton Accords, chapter 11 of the present volume, by Rohan Maxwell, contains evidence in spades of mobilization and conflict along communal lines—and thus of the failure of a Bosnian civic identity to emerge.) The next column of the table presents the chapter author's judgment about whether the success or failure of military integration was at least partly responsible for—that is, at least a reasonably substantial contributing cause of—the postwar peace, or lack thereof. My own assessment (defended in the discussion below), as well as possible alternative causes drawn from the empirical narratives, follow.

If military integration were merely a problem of engineering, we would be able to observe clear systematic associations between certain design features in table 15.1 and the conflict outcomes in table 15.2. But no such associations emerge from the data; neither a detailed agreement regarding military integration, nor the involvement of outside actors, nor the construction of a fully national army, is correlated with a stable peace, let alone with the transcendence of communal conflict. Those who supposedly possess the secrets of military integration—international actors, representatives of both nation-states and multinational organizations—have ranged from being the crucial drivers of the process to being tangential to it, again with no clear impact on outcomes. A first glance at this volume's comparative findings suggests that the emphasis, among both practitioners and scholars, on military integration after civil war may be misplaced.

The rest of this chapter proceeds in two sections to sustain my skepticism. First, I critically assess the possible theoretical logics and causal mechanisms mediating between military integration and the durability of peace. I argue that they are either theoretically implausible, or their scope is likely to be narrow, or they are particularly ill suited to post–civil war contexts. Second, I reverse this volume's causal arrow and inquire into the sources of military integration. Drawing on this volume's eleven finely drawn case studies, I conclude—admittedly sometimes at odds with the authors themselves—that military integration's shape, its success, and its failure are more the product of politics than their cause. The result is to cast grave doubt on the policy community's faith that military integration is crucial to the fate of peace after civil war.

THINKING THEORETICALLY—AND CRITICALLY—ABOUT MILITARY INTEGRATION

The introductory chapter presents a number of causal pathways through which military integration can lead to peace—by serving as a costly signal that reveals one's true intentions (or type, in the jargon) or that ties one's hands, by providing security, by employing ex-combatants who might otherwise threaten the peace, by serving as a symbol of the unified nation, and by fostering trust via the process of negotiation. To transform this catalogue of mechanisms into testable hypotheses, one must add scope conditions. After doing so, I assess their logic and plausibility.

1. *When* is military integration a strong signal that the parties are committed to postwar arrangements and will not take up arms to revise those arrangements?

Hypotheses
a. When it is costly, requiring the former rivals to overcome domestic opposition. Ironically, too smooth a process of military integration—when the parties readily accept it, without hesitation, as in Mozambique or the Philippines—may undercut that integration's signaling power. Excluding those that most require reassurance, such as Maronite militias in Lebanon, weakens the signal.
b. When a communally dominated military has long been at issue, and thus the military is, compared with other national institutions, especially salient. This apparently was the case in Burundi, where the army, perceived as the *armée mono-ethnique* and a crucial instrument of southern Tutsi domination, had long been on the agenda for Hutu activists.
c. When military integration is rendered irreversible. If reversing institutional innovations is not especially costly, the signal inherently lacks credibility.

Analysis
The signaling mechanism is plausible, but there is reason for skepticism. First, the signaling hypotheses presume that all observers can clearly identify when the deal has been upheld or violated. But in fact there is often substantial disagreement about whether the state has actually upheld its end of the bargain—as in Sudan and the Philippines, for instance. The problem is a difficult one because vagueness is often what enables the deal to go through in the first place; incomplete contracting has real virtues in uncertain environments (Cooley and Spruyt 2009).

Second, the informational logic of signaling presumes that leaders are sincere and just need to find a way to communicate that sincerity. But it is not clear that the cases are generally well characterized as commitment problems. Rather, local actors often see a negotiated settlement as merely a temporary arrangement, at odds with their long-term objectives. In other words, for many local actors, peace is but a short-term exigency—which the lack of military integration reflects.

Third, it is commonly thought that international actors are crucial to achieving military integration, because they can provide side payments to reluctant parties, offer reassurance, or knock heads together as necessary. Although the cases in this volume do not suggest that outside involvement is a necessary condition for either military integration or a stable peace, it has certainly been prominent in some cases. However, if the value of military integration lies in signaling the parties' political will, the more important international actors are as drivers of military integration—that is, the more they provide carrots to soften the blow and induce compliance or beat the recalcitrant with sticks to coerce their good behavior—the less reliable military integration is as an index to the parties' commitment to peace and thus as a predictor of their future behavior once outside actors leave or lose interest.[3]

2. *When* does military integration, and more generally security-sector reform, keep the discontented off the streets?

Hypothesis
a. When the new military employs large numbers of the demobilized.

Analysis
The hypothesized scope condition, regarding the size of the postwar military, is intuitive, yet it is also obvious how rarely this mechanism has operated. In every case in this volume the postwar military was ultimately smaller—much smaller—than the numbers under arms during the civil war. Sustaining a large military is often beyond the fiscal capacity of postwar states in the developing world, especially given their great needs. Only a fraction of former armed combatants normally find places in the integrated military. In fact, insofar as security-sector reform includes "right-sizing" the military, it exacerbates the challenge of demobilization—that is, the need to keep idle hands busy.

3. *When* can a unified military serve as a symbol of national unity?

Hypotheses
a. When the dominant citizenship discourse is "civic republican"—promising equal rights in exchange for contributions to the common good, with military service held up as the epitome of civic virtue—as opposed to "liberal" or "ethnonational" (Krebs 2006; for skepticism about other mechanisms mediating between military service and national identity, see Krebs 2004).
b. When a communally dominated military has long been at issue, and thus the military is, compared with other national institutions, especially salient.
c. When the new postwar military is communally balanced across the ranks and fully integrated—serving as a living, breathing symbol of the nation. It is also essential that the military's communal balance be clear to all and desired by all.

Analysis
Symbolic politics is extraordinarily important, as Murray Edelman's classic work on the subject reminds us. But that phrase—*symbolic politics*—suggests that the political is as important as the symbolic (Edelman 1964, 1971, 1988). After all, symbolic constellations are normally sufficiently varied and contradictory that political entrepreneurs can draw on either the failings of military integration—of which there are sure to be some—or continued discrimination in other institutions to mobilize the masses to undermine the peace. The integrated military is only one among many conceivable symbols, and the key question is which account of the nation, if any, dominates the public sphere. Moreover, the symbolic prominence of the military derives in part from its historical centrality to processes of state and nation building. This was true in the European experience, where, as Charles Tilly (1975, 42) writes, "war made the state, and the state made war"; where, Max Weber (1946, 261) observes, "the discipline of the army [gave] birth to all discipline"; where militaries were key institutions for the labeling and transmission of social values; and where, by the second half of the nineteenth century, the concept of the military as a "school for the nation" had face validity (Feld 1977). But interstate war has, in many postcolonial regions, been rare; postcolonial states have remained weak (Herbst 2000, but also see Chowdhury 2009), and their militaries have typically been dominated by the residents of particular regions and the members of certain tribes and ethnic groups. The armed forces in these states have lacked the historical and cultural prominence that militaries have had in Europe, and they have often been entirely at odds with an inclusive conception of the national political community. All this casts doubt on the premise that militaries in, say, Sub-Saharan Africa are likely to be powerful symbolic resources for representing the nation and thereby for reweaving torn national threads and stabilizing the boundaries of the political community.

In short, there are good theoretical and empirical reasons to be skeptical that military integration works to produce and reinforce peace in post–civil war environments. But the foregoing analysis has not addressed an even more fundamental question:

Is military integration a causally independent variable, or is it more the product of deeper causal forces that themselves drive the resumption of violence? What happens if we turn the question around and treat military integration as the explanandum? This volume's chapters are a rich trove that can be mined for answers. Their findings do not allow us to reject the null hypothesis—that military integration has little independent impact on the duration or quality of post–civil war peace and stability.

MILITARY INTEGRATION AND THE PRIMACY OF POLITICS

One reason we cannot reject the null hypothesis is that the many moving pieces confound efforts to isolate military integration's impact on postwar peace. Military integration takes place either as part of a larger package in a negotiated settlement or subsequent to that settlement. Ideally, scholars might examine cases in which there was a strong commitment to a peace deal and otherwise auspicious structural conditions, but in which security sector reform was, for various reasons, badly handled—and voilà, peace collapsed. Or they might look at cases in which, although the commitment to peace was weak, security-sector reform was, due to extraordinary leadership, handled well—and voilà, peace was sustained. Alas, the real world is not so (social-scientifically) tidy. Where the commitment to peace was weak, military integration tended to be incomplete or short-lived; fighting did not always resume, but communal conflict persisted, sometimes with great intensity. This is the story of Bosnia-Herzegovina, Congo, and the Philippines. Where the commitment to peace was strong, military integration tended to be complete and durable; not only fighting, but even intense communal divides, became a thing of the past. This is the story of South Africa, Mozambique, to some extent Burundi, and perhaps (at least according to the chapter in this volume) Rwanda. It is then necessarily very difficult to isolate the effects of military integration, and thus we cannot attribute outcomes, good or bad, to integration's success or failure. This volume does not include instances of poor military integration, despite pro-peace underlying conditions. It does record (disturbing) evidence of relatively complete military integration without the hypothesized effects on communal conflict and nation building—with Lebanon, Sierra Leone, Sudan, and Zimbabwe as the salient examples; in some cases an initially professional force regressed to become the plaything of communal politics.[4]

There is a second, more important reason we cannot reject the null hypothesis: The chapters in this volume present substantial evidence that military integration is more the product than it is the cause of wartime and postwar political dynamics and structures. The authors of the chapters on Bosnia, the Philippines, Sierra Leone, and Zimbabwe themselves conclude that security-sector reform cannot be credited with the relative stability or blamed for the resumption of violence in those countries. Although the other authors in the volume see military integration as more important, they present little evidence to substantiate their claims. Cyrus Samii, for instance, concludes in chapter 13 that "the apparent success of Burundi's military integration process ha[s] nonetheless made the resumption of large-scale violence less likely," thanks especially to ethnic quotas that have served as a commitment device to signal moderation to nervous Tutsis. Yet this remains an unsubstantiated assertion in his otherwise data-rich chapter. So, too, in chapter 5 Florence Gaub credits the multiconfessional and professional Lebanese army—"the one successful institution in the country that unites all ethnic groups," with its "fully integrated units and carefully balanced posts"—with "considerable" symbolic impact that has "contributed to postconflict stabilization" by "reestablishing trust among politicians, army personnel, and ultimately civil society." Again, a ringing endorsement, but without any evidence backing it up.

In contrast, the cases contain much evidence to the contrary. This is explicit in the chapters on Bosnia, Sierra Leone, and Zimbabwe, and it is implicit in other chapters as well. In Burundi, for instance, as long as the Forces pour la défense de la démocratie (FDD) held out hope for complete victory, it was not willing to accept anything short of the "dismantling" of the army. As the FDD's confidence in its ability to vanquish government forces and retain popular support faded, it became more open to true integration. The power-sharing agreements that created an integrated military reflected the new distribution of power, Samii concludes. Can we therefore credit the new integrated military with keeping the peace? Or, rather, did neither Hutu nor Tutsi see any advantage in returning to war because neither foresaw the underlying distribution of power changing imminently and neither believed that exhausted populations could again be mobilized? Samii himself suggests that the real source of peace was popular exhaustion with the conflict. Thus he worries that "war-weariness may fade," and that, if it does, "tension over ethnic balance in the military could reemerge as part of a broader resource struggle fought again across ethnic lines." The post–civil war Burundian military, in Samii's account, gets it right, but even he has little faith in its capacity to weave together a durable nation.

In Congo, Judith Verweijen shows in chapter 9, the post–civil war military did not get it right at all. A "half-brewed" military emerged from the civil war, and she blames it for "entrench[ing] incentive structures rewarding violence" and for "enabl[ing] wartime power networks to remain partially intact and allow[ing] for the manipulation of the military for private ends," thus "block[ing] a transformation of the war economy." Her analysis of the DRC's postwar political economy is insightful, but the causal analysis is problematic in that it places substantial (though by no means exclusive) emphasis on the armed forces. The failures of the military integration process quite clearly, in her account, are symptomatic of the failures of the "1 + 4" power-sharing formula that the public derides as "1 + 4 = 0." As she observes, the "underlying reason" for failure of military integration in the DRC is that the ex-belligerents were "anxious to keep open the option of a return to violence." In other words, lingering insecurity was not the result of the failures of military integration, but rather the chief cause of the various parties' "halfhearted commitment to military integration." The military became yet another forum in which to prosecute their war, another institution that could be bent to their parochial ends. It is tempting to say that a flawed military integration accounts for local actors' lingering insecurities, but it is clear that those insecurities, or perhaps the actors' greed, account for the flawed military integration.

Mozambique comes across as a remarkable success story. It rings true that Mozambique's thriving despite adverse conditions "would not have been possible without the merging of the military forces that fought for more than sixteen years a very bloody and violent civil war," as Andrea Bartoli and Martha Mutisi write in chapter 10. But is it the case that "the readiness, quality and execution of the merging by the militaries themselves had huge effects in establishing the peace process viability, its longevity and effectiveness," as they claim? Although they endorse several of the introduction's hypothesized causal mechanisms, they fail to provide any evidence. There is, moreover, a more plausible alternative: that the very factors that brought Renamo and Frelimo to the negotiating table—a battlefield stalemate and war exhaustion—both compelled them to resolve their differences over military integration and created the circumstances conducive to peace. In fact, the military has not been a prominent institution in postwar Mozambique; as the authors note, filling the ranks has been a challenge, because most ex-combatants and Mozambicans in general want nothing more than to get on with their lives and leave the war behind. It is hard to envision failed military integration given the factors that drove Mozambicans to end their civil war, and that in itself should be telling.

Much the same might be said about South Africa. "South Africa has a fair claim to be the poster child of negotiated settlements of civil war," Roy Licklider writes in chapter 8, and it comes across also as the exemplar of post–civil war military integration. The process in South Africa was inclusive, professional, and, perhaps not coincidentally, generally successful. Although Licklider offers no extravagant claims about the impact of the South African National Defence Force (SANDF) on the stability of postapartheid South Africa, he does endorse some of the hypothesized causal mechanisms (albeit with little supporting evidence). In particular, though he acknowledges that integration "began fairly late so its progress probably did not directly influence similar activities in other areas of society," he argues that "its failure certainly would have done so." But, as in Mozambique, it is hard to imagine military integration failing dramatically here. Once F. W. de Klerk and the ruling regime had made the strategic decision to end apartheid, there was no real going back in any major arena or institution. The protracted negotiations were successful not primarily because of the process' inclusivity and professionalism—valuable though those were—but because the basic outcome—the end of the apartheid regime—was a foregone conclusion and because the domestic distribution of power—heavily favoring the African National Congress—was known to all. As a result, Nelson Mandela (but by no means all his ANC colleagues) understood that black South Africans could afford to be magnanimous in victory; the ANC's stance during the military integration process, and the outcome—a SANDF that "adopt[ed] the SADF [apartheid-era South African Defence Force] model in many ways"—reflected that wisdom.

Finally, consider the case of Sudan. In deeply researched chapter 3, Matthew LeRiche argues that "the failure of the attempt to integrate the South Sudanese Anyanya ("snake poison") rebel fighters and the Sudanese military, a major requirement of the 1972 Addis Ababa Peace Agreement, made a return to war in Sudan almost inevitable." Specifically, he identifies the decision by President Nimeiri to fully integrate former rebel fighters into the armed forces and assign them to units stationed far from the South as "a major catalyst" in the return to open civil war. One might interpret these developments in Sudan in at least two ways, but in neither was military integration anything more than a trigger. Through one optic, the resumption of civil war was the product not of poorly done military integration, as LeRiche suggests, but rather of Nimeiri's ambitious state-building agenda. This agenda, reflected not only in Nimeiri's assignment of former rebels to geographically dispersed units but also in his abrogation of autonomy provisions, rendered the costs of peace too great for Southerners to bear.

Given Nimeiri's truly national vision, it is hard to see how any meaningful military integration would have kept Sudan together. A second optic, which appears as a skein throughout LeRiche's chapter, draws attention not to what Nimeiri did but to what he did *not* do—specifically, improve the socioeconomic circumstances of the South. The Sudanese state had a limited window in which to demonstrate to Southerners what they could gain from being part of a unified Sudan. The state's failure to deliver on that promise, not the failures of military integration or even Nimeiri's state-building agenda, doomed Sudan to renewed civil war. Only in that context did Southerners feel that they needed "insurance" against Northern aggression, and only in that context was the assignment of Southern forces to distant regions significant. In fact, that context seems to help explain why Nimeiri sought to disperse the Absorbed Forces in the first place in an effort, LeRiche records, to preempt organized revolt. It is rather strange to treat core issues like the introduction of sharia law and the lack of regional development as secondary and to treat military integration as primary. The deployment of Southern forces beyond the South may have been the straw that broke the camel's back, but it was not a fundamental driver of renewed violence.

CONCLUSION: MILITARY INTEGRATION AND THE MALADIES OF NATION BUILDING

Military integration is, first and foremost, a local *political* problem, not a matter of *technical* expertise that local actors lack and whose secrets international actors must therefore impart. The post–civil war military operates within, and reflects, a domestic sphere in which the issues that drove the war still matter (or not) and in which publics remain willing to pay the price of war (or, exhausted, are not). No matter what the formal rules say about the military's staying out of politics, post–civil war armies, more often than not, wallow in communal politics. We cannot expect militaries to have escaped the grip of intercommunal competition when their surrounding societies have failed to do so.[5] Military integration may be a prominent failing in postconflict states, but it may also be the canary in the coal mine—an important early sign that the peace is fragile. Local actors' resistance to military integration, their retention of the means of coercion, their hedging their bets against the breakdown of peace—these are the symptoms of the malady, not the malady itself.[6]

This volume's rich chapters suggest that post–civil war military integration is more product than cause and that it reflects more than it shapes the incentives confronting political actors. It would be premature, based on this volume alone, to conclude that military integration does not matter all that much to the durability of postwar peace. It would be premature to recommend that scholars concerned with the challenges of sustaining peace and stability in war-torn states stop researching military integration. It would certainly be premature to advise policymakers that they should put their energies and resources elsewhere. Perhaps military integration achieves beneficial long-term effects through mechanisms operating at the level of individual soldiers and veterans.[7] Perhaps the value of military integration lies in the way it enmeshes local actors in processes from which they cannot easily extricate themselves, altering their preferences over strategies and ultimately their preferences over outcomes.[8] But this volume should give pause to the true believers, direct scholars to a more careful examination of the causal mechanisms all along the chain, and caution policymakers against putting too many eggs in this single basket, which now appears more fragile.

The chapters in this volume, for all their richness, do not allow dispositive answers, in part because they do not ask all the right questions. Information on nations' post–civil war military integration processes, which the chapters contain in spades, is not only important; it is essential. But it is also only the starting point for any deeper analysis of how military integration shapes postwar politics. If one hypothesizes that military integration is a crucial signal of the actors' intent, then one would want to see at least suggestive evidence that the actors themselves perceived military integration negotiations in this light and interpreted the signal as hypothesized. Such evidence is not always available, but at the very least one would want to analyze carefully the costliness, irreversibility, and symbolic meaning of military manpower policy in a particular environment—all features of a signal that are hypothesized to conduce to its impact. Similarly, before concluding that the postwar military has served to stitch the nation together, sustaining its fragile unity, one would want to see evidence of national leaders' invoking to mass audiences the military as a living symbol of the nation; one would also want to see evidence of such arguments' positive reception in the mass media and mass culture.

Even if the right questions had been asked and spot-on data were provided, however, the chapter might not allow dispositive answers because of endogeneity concerns, which have featured prominently in this chapter. Individual cases may not permit untangling the complex causal chains and isolating the effects of military integration. The most persuasive cases, in which endogeneity concerns are most easily addressed, are those in which integration processes are at odds with other trends—for instance, when political matters are progressing well until military integration throws a wrench in the works, or when successful integration, conducted among professionals and relatively insulated, greases the wheels of frozen talks or silences charges of deception and revanchism. Such cases, which might powerfully establish the effects of military integration, are missing from this volume but may exist in the broader universe of cases.

Perhaps, however, military integration is of great utility—even when it fails. To pursue the metaphor, dead canaries are awfully useful, too: They save lives, by giving early warning to vulnerable miners.[9] But military integration is a wrenching process, even when political will is present. To use it as an early-warning system is a bit like replacing the canary with an endangered and expensive African parrot. The parrot's death will still help the miners, but there are far cheaper ways of achieving the same end.

Even more important, if my concerns have merit, military integration is a distraction from the more consequential effort to figure out the real causes of the resumption of violence and to identify the mechanisms and institutions, both domestic and international, that can forestall or prevent it. After all, it is the noxious fumes that are dangerous, to canaries and coal miners alike.

ABBREVIATIONS

FDD   Forces pour la défense de la démocratie—largest Hutu rebel group in Burundi

Frelimo	Frente de Libertação de Moçambique—government party in Mozambique
Renamo	Resistência Nacional Moçambicana—rebel group in Mozambique
SADF	South African Defence Force—apartheid-era military
SANDF	South African National Defence Force—postapartheid military

ACKNOWLEDGMENT

Thanks to Paul Jackson, Roy Licklider, and Cyrus Samii for comments on an earlier version of this chapter.

NOTES

1. This has, however, entailed some judgment calls, because the authors do not always present final judgments clearly, nor do they use precisely the same terminology. When the chapter did not contain sufficient information to answer a question, I consulted with the author, as indicated in the tables.

2. In those cases in which there was a peace settlement. I have recorded as "No" both cases in which there was no peace settlement (e.g., Rwanda) as well as those in which there was a peace settlement but military integration was not a key issue.

3. I use *index* here in the way done by Robert Jervis (1989).

4. As a result, I cannot exclude the possibility that "good" military integration is a necessary, but not sufficient, condition of nation building and communal peace, and I cannot, based on the cases presented in this volume, definitively conclude that military integration is irrelevant to postwar outcomes. Due to my own ignorance, I cannot say whether these limitations in the volume's universe of cases reflect patterns in the full universe of cases.

5. Militaries of course rarely perfectly reflect their societies—their values, their norms of behavior, their demographic and class composition. History certainly offers examples of militaries whose basic allegiances and identity departed radically from those of the surrounding society. The officer corps of the Austro-Hungarian military, for instance, was perhaps the only segment in the empire that truly bought into the imperial vision. But the circumstances in which the Habsburg officer corps transcended intensifying intercommunal conflict were very different from the post–civil war cases that are this volume's concern.

6. My argument here parallels critiques of the voluminous literature attributing war to arms-racing behavior.

7. Elsewhere, I have expressed skepticism about such mechanisms in a different context (Krebs 2004), due especially to concerns about how such individual-level changes, wrought in the military, are sustained in different social environments and how they aggregate to social and collective outcomes. I remain skeptical in this context as well. Thanks to Cyrus Samii for pressing me along these lines.

8. This is an empirical question, and the findings in this volume are not encouraging. Thanks to Roy Licklider for suggesting this.

9. Thanks to Paul Jackson for bringing this to my attention.

Chapter 16

SO WHAT?

Roy Licklider

So what does all this add up to? The cases are all interesting, but we are not historians; we are interested in them for what they may tell us about current and future situations. Each is obviously different, but do any common elements emerge?

Perhaps the most counterintuitive conclusion to be drawn from these studies is that it is in fact possible under a variety of circumstances to integrate personnel from competing military groups after civil wars. Obviously the outcomes varied a lot, but none of the integrations failed because of violence within integrated forces. In many instances there was intermittent, low-level violence, and some cases there were close calls, but the overall record is quite positive. Forces were successfully integrated in countries where there was intensive international involvement, like Bosnia-Herzegovina and Sierra Leone, and where there was not, like South Africa and Zimbabwe. They were created when integration was linked to negotiated settlements, as in Burundi and Sudan, and when it followed a military victory, as in Rwanda and Sierra Leone. They were created where military integration had substantial local political support, such as South Africa and Mozambique, and where it did not, such as the Democratic Republic of the Congo and Bosnia-Herzegovina. Gaub (2011, 133–37) attributes this to the military culture, which stresses hierarchy. Zirker, Danopoulos, and Simpson (2008) argue that militaries in developing countries may develop an identity so strong that it is equivalent to a separate ethnicity; something like this seems to have happened in some of these cases.

A second generalization is that the military capabilities of the new forces were often irrelevant to their successes. New armies in South Africa and Mozambique were weaker than their predecessors; those in Bosnia-Herzegovina and Lebanon were weaker than potential adversaries. But all seem to have had some impact with the example of integration that they set for their societies—although, as Krebs points out in chapter 15, these conclusions are based on anecdotes rather than precise evidence from process-tracing exercises. This was true where there were no obvious external enemies (South Africa and Mozambique) and where there were (Lebanon and Bosnia-Herzegovina).

I expected that the most extreme test of the new militaries would be fighting other members of groups that had been integrated. The new armies were asked to do this in Rwanda, Burundi, Sierra Leone, and the Philippines, and in all these cases it seems to have strengthened rather than weakened the army's cohesion. The general point may be that integration is facilitated when it has an immediate task. A substantial portion of the new forces of the Democratic Republic of the Congo collapsed almost immediately when asked to do this, but this unfortunate group had so many problems that it is hard to isolate the importance of this particular issue.

At the same time that the militaries were being integrated, their numbers were usually also being drastically reduced as a result of the end of the civil war, the lack of an obvious security threat (in many cases), and a desire to get a peace dividend to bolster the economy. This meant that, in the end, relatively few people were employed (one reason there was not much support for the causal model about employing former fighters), but this may sometimes have allowed more selectivity in recruiting (Gaub 2011, 129–33), which presumably increased the chances of success.

WHEN WAS MILITARY INTEGRATION ATTEMPTED?

Military integration is often linked to negotiated settlements of civil wars. But this project shows that it can also be employed after military victories. The obvious case is Rwanda, where a group that had been victims of a genocide won the civil war (itself an unprecedented event) and then created a new military with a majority from the losing ethnic group, including former *genocideiares*. In Sierra Leone the British intervention essentially created a military victory, and in the Philippines the MNLF seems to have been mostly defeated before it agreed to be integrated into the Philippine army. All these cases look like successes to one degree or another. This may just be because civil wars that end in military victories are less likely to resume than those that end in negotiated settlements (Licklider 1995; Toft 2010), although that issue remains open to debate.

WHAT STRATEGIES WERE MOST EFFECTIVE?

Former adversaries were integrated as individuals rather than units in nine of the eleven cases in this volume; only in Sudan and Bosnia-Herzegovina were the basic units segregated. Successfully mixing the groups at the lowest level seems likely to ultimately improve cohesion (Gaub 2011, 126–29), so it is encouraging that this rather risky strategy seems to have worked fairly well.

International assistance is often helpful, though it is not necessary for success (South Africa, Rwanda) and is not sufficient to prevent failure (Democratic Republic of the Congo). This is particularly significant since Caroline Hartzell shows in chapter 2 that military integration is often implemented because of pressure from the international community, sometimes against considerable local resistance (Bosnia-Herzegovina is a case in point), which raises significant ethical issues to which I will return.

Human rights violators were often not excluded from the new armies. It sounds plausible to recommend that each individual be vetted, but this is actually quite difficult and expensive. Moreover, negotiated settlements to civil war often involve some sort of formal or informal amnesty. Interestingly, even forces whose members were not screened often did fairly well in terms of human rights violations; training and environment may be more important than past behavior.

Quotas were often used and were generally quite successful. Simple formulas (50/50 in cases with two groups, such as Burundi; 33/33/33 with three groups, as for senior appointments in Bosnia-Herzegovina) often were more useful, even if less obviously fair, than ratios based on population or other figures that might be unclear and disputed (Samii 2013; for a sophisticated discussion of quotas, see Gaub 2011, 122–26).

I had assumed that military integration would be a single process, but the chapter authors considered examples of what might be called serial integration, with some groups being integrated at different times. In the DRC this became routine and was clearly destructive. In Burundi, conversely, the successful integration of a small rebel group seems to have facilitated the integration of the larger group a few years later. The opposite seems to have occurred in the Philippines, where the successful integration of members of the MNLF seems to have persuaded the MILF that this was not a good way to end the war.

DOES MILITARY INTEGRATION MAKE CIVIL WAR RESUMPTION LESS LIKELY?

In chapter 15, Ronald Krebs makes a strong argument that military intervention is a political process rather than a technocratic solution that outsiders can use to work around local political divisions. In other words, the success of military integration depends on the political will of the local elites (cf. Burgess 2008). We have at least five cross-national studies of civil war resumption that compare countries with and without military integration and postwar peace (Walter 2002, 63–86; Hoddie and Hartzell 2003; Glassmyer and Sambanis 2008; DeRouen, Lea, and Wallensteen 2009; Toft 2010, 12–13, 58–60). Two general conclusions emerge from these studies. First, four of the five show an association between military integration and length of postwar peace, and the fifth says it cannot exclude the possibility that effective integration might have this effect (Glassmyer and Sambanis 2008). The studies use different measures of both dependent and independent variables and cover different time spans; the findings can be fairly called robust. Second, none of these studies disproves Krebs's argument that the causal arrow is reversed and that successful military integration is the result of peace or, more precisely, that both military integration and peace are the results of political will of the parties. These studies all focus on whether the parties *agreed* to military integration, not whether it was actually implemented. Moreover, we have no indicator for political will other than the actions that it is supposed to produce; it is simply a construct of our minds to explain behavior, and its existence and strength are strictly a function of analysts' opinions (rather similar to our case authors' opinions). Unfortunately, this does not mean that it or something like it does not exist or is not important.

This study excluded cases in which military integration was not attempted. As a result we also cannot make strong claims about whether or not military integration makes renewed civil war less likely. Instead, we asked the authors to give their opinion on the question, and suggested a number of possible causal paths, but, as Krebs correctly points out, these are just opinions. On one hand, they are the opinions of people who know a lot more about the cases than Krebs or I do. On the other hand, they were asked to write about military integration and may have assigned it greater importance than if they had not been sensitized to this issue. For what it is worth,

TABLE 16.1 Causal Links to a Reduced Probability of Civil War, by Case

Case	Renewed Civil War	Costly Commitment	Provide Security	Employ Fighters	Example for Society	Trust from Negotiations
Military integration successes						
Lebanon	No	No	No	Some (25% of militias)	Yes	No
Philippines	Continued war with different groups	No	No	No	Yes (?)	No
South Africa	No	No	No	No	Yes	No
Sierra Leone	No	Somewhat	Somewhat	Somewhat	Yes	No
Mozambique	No	Yes	Somewhat	Yes	Yes	Yes
Bosnia-Herzegovina	No	No	No	A few	Yes	No
Burundi	Continued war with different groups	Yes	Yes	?	Slowly	
Rwanda II	No	Yes	Yes	Yes	Yes	Yes
Failures						
Sudan 1971 (after 11 years)	Yes	Yes	No	Yes	No	No
Zimbabwe	Yes	Yes	Protect autocracy	Somewhat	No	No
Rwanda I	Yes	No	No	No	No	No
DRC	Not among parties to settlement	Somewhat	No	Yes	No	No

opinions varied; some authors believed that military integration was an important part of the process in the cases they studied, while others said that it was not especially important. The good news is that no one seemed to think it actually made renewed war more likely (although the evidence of the Arusha agreements in Rwanda suggests an example). Table 16.1 summarizes the judgments of the case authors.

Of the five causal paths, the symbolic appeal of an integrated institution was selected most often by the analysts, although admittedly determining the policy impact of symbolic actions is very much a work of art at this point. No one explicitly credited the new military's fighting prowess, either actual or potential, with establishing and keeping the peace, although the new militaries in Rwanda, Burundi, and Sierra Leone did play a major part in winning the respective continuing wars, which suggests that military efficacy was important in some cases. No one seems to have been impressed with the argument that providing security built trust and allowed political and economic development.

The idea that peace is more likely if the participants make costly commitments is currently in vogue in political science (Fearon 1998), but no one saw much evidence that this was the case. Similarly, employing enough fighters to seriously reduce the risk of future violence did not seem to play a role in these cases, except in the Philippines, where the MNLF saw integration as a means of providing employment for some of its members. The argument that successful negotiation of security issues led to more trust is mentioned only in the Mozambique case.

Burgess and Krebs argue that political will is necessary for military integration to succeed. At one level this is clearly true, and chapter 9, on the Democratic Republic of the Congo, is a powerful example of what happens when this political will is lacking. However, in practice this argument is often tautological; because we have no independent measure of political will, there is a tendency to say that political will existed if there was success and that it did not if there was not. I do not agree that there was overwhelming political support in South Africa, where the threat of continued violence by the military forced a universal amnesty (Samuels 2009), or in Burundi, where it took several years of war after the Arusha accords to bring the Forces pour la défense de la démocratie to an agreement. I also do not think it is fair to say that there was no political will for integration in Sudan, where the settlement lasted for eleven years before being deliberately broken by Nimieri in a desperate and futile attempt to save his regime from northern Islamists.

More important, we need to see the political wills of various actors as variables rather than constants. Military integration will reduce the probability of renewed civil war only if it changes the political will of some important actors. Surely this is what we mean when we say that military integration has had an impact by showing that people who have been killing one another can cooperate, implying that this might be possible in other areas as well, and that people can go about their business with less far of repression by the military because some of their compatriots are in it. After all, few participants start civil wars with a goal of military integration (although Burundi seems to have been an exception). Each side wants to win; military integration, like negotiated settlements in general, is usually a second-best solution when it becomes clear that it cannot win. The decision to participate in military integration reflects some sort of change in political behavior; the issue is whether, if it is successful, this will result in changes in political will among important actors. And though Krebs is correct that there is no extensive historical military tradition in many of these countries, the fact that they are emerging from civil war might reasonably make effective cooperation among competing militaries resonate through society. Having said this, the authors agree that this project has not gotten down to the micro level in the various societies to show whether this is true or not and so the issue remains unsettled. Some of the issues involved in such research are discussed below.

I think it is fair to say that no contributor to this project believes that military integration by itself will make renewed civil war impossible. Like all the other behaviors linked to liberal peace building (international peacekeepers, effective public institutions, democracy, rule of law, reducing corruption, improving the economy, reducing inequality, transitional justice, etc.), the argument in its favor is that it will help reduce the chances of such conflict. And like these other behaviors, it is extraordinarily difficult to demonstrate empirically that this is true. Indeed, we are reasonably certain that these strategies, individually or collectively, do not work in all cases, but that some do work sometimes; we are still trying to find out why this is true and what conditions are necessary for success. Our hope is that the detailed cases in the volume will lay down a basis on which further work can be done.

Caroline Hartzell's conclusion in chapter 2 that military integration is often introduced and supported by the international community reveals an obvious paradox. Two decades of state-building efforts seems to have produced at least two generally accepted conclusions: that security is essential for peace, and that local ownership of the new institutions is essential for long-term peace. But how is local ownership supposed to develop for an idea being brought in from the outside, often over local opposition? Indeed, as Krebs notes, the fact that it came from the outside reduces the cost of adopting it, which in turn makes it less useful in signaling peaceful intent to domestic opponents.

The strategy presumably depends on two assumptions: that locals will change their minds and accept military integration as a good thing once it is in place, and that this acceptance will make the resumption of civil war less likely. These are not crazy ideas; we have seen that some of our authors supported rather similar ones. But they are also very controversial. As noted above, if the value of military integration is in creating a symbol, it will be very hard to develop empirical analyses to establish that this has actually had the predicted effect.

Judith Verweijen's analysis of the Democratic Republic of the Congo in chapter 9 raises a quite different possibility—that a *failure* of military integration (or at least a failure to establish a strong central military) may be necessary for the achievement of at least some negotiated settlements. She argues that in the DRC, conflicting elites with their separate networks would only accept a settlement involving a weak central government, and that as a result the integration process was deliberately sabotaged.

Indeed, as the DRC example suggests, peace may not be the ultimate objective of some of the players. The DRC is hopefully an extreme case, but it illustrates the general point that in any postwar country, some political actors will initially see a stable, effective government as useful and others will disagree; the gamble of any negotiated settlement is that over time some people who originally opposed a settlement will see it in their interests to come to terms with it and that an integrated military may make that more likely. It is hardly surprising that some settlements fail; what is remarkable is that any succeed.

POSTWAR PROBLEMS

Even if we assume that military integration can help prevent renewed civil wars in some situations, we may decide it is not a good idea. A similar debate about the utility of political power sharing as a way of ending civil wars seems to have concluded that this may be the only way to end some wars but that after a few years the resulting government often becomes unable to respond appropriately to its many challenges (Rothchild and Roeder 2005; Hoddie and Hartzell 2005). A possible parallel may be the shift of the United States from the Articles of Confederation to the Constitution after the American Revolution.

Our cases suggest at least two such long-term problems. Is the resulting military really appropriate for the country in the long run? What is the impact of a strong military on political democracy?

There often does not seem to be real need for a strong military after a civil war because there is little real outside threat. Military integration frequently occurs at the same time as the total force is being reduced. Many of the new militaries search for a mission to justify expenditures that loom large in relatively poor countries. As is noted by David Laitin in chapter 14, peacekeeping is one popular choice; it employs soldiers, gains some prestige for the country, and can be self-supporting, but it seems perverse to create a military just for peacekeeping.

This in turn raises the question of what sort of military the country really needs. Obviously, it varies from case to case—South Africa has no obvious outside enemies, for example, while the Democratic Republic of the Congo literally cannot control its own territory in part because of encroachments by its neighbors. Outsiders are sometimes accused of forcing other countries to adopt their own military models, not without some reason. But South Africa is a relatively advanced country with no obvious external foes and no foreign involvement in its decision making whose military wants to sustain an expensive, mechanized, land-based force even though its major tasks seem likely to be counterinsurgency and peacekeeping on land and monitoring of coastal waters.

At a minimum it seems plausible that locals should decide on the kind of military that is required because they will pay the penalty for any errors in the consequential decision. But internationals should try to avoid creating a military that is not financially and politically viable after the war and work with locals to develop plans for a military that is sustainable in the long run.

However, there is a more fundamental question: Should we really be doing all this? Military integration is attractive to outsiders because it promises to fulfill a real need for security in the postconflict society. The importance of security was overlooked by many of those in peacekeeping for a long time, although it was highlighted in one of the first systematic studies of the field (Stedman 2002, 668), but it is now seen as critical. Military integration is also relatively easy to do by outsiders. It involves training a fairly small group of people to do things that they are generally interested in doing, as compared with, for example, setting up a functioning justice system, which would usually be much more useful in establishing security but may well require changing the culture of the society as a whole. Moreover, it fits the skill sets of the international community as a whole; we can readily deploy substantial numbers of military trainers, but, as shown in Iraq and Afghanistan, we simply do not have the organizations or personnel to do the same for important civilian activities.

Unfortunately, it is not entirely clear that it is a good idea to strengthen the military in post–civil war countries. It is much easier to create a reasonable military from the outside than, for example, to reform the police force and the judiciary, which are deeply embedded in the local power structures of everyday life. A strong security apparatus inside a weak and ineffective government creates a temptation for military domination or coup: "It can be posited that the military replaces a civilian regime when the military is the most cohesive and *politically* the best organized group at a given time in a political system; the military intervenes and succeeds in the *absence* of relatively more powerful opposition" (Perlmutter and Bennett 1980, 16; emphasis in the original).

Between the world wars, the United States intervened in a number of Central American countries and established strong security forces that undermined the civilian governments and produced a series of military dictatorships. Paul Jackson's analysis of Zimbabwe in chapter 4 illustrates a similar situation: an alliance between some civilian politicians and military leaders to eliminate democracy (and military integration) in favor of authoritarian rule. Similar tendencies can be seen in Rwanda and Uganda, and when in South Africa I heard references to a "Zulu security sector." I am on record in favor of policies that will produce short-term peace, even at the possible risk of longer-term problems, such as amnesty, power-sharing governments, and military integration; my preference is to save as many lives as we can now and worry about the consequences later (Licklider 2008). But ignoring the risks involved is not simply oversight; it is negligence. Ultimately, of course, these decisions will be made by some of the locals, but it is likely that those with guns will have a disproportionate influence on the choices. So outsiders have an obligation to stress not only the abstract values of civilian control of the military but also the evidence that military governments do not bring economic development or political democracy and often result in the eventual weakening of the military itself.

There are serious ethical issues when outsiders pressure local people to follow particular policies in their own countries. If these policies backfire, the locals will pay the price while the outsiders go home. In this situation outsiders have multiple ethical obligations: to learn more about whether military integration works to help keep the peace; to learn more about the "best" way to go about integrating militaries under different sets of circumstances; and to give more thought to the implications of all of this beyond outcomes such as keeping the peace (e.g., effects on democracy, human rights). Hopefully, this book will add to our knowledge in these areas. But this in turn suggests another obligation: to be as candid as we can with the locals about these costs and benefits, not withholding our own inevitable uncertainties, even if we are convinced that some of the locals will use this information for ends that we would deplore (Hartzell 2012).

The current mantra of the international community is local ownership, and this sounds like another version of it. But in fact we have other ethical obligations as well. Outsiders have influence because they bring resources to the table. One risk is that they will use those resources to persuade or coerce locals into accepting policies that make more sense to outsiders than to locals. But we also have a responsibility to the people who have given us these resources. If a local government says it is upholding its cultural traditions by trying to wipe out a local minority, this does not settle the issue; we have an obligation to the people who supply our resources not to wind up supporting organizations whose actions they deplore. After all, one of the advantages of being an outsider is that exit is always a possibility and it may become an obligation.

So what have we learned about the costs and benefits of military integration? First, we know that it can be done. This is not just another academic idea; it has actually been carried out in a number of different countries with varying degrees of success. But we also know that it is not a shortcut for the long and difficult task of creating a strong and hopefully democratic government, although it may help. The integration process can be sabotaged or taken over by local political leaders. Even if the integration itself is successful, it may result in a military that has undue influence on a weak government, which makes democracy problematic. Like the military itself, it is an instrument that is sometimes powerful but potentially dangerous. In order to reduce this danger, we should alert the locals to these issues and continue to monitor individual countries. And we need to increase our understanding about the complex relationships involved in our theories, in the ways suggested below.

FUTURE RESEARCH

One general strategy would be to move below the fairly broad approach of our case chapters. The causal path that was most popular with our authors has the integrated military acting as a symbol, presumably changing people's minds about the value of integrating other institutions and making them more likely to resist the temptation to return to civil war. Demonstrating the empirical impact of symbols is very much a work of art at this point, but one strategy would be to use process tracing of individuals and groups within a particular elite over time. We did not have the time or expertise to do this microanalysis on any one country, much less all of them, but it remains a potentially important path for future research.

We also need more knowledge about how integration is actually carried out on the ground; getting it will require more micro-level research. Practitioners need something like a handbook, drawn from different experiences and noting what they have in common and what they do not. Despite the fact that the international community has been deeply involved in the issues of military integration for several decades, there is remarkably little institutional memory of what works and what does not, even within individual countries like Britain and the United States, much less the larger group of states and organizations that are intermittently involved in integration activities. In particular, a history of the activities of the various British military assistance teams would be very helpful.

Some of the most interesting quantitative research on the efficacy of various peace strategies has involved calculating the probability of renewed war in a number of cases and then using this degree of difficulty to sort out whether the strategies are used in more-or-less difficult cases and then looking about outputs. One impressive example looks at multidimensional peacekeeping and suggests that it is associated with more failures than successes; but it also shows that, if one controls for the degree of difficulty, it is actually associated with success more than comparable cases without peacekeeping (Fortna 2008). This strategy might be applied to military integration as well.

At another level, how have settlements and victories that did not involve military intervention handled the security problem? In chapter 7, Rosalie Arcala Hall notes that the problems of military integration of the MNLF have persuaded the remaining rebel group, the MILF, to work for regional autonomy instead, and there is substantial evidence that this approach has been helpful in preserving peace in ethnopolitical conflicts (Marshall and Gurr 2005). Are there other approaches which have proven useful?

Finally, we need more work at the macro level. In chapter 14, David Laitin provides an excellent example of how to look at a social process similar in outline to military integration, teasing out similarities and differences, and using both to suggest potential problem areas and to give some idea of the probability of success. Expanding our knowledge by this important strategy is an ethical imperative as well as a research opportunity.

More broadly still, we need to know how military integration relates to other standard strategies in the state-building toolbox, like transitional justice, power sharing, the rule of law, reform of the justice system, democratization, and economic development. If our goal is to encourage more integrated nations, military integration may be very useful, but it cannot accomplish much by itself.

REFERENCES

Abasama, Datukan. 2009. Interview, March 5. Former Moro National Liberation Front combatant and corporal, Philippine army. Camp Siongco, Parang, Maguindanao.

Abdullah, Ibrahim. 1998. Bush Path to Destruction: The Origin and Character of the Revolutionary United Front/Sierra Leone. *Journal of Modern African Studies* 36, no. 2: 203–35.

Abinales, Patricio. 2000. *Making Mindanao: Cotabato and Davao in the Formation of the Philippine State*. Quezon City: Ateneo de Manila Press.

Abraham, Arthur. 2004. The Elusive Quest for Peace: From Abidjan to Lomé. In *Between Democracy and Terror: The Sierra Leone Civil War*, edited by I. Abdullah. Dakar: CODESRIA.

Adebajo, Adekeye. 2002. *Building Peace in West Africa: Liberia, Sierra Leone and Guinea-Bissau*. Boulder, CO: Lynne Rienner.

Adolfo, Eldridge. 2010. *Sierra Leone: Splitting under the Strain of Elections?* Stockholm: FOI, Swedish Defence Research Agency.

Africa Research Bulletin. 2000a. Volume 37, issue 4.

———. 2000b. Volume 37, issue 5.

Africa, Sandy. 2008. South Africa: SSR after Apartheid. In *Local Ownership and Security Sector Reform*, edited by Timothy Donais. Geneva: Geneva Center for the Democratic Control of Armed Forces.

Agence France-Press. 2007. Burundi Accused of Favouring Hutus in Security Forces. August 8.

Akbar, Johnny. 2009. Interview, April 6. Former Moro National Liberation Front military commander. General Santos City.

Akerlof, G. A., and R. E. Kranton. 2000. Economics of Identity. *Quarterly Journal of Economics* 115, no. 3: 715–53.

Albrecht, Peter, and Paul Jackson. 2009. Security System Transformation in Sierra Leone, 1997–2007. Global Facilitation Network for Security Sector Reform (GFN-SSR) and International Alert.

Alexander, J., J. McGregor, and T. Ranger. 2000. *Violence and Memory: One Hundred Years in the "Dark Forests" of Matabeleland*. Oxford: James Currey.

Alier, Abel. 1992. *Southern Sudan: Too Many Agreements Dishonoured*, Second Edition. Reading, U.K.: Ithaca Press.

All Africa. 2007. Army Positions Create Insecurity in Kabezi. October 26.

Allen, Tim. 1991. Full Circle? An Overview of Sudan's "Southern Problem" since Independence. *Northeast African Studies* 13, nos. 2–3: 191–215.

REFERENCES

Allport, R. N.d. Operation Quartz: Rhodesia 1980. www.rhodesia.nl/quartz.htm.
Amnesty International, 2007. *Democratic Republic of Congo: Disarmament, Demobilization and Reintegration (DDR) and Reform of the Army*. London: Amnesty International.
Associated Press. 2005. UN Suspects Burundian Army behind Election Violence. June 3.
Ayao, Khanappi. 2009. Interview, March 2. Former Moro National Liberation Front military commander. Cotabato City.
Azar, Fabiola, and Etienne Mullet. 2002. Muslims and Christians in Lebanon: Common Views on Political Issues. *Journal of Peace Research* 11.
Azurin, Arnold. 1996. *Beyond the Cult of Dissidence in Southern Philippines and War-Torn Zones in the Global Village*. Manila: University of the Philippines Press and UP Center for Integrative and Development Studies.
Bairstow, Major Timothy M., USMC. 2012. *Border Interdiction in Counterinsurgency: A Look at Algeria, Rhodesia and Iraq*. Biblio Scholar.
Baker, Deane-Peter. 2009. *New Partnerships for a New Era: Enhancing the South African Army's Stabilization Role in Africa*. Carlisle, PA: Strategic Studies Institute, US Army War College.
Baker, Deane-Peter, and Evert Jordaan. 2010. *South Africa and Contemporary Counterinsurgency: Roots, Practices, Prospects*. Claremont, South Africa: UCT Press.
Bangura, Yusef. 1997. Reflections on the Abidjan Peace Accord. *Africa Development* 22, nos. 3–4: 217–41.
———. 2000. Strategic Policy Failure and Governance in Sierra Leone. *Journal of Modern African Studies* 38, no. 4: 551–77.
BBC. 2010. Burundi Peacekeepers on Trial over Somalia Mutiny. February 5.
BBC Monitoring. 2005. Former Fighters Differ over Ranks in the New Burundi Army. Via Radio Bonesha FM, March 22.
———. 2008. Burundi: Donors Demand Reduction of Army Size. Via Agence Burundaise de Presse, March 27.
Beber, Bernd. 2009. The Effect of International Mediation on War Settlement: An Instrumental Variables Approach. Unpublished manuscript. http://homepages.nyu.edu/~bb89/files/Beber_MediationIV.pdf.
Bell, J. Bowyer. 1972. Societal Patterns and Lessons: The Irish Case. In *Civil Wars in the Twentieth Century*, edited by Robin Hingham. Lexington: University Press of Kentucky.
Bennett, Andrew, and Alexander George. 2005. *Case Studies and Theory Development in the Social Sciences*. Cambridge, MA: MIT Press.
Bermeo, Nancy. 2003. What the Democratization Literature Says—or Doesn't Say—about Postwar Democratization." *Global Governance* 9:159–77.
Bertrand, Jacques. 2000. Peace and Conflict in Southern Philippines: Why the 1996 Peace Agreement Is Fragile. *Pacific Affairs* 73, no. 1: 37–54.
Blanchetti-Revelli, Lafranco. 2003. Moro, Muslim or Filipino? Cultural Citizenship as Practice and Process. In *Cultural Citizenship in Island Southeast Asia: Nation and Belonging in the Hinterlands*, edited by Renato Rosaldo. Berkeley: University of California Press.

Boshoff, H. 2003. Burundi: The African Union's First Mission. Situation Report, African Security and Analysis Programme, Institute for Security Studies, Pretoria, June 10.

———. 2005. *Update on the Status of Army Integration in the DRC*. Pretoria: Institute for Security Studies.

Boshoff, H., and J. M. Gasana. 2003. Mapping the Road to Peace in Burundi: The Pretoria Sessions. Situation Report, African Security and Analysis Programme, Institute for Security Studies, Pretoria, November 24.

Boshoff, H., and H. Hoebeke. 2008. *Peace in the Kivus? An Analysis of the Nairobi and Goma Agreements*. Pretoria: Institute for Security Studies.

Bosnia and Herzegovina, Defence Reform Commission Secretariat. 2004. *The Path to Partnership for Peace. Report of the Defence Reform Commission*. Sarajevo: Government of Bosnia and Herzegovina.

Bosnia and Herzegovina, Presidency of Bosnia and Herzegovina. 2006. *The Size, Structure and Locations of the Armed Forces of Bosnia and Herzegovina*. Sarajevo: Government of Bosnia and Herzegovina.

Bratton, M., and N. Van de Walle. 1994. Neopatrimonial Regimes and Political Transitions in Africa. *World Politics* 46, no. 4: 453–89.

Bucyalimwe, Mararo, S. 2007. "Les élections de 2006 et l'ordre post-transition au Kivu: Changements et Continuités." In *L'Afrique des grands Lacs: Annuaire 2006–2007*, edited by F. Reyntjens and S Marysse. Paris: L'Harmattan.

Burgess, Stephen F. 2008. Fashioning Integrated Security Forces after Conflict. *African Security* 1, no. 2: 69–91.

Butler, Christopher, and Scott Gates. 2009. Asymmetry, Parity, and (Civil) War: Can International Theories of Power Help Us Understand Civil War? *International Interactions* 35, no. 3: 330–40.

Call, Charles, and William Deane Stanley. 2003. Military and Police Reform after Civil Wars. In *Contemporary Peacemaking: Conflict, Violence and Peace Processes*, edited by John Darby and Roger MacGinty. New York: Palgrave Macmillan.

Callaghy. T. 1984. *The State–Society Struggle: Zaire in Comparative Perspective*. New York: Columbia University Press.

Capron, Laurence. 1999. The Long-Term Performance of Horizontal Acquisitions. *Strategic Management Journal* 20, no. 11: 987–1018.

Catholic Commission for Justice and Peace in Zimbabwe. 1997. Report on the 1980s Disturbances in Matabeleland and the Midlands. www.newzimbabwe.com/pages/gukurahundi.html.

Cawthra, Gavin. 1997. *Securing South Africa's Democracy: Defence, Development and Security in Transition*. New York: St. Martin's Press.

———. 1999. *From "Total Strategy" to "Human Security": The Making of South Africa's Defence Policy 1990–98*. Working Paper 8. Copenhagen: Copenhagen Peace Research Institute.

———. 2003. Security Transformation in Post-Apartheid South Africa. In *Governing Insecurity: Democratic Control of Military and Security Establishments in Transitional Democracies*, edited by Gavin Cawthra and Robin Luckham. London: Zed Books.

———. 2012a. Director of the Centre for Defence and Security Management, Uni-

versity of Witwatersrand. Personal interview by Roy Licklider, Park Town, South Africa, June 13.
———. 2012b. Political and Security Negotiations and Security Sector Transformation in South Africa. In *Post-War Security Transitions: Participatory Peacebuilding after Asymmetric Conflicts*, edited by Véronique Dudouet, Hans J. Giessmann, and Katrin Planta. London: Routledge.
Cayton, Alfredo. 2009. Interview, March 5. Division commander, Sixth Infantry Division, Philippine Army. Camp Siongco, Parang, Maguindanao.
Central Intelligence Agency. 2010. *CIA World Factbook: Lebanon.* www.cia.gov/library/publications/the-world-factbook/geos/le.html.
Chartouni, Charles. 1995. *Conflict Resolution in Lebanon: Myth and Reality.* Beirut: Foundation for Human and Humanitarian Rights.
Che Man, W. K. 1990. *Muslim Separatism: The Moros of Southern Philippines and the Malays of Southern Thailand.* Singapore: Oxford University Press.
Chitoyo, K. 2009. *The Case for Security Sector Reform in Zimbabwe.* Occasional Paper. London: Royal United Services Institute.
Chitiyo, K., and M. Rupiya. 2005. Tracking Zimbabwe's Political History: The Zimbabwe Defence Force from 1980 to 2005. In *Evolutions & Revolutions: A Contemporary History of Militaries in Southern Africa*, edited by M. Rupiya. Pretoria: Institute for Security Studies.
Chowdhury, Arjun. 2009. Expectations of Order: State Failure in Historical Context. Ph.D. diss., University of Minnesota.
Chrétien, Jean-Pierre, André Guichaoua, and Gabriel Le Jeune. 1989. *La crise d'aout 1988 au Burundi.* Paris: Centre d'études africaines.
Chuter, David. 2006. Understanding Security Sector Reform. *Journal of Security Sector Management* 4, no. 2 (April): 1–22.
Cilliers, J. K. 1985. *Counterinsurgency in Rhodesia.* Dover, DE: Croom Helm.
Cilliers, Jakkie. 2012. Executive Director, Institute for Security Studies, Pretoria, South Africa. Telephone interview by Roy Licklider, June 14.
Colletta, Nat J., and Robert Muggah. 2009. Context Matters: Interim Stabilisation and Second Generation Approaches to Security Promotion. *Conflict, Security & Development* 9, no. 4 (December): 425–53.
Collier, Paul, and Anke Hoeffler. 2006. Military Expenditure in Post-Conflict Societies. *Economics of Governance* 7, no. 1: 89–107.
Cooley, Alexander, and Hendrik Spruyt. 2009. *Contracting States: Sovereign Transfers in International Relations.* Princeton, NJ: Princeton University Press.
Corm, Georges. 1991, Liban: Hégémonie milicienne et problème de rétablissement de l'Etat. *Monde arabe, Maghreb Machrek* 131, January–March.
Cragin, Kim, and Peter Chalk. 2003. *Terrorism & Development: Using Social and Economic Development to Inhibit a Resurgence of Terrorism.* Santa Monica, CA: RAND Corporation.
Cunningham, David E. 2011. Who Gets What in Peace Agreements? In *To Block the Slippery Slope: Reducing Identity Conflicts and Preventing Genocide*, edited by Mark Anstey, Paul Meerts, and I. William Zartman. New York: Oxford University Press.
Cunningham, David E., Kristian Skrede Gleditsch, and Idean Salehyan. 2009. It

Takes Two: A Dyadic Analysis of Civil War Duration and Outcome. *Journal of Conflict Resolution* 53, no. 4: 570–97.

Daily Star 2006. Hizbullah Rejects UN Call for Integration into Armed Forces. April 20.

———. 2009. Lebanon's Defense Budget to Rise 22 per Cent in 2009. September 14.

De Goede, M. 2007. The Price for Peace in the Congo: The Incorporation of the Political Economy of War in the State and Governance System. Saint Andrews: Centre for Peace and Conflict Studies

De Goede, M., and C. Van Der Borgh. 2008. A Role for Diplomats in Post-War Transitions? The Case of the International Committee in Support of the Transition in the Democratic Republic of Congo. *African Security* 1, no. 2: 115–33.

Dempsey, Thomas A. 2009. The Transformation of African Militaries. In *Understanding Africa: A Geographic Approach*, edited by Amy Richmond Krakowka and Laurel J. Hummel. Carlisle, PA: Center for Strategic Leadership, US Army War College.

Dennis, Major General A. W., CB, OBE (Retired). 1992. The Integration of Guerrilla Armies into Conventional Forces: Lessons Learnt from BMATT in Africa. *South African Defence Review*, issue 5. www.iss.org.za/Pubs/ASR/SADR5/Dennis.html.

DeRouen, Karl, Jr., Jacob Bercovitch, and Paulina Pospieszna. 2011. Introducing the Civil Wars Mediation (CWM) Dataset. *Journal of Peace Research* 48, no. 5: 663–72.

DeRouen, Karl, Jr., Jenna Lea, and Peter Wallensteen. 2009. The Duration of Civil War Peace Agreements. *Conflict Management and Peace Science* 26, no. 4: 367–87.

Dietz, Henry, Jerrold Elkin, and Maurice Roumani. 1991. *Ethnicity, Integration and the Military*. Boulder, CO: Westview Press.

Dorman, Andrew M. 2009. *Blair's Successful War: British Military Intervention in Sierra Leone*. Aldershot, UK: Ashgate.

Drewienkiewicz, John. 2003. Budgets as Arms Control: The Bosnian Experience. *RUSI Journal* 148 (April): 30–35.

Ebenga, J., and T. N'Landu. 2005. The Congolese National Army in Search of an Identity. In *Evolutions and Revolutions: A Contemporary History of Militaries in Southern Africa*, edited by Martin Rupyia. Pretoria: Institute for Security Studies.

Ebo, Adedeji. 2006. The Challenges and Lessons of Security Sector Reform in Post-Conflict Sierra Leone. *Conflict, Security and Development* 6, no. 4: 481–501.

Edelman, Murray. 1964. *The Symbolic Uses of Politics*. Urbana: University of Illinois Press.

———. 1971. *Politics as Symbolic Action: Mass Arousal and Quiescence*. Chicago: Markham.

———. 1988. *Constructing the Political Spectacle*. Chicago: University of Chicago Press.

Edwards, Martin, Greg Mills, and Terence McNamee. 2009. Disarmament, Demobilization, and Reintegration and Local Ownership in Great Lakes: The Experience of Rwanda, Burundi, and the Democratic Republic of Congo. *African Security* 2:29–58.

El-Tom, Abdullahi Osman. 2011. *Darfur, JEM and the Khalil Ibrahim Story*. London: Red Sea Press.

REFERENCES

Endrew, Jürgen, 2000. Vom "Monopoly" privatisierter Gewalt zum Gewaltmonopol? Formen der Gewaltordnung im Libanon nach 1975. *Leviathan Zeitschrift für Sozialwissenschaft* 28, no. 2 (June).

Englebert, P., and D. Tull. 2008. Postconflict Reconstruction in Africa: Flawed Ideas about Failed States. *International Security* 32, no. 4: 106–39.

Eriksson Baaz, M., and M. Stern. 2010. *The Complexity of Violence: A Critical Analysis of Sexual Violence in the Democratic Republic of Congo (DRC).* Stockholm: Swedish International Development Cooperation Agency.

Eriksson Baaz, M., and J. Verweijen. 2013. The Volatility of a Half-Cooked Bouillabaisse: Rebel–Military Integration and Conflict Dynamics in Eastern DRC. *African Affairs*, July.

Esterhuyse, Abel. 2012. Associate professor of strategy, Faculty of Military Science, Stellenbosch University. Telephone interview by Roy Licklider, Saldanha, South Africa, June 14.

Farrell, Joseph, and Carl Shapiro. 1990. Horizontal Mergers: An Equilibrium Analysis. *American Economic Review* 80, no. 1: 107–26. www.jstor.org/stable/2006737.

Fearon, James D. 1998. Commitment Problems and the Spread of Ethnic Conflict. In *The International Spread of Ethnic Conflict: Fear, Diffusion, and Escalation*, edited by David A. Lake and Donald Rothchild. Princeton, NJ: Princeton University Press.

Fearon, James D., and David Laitin. 2003. Ethnicity, Insurgency, and Civil War. *American Political Science Review* 97, no. 1: 75–90.

———. 2004. Neotrusteeship and the Problem of Weak States. *International Security* 28, no. 4: 5–43.

Fegley, Randall. 2011. *Beyond Khartoum: A History of Subnational Government in Sudan.* Asmara, Eritrea, and Trenton, NJ: Red Sea Press.

Feld, Maury. 1977. *The Structure of Violence: Armed Forces as Social Systems.* Beverly Hills, CA: Sage.

Ferrer, Mirriam C. 1999. Integration of the MNLF Forces into the PNP and AFP: Integration without Demobilization and Disarmament. University of the Philippines Project on Assessment of the Implementation of the GRP-MNLF Peace Agreement, Phase I, University of the Philippines Center for Integrative and Development Studies.

Figuié, Gerard. 2000. *Le point sur le Liban 2000.* Paris: Maisonneuve & Larose.

Flower, K. 1987. *Serving Secretly: An Intelligence Chief on Record: Rhodesia into Zimbabwe, 1964 to 1981.* London: John Murray.

Fortna, Virginia Page. 2008. *Does Peacekeeping Work? Shaping Belligerents' Choices after Civil War.* Princeton, NJ: Princeton University Press.

Francis, David J. 2000. Torturous Path to Peace. The Lomé Accord and Postwar Peacebuilding in Sierra Leone. *Security Dialogue* 31, no. 3: 357–73.

Frankel, Philip. 2000. *Soldiers in a Storm: The Armed Forces in South Africa's Democratic Transition.* Boulder, CO: Westview Press.

Fyfield, J. A. 1982. *Re-Educating Chinese Anti-Communists.* New York: St. Martin's Press.

Gacis, Feliciano. 2010. Interview, March 18. Former member for the government of the Republic of the Philippines panel in negotiations with the MNLF. University of the Philippines–Diliman Asian Center, Quezon City.

Ga'le, Severino Fuli Boki Tombe. 2002. *Shaping a Free Southern Sudan: Memoirs of Our Struggle 1934–1985.* Limuru, Kenya: Loa Catholic Mission Council and Pauline Publications Africa.

Gartner, Scott Sigmund. 2011. Signs of Trouble: Regional Organization Mediation and Civil War Agreement Durability. *Journal of Politics* 73, no. 2: 380–90.

Gaub, Florence. 2007. Multiethnic Armies in the Aftermath of Civil War: Lessons Learned from Lebanon. *Defense Studies* 7, no. 1: 5–20.

———. 2011. *Military Integration after Civil Wars: Multiethnic Armies, Identity and Post-Conflict Reconstruction.* London: Routledge.

Gberie, Lansana. 2005. *A Dirty War in West Africa: The RUF and the Destruction of Sierra Leone.* London: Hurts & Company.

Gbla, Osman. 2006. Security Sector Reform under International Tutelage in Sierra Leone. *International Peacekeeping* 13, no. 1: 78–80.

Gear, Sasha. 2002. *Now That the War Is Over: Ex-Combatants, Transition and the Question of Violence: A Literature Review.* Violence and Transition Series. Johannesburg: Center for the Study of Violence and Reconciliation. www.csvr.org.za/wits/papers/papvtp8a.htm.

Gent, Stephen E. 2011. Relative Rebel Strength and Power Sharing in Intrastate Conflicts. *International Interactions* 27, no. 2: 215–28.

Gilligan M., E. Mvukiyehe, and C. Samii. 2013. Reintegrating Rebels into Civilian Life: Quasi-Experimental Evidence from Burundi. *Journal of Conflict Resolution* 57, no. 4.

Ginifer, Jeremy. 2006. The Challenges of Security Sector and Security Reform Processes in Democratic Transitions: The Case of Sierra Leone. *Democratization* 13, no. 5: 791–810.

Gladwell, Malcolm. 2008. *Outliers.* New York: Little, Brown.

Glassmyer, Katherine, and Nicholas Sambanis. 2008. Rebel Military Integration and Civil War Termination. *Journal of Peace Research* 45:365–84.

Goodwin, P., and J. Hancock. 1993. *Rhodesians Never Die: The Impact of War and Political Change on White Rhodesia, 1979–1980.* Oxford: Oxford University Press.

Gregorian, Raffi. 2005a. *Agreed Principles for the Way Ahead on Defence Reform.* Sarajevo: Defence Reform Commission Secretariat.

———. 2005b. Letter to Entity Ministers of Defence. Defence Reform Commission Secretariat, Sarajevo, June 7.

Grundy, Kenneth W. 1983. *Soldiers without Politics: Blacks in the South African Armed Forces.* Berkeley: University of California Press.

Gumampangan, Salib. 2009. Interview, March 4. Former Moro National Liberation Front commander and captain, Philippine Army. Camp Siongco, Parang, Maguindanao.

Gutierrez, Eric, and Marites Danguilan-Vitug. 1999. ARMM after the Peace Agreement: An Assessment of Local Government Capability in the ARMM. In *Rebels, Warlords and Ulama: A Reader in Philippines Separatism and the War in Southern Philippines.* Quezon City: Institute for Popular Democracy.

Habib, Rana, 2002. La vocation des militaires au sein de l'exception démocratique libanaise. Master's thesis, University Saint Joseph, Beirut.

REFERENCES

Hadji Ebrahim, Abdullawi. 2009. Interview, March 3. Former Moro National Liberation Front military commander. Cotabato City.

Hamel, G., and C. K. Prahalad. 1994. *Competing for the Future*. Cambridge, MA: Harvard Business School Press.

Hanson-Alp, Rosalind. 2010. Civil Society's Role in Sierra Leone's Security Sector Reform Process: Experiences from Conciliation Resources West Africa Programme." In *Security Sector Reform in Sierra Leone 1997–2007: Views from the Front Line*, edited by Paul Jackson and Peter Albrecht. Geneva Centre for the Democratic Control of Armed Forces. Berlin: LIT Verlag.

Hariri, Ahmad, 1990. L'Armée et le pouvoir politique au Liban. PhD diss., University Paris I Panthéon Sorbonne.

Hartzell, Caroline. 2009. Settling Civil Wars: Armed Opponents' Fates and the Duration of the Peace. *Conflict Management and Peace Science* 26, no. 4: 347–65.

———. 2012. Personal communication.

Hartzell, Caroline, and Matthew Hoddie. 2003. Institutionalizing Peace: Power Sharing and Post–Civil War Conflict Management. *American Journal of Political Science* 47, no. 2: 318–32.

———. 2007. *Crafting Peace: Power-Sharing Institutions and the Negotiated Settlement of Civil Wars*. University Park: Pennsylvania State University Press.

Haupt, Christian, and Jeff Fitzgerald. 2004. Negotiations on Defence Reform in Bosnia and Herzegovina. In *From Peace Making to Self Sustaining Peace: International Presence in South East Europe at a Crossroads?* edited by Predrag Jurekovic and Frederic Labarre. Report of the Eighth Workshop of the Study Group Regional Stability in South East Europe. Vienna: National Defence Academy.

Haupt, Christian, and Daniel Saracino. 2005. Defense Reform in Bosnia and Herzegovina. In *Defence Reform Initiative for Bosnia and Herzegovina/Serbia and Montenegro. The DRINA Project*, edited by Jos Boonstra. Groningen: Centre for European Security Studies.

Heinecken, Lindy. 2009. A Diverse Society, a Representative Military? The Complexity of Managing Diversity in the South African Armed Forces. *Scientia Militaria: South African Journal of Military Studies* 37, no. 1: 25–49.

———. 2012. Professor of sociology, Stellenbosch University. Interview by Roy Licklider, Stellenbosch, South Africa, June 11.

Heinecken, Lindy, and No ll Van Der Waag-Cowling. 2011. The Politics of Race and Gender in the South African Armed Forces: Issues, Challenges, Lessons. In *Defending Democracy and Securing Diversity*, edited by Christian Leuprecht. London: Routledge.

Heinecken, Lindy and Rialize Ferreira. 2012a. "Fighting for Peace": South Africa's Role in Peacekeeping Operations in Africa, Part I. *African Security Review* 21, no. 2 (June): 20–35.

———. 2012b. "Fighting for Peace": South Africa's Role in Peacekeeping Operations in Africa, Part II. *African Security Review* 21, no. 2 (June): 36–49.

———. 2012c. "Fighting for Peace": South Africa's Role in Peacekeeping Operations in Africa, Part III. *African Security Review* 21, no. 2 (June): 50–60.

Hendrickson, Dylan, and Andrzej Karkoszka. 2005. Security Sector Reform and Do-

nor Policies. In *Security Sector Reform and Post-Conflict Peacebuilding*, edited by Albrecht Schnabel and Hans-Georg Ehrhart. New York: United Nations Press.
Hendrix, Cullen S. 2010. Measuring State Capacity: Theoretical and Empirical Implications for the Study of Civil Conflict. *Journal of Peace Research* 47, no. 3: 273–85.
Herbst, Jeffrey. 2000. *States and Power in Africa: Comparative Lessons in Authority and Control.* Princeton, NJ: Princeton University Press.
Heston, Alan, Robert Summers, and Bettina Aten. May 2011. "Penn World Tables, Version 7.0." Center for International Comparisons of Production, Income, and Prices at University of Pennsylvania. http://pwt.econ.upenn.edu/php_site/pwt70/pwt70_form.php.
Higgs, James A. 2000. Creating the South African National Defence Force. *Joint Force Quarterly* 25 (Summer): 45–50.
Hironaka, Ann. 2005. *Neverending Wars: The International Community, Weak States, and the Perpetuation of Civil War.* Cambridge, MA: Harvard University Press.
Hirsch, John L. 2001. *Sierra Leone: Diamonds and the Struggle for Democracy.* Boulder, CO: Lynne Rienner.
Hoddie, Matthew, and Caroline Hartzell. 2003. Civil War Settlements and the Implementation of Military Power-Sharing Arrangements. *Journal of Peace Research* 40, no. 3 (May): 303–20.
———. 2005. Power Sharing in Peace Settlements: Initiating the Transition from Civil War. In *Sustainable Peace: Power and Democracy after Civil Wars*, edited by Philip G. Roeder and Donald Rothchild. Ithaca, NY: Cornell University Press.
Hoebeke, H., H. Boshoff, and K. Vlassenroot. 2008. *Assessing Security Sector Reform and Its Impact on the Kivu Provinces.* Pretoria: Institute for Security Studies.
Hoffman, B., J. Taw, and D. Arnold. 1991. *Lessons for Contemporary Counterinsurgencies: The Rhodesian Experience.* Arroyo, CA: RAND Corporation. www.rand.org/pubs/reports/R3998.htm.
Hoffmann, Stanley. 1968. *Gulliver's Troubles, or the Setting of American Foreign Policy.* New York: McGraw-Hill.
Horn, Adrian, Funmi Olonisakin, and Gordon Peake. 2006. "United Kingdom–Led Security Sector Reform in Sierra Leone. *Civil Wars* 8, no. 2: 109–23.
Horowitz, D. 1985. *Ethnic Groups in Conflict.* Berkeley: University of California Press.
Horvatich, Patricia. 2003. The Martyr and the Mayor: On the Politics of Identity in Southern Philippines. In *Cultural Citizenship in Island Southeast Asia: Nation and Belonging in the Hinterlands*, edited by Renato Rosaldo. Berkeley: University of California Press.
Howe, H. 2001. *Ambiguous Order: Military Forces in African States.* Boulder, CO: Lynne Rienner.
Hughes, Gordon. 2007. South Africa Defence Transformation: A Project Still in Progress. Keynotes Series, Center for Security Studies Management, Cranfield University, Shrivenham, UK (August).
Hull, Cecilia, and Emma Svensson. 2008. *African Union Mission in Somalia (AMISOM): Exemplifying African Union Peacekeeping Challenges.* Stockholm: Swedish Defense Research Agency.
L'Humanité. 1991. Désarmement des milices libanaises: Les Forces libanaises. April 30.

———. 1994. Beyrouth: Dissolution du parti des Forces Libanaises et information sous contrôle. March 24.
Human Rights Watch. 2006. *"We Flee When We See Them": Abuses with Impunity at the National Intelligence Service in Burundi.* New York: Human Rights Watch.
———. 2007. Burundi: Bring Muyinga Massacre Suspects to Trial. Press release, September 25.
———. 2008. Burundi: Muyinga Massacre Convictions a Victory. Press release, October 24.
———. 2009a. Diamonds in the Rough: Human Rights Abuses in the Marange Diamond Fields of Zimbabwe. www.hrw.org/reports/2009/06/26/diamonds-rough-0.
———. 2009b. *Pursuit of Power: Political Violence and Repression in Burundi.* New York: Human Rights Watch.
———. 2010a. *Closing Doors? The Narrowing of Political Space in Burundi.* New York: Human Rights Watch.
———. 2010b. *We'll Tie You Up and Shoot You: Lack of Accountability and Political Violence in Burundi.* New York: Human Rights Watch.
Humphreys, Macartan, and Jeremy M. Weinstein. 2007. Demobilization and Reintegration. *Journal of Conflict Resolution* 51:531–67.
Huntington, S. 1957. *The Soldier and the State: The Theory and Politics of Civil–Military Relationships.* Cambridge, MA: Belknap/Harvard University Press.
Hussin, Parouk. 2005. Excerpts from the presentation as ARMM governor. Meeting with the Committee of Eight. August 15. Indonesian Embassy, Makati, Philippines.
Iklé, Fred C. 1971. *Every War Must End.* New York: Columbia University Press.
International Crisis Group. 1999a. *The Agreement on a Cease-Fire in the Democratic Republic of Congo: An Analysis of the Agreement and Prospects for Peace.* Brussels: International Crisis Group.
———. 1999b. *How Kabila Lost His Way. The Performance of Laurent Désiré Kabila's Government.* Brussels: International Crisis Group.
———. 2000a. *The Mandela Effect: Prospects for Peace in Burundi.* Central Africa Report 13. Brussels: International Crisis Group.
———. 2000b. *Scramble for the Congo. Anatomy of an Ugly War.* Brussels: International Crisis Group.
———. 2001a. *Sierra Leone: Managing Uncertainty.* Africa Report 35. Freetown: International Crisis Group.
———. 2001b. *Time for a New Military and Political Strategy.* Africa Report 28. Freetown: International Crisis Group.
———. 2002. *Sierra Leone: Politics as Usual.* Africa Report 49. Freetown: International Crisis Group.
———. 2003. *Sierra Leone: The State of Security and Governance.* Africa Report 67. Freetown: International Crisis Group.
———. 2004. *Liberia and Sierra Leone: Rebuilding Failed States.* Africa Report 87. Dakar: International Crisis Group.
———. 2005. *The Congo's Transition Is Failing: Crisis in the Kivus.* Brussels: International Crisis Group.

———. 2006a. *Escaping the Conflict Trap. Promoting Good Governance in the Congo.* Brussels: International Crisis Group.

———. 2006b. *Security Sector Reform in the Congo.* Brussels: International Crisis Group.

———. 2007. *Sierra Leone: The Election Opportunity.* Africa Report 129. Freetown: International Crisis Group.

International Institute for Strategic Studies. 1975. *Military Balance 1974–1975.* London: International Institute for Strategic Studies.

———. 2010. *The Military Balance 2010.* London: Routledge.

Ismail, Aboobaker. 2012. South Africa's Experience of Military Integration through Interactive Negotiations and Planning. In *Post-War Security Transitions: Participatory Peacebuilding after Asymmetric Conflicts*, edited by Véronique Dudouet, Hans J. Giessmann, and Katrin Planta. London: Routledge.

Jabbour, Milia. 1989. L'armée libanaise: Entre le professionalisme et le destin national 1945–1989. Master's thesis, University Saint Joseph, Beirut.

Jacildo, Nerlyne C. 2003. Experiences of MNLF Integrees in Basilan and Zamboanga: Issues and Problems. Master's thesis, University of the Philippines–Diliman, Quezon City.

Jackson, Paul, and Peter Albrecht. 2010. *Reconstructing Security after Conflict: Security Sector Reform in Sierra Leone.* Basingstoke, UK: Palgrave Macmillan.

Jackson, S. 2006. The United Nations Operation in Burundi (ONUB): Political and Strategic Lessons Learned. Best Practices Unit, Department of Peacekeeping Operations, United Nations.

Jackson, Terence, and Elize Kotze. 2005. Management and Change in the South African National Defense Force: A Cross-Cultural Study. *Administration and Society* 37, no. 2: 168–98.

Jarstad, Anna K., and Desireé Nilsson. 2008. From Words to Deeds: The Implementation of Power-Sharing Pacts in Peace Accords. *Conflict Management and Peace Science* 25, no. 3: 206–23.

Jensen, M. 1986. Agency Costs of Free Cash Flow, Corporate Finance, and Takeovers. *American Economic Review* 76:323–29.

Jervis, David T. 2005. Miracle or Model? South Africa's Transition to Democracy. *International Third World Studies Journal and Review* 16:37–46.

Jervis, Robert. 1989. *The Logic of Images in International Relations.* New York: Columbia University Press. Orig. pub. 1970.

Jikiri, Yusuf. 2009. Interview, June 30. Former deputy commander of AFP Southern Command and highest-ranking MNLF military officer. House of Philippine Representatives, Quezon City.

Johnson, P. and Martin, D. 1986. *Zimbabwe: Apartheid's Dilemma, Destructive Engagement: Southern Africa at War.* Harare: Zimbabwe Publishing House.

Jones, Seth G., Jeremy M. Wilson, Andrew Rathmell, and K. Jack Riley. 2005. *Establishing Law and Order after Conflict.* Santa Monica, CA: RAND Corporation.

Jordaan, Evert. 2004. South African Defence since 1994: A Study of Policy-Making. Master's of military science thesis, University of Stellenbosch.

Jundam, Mashur Bin-Ghalib. 2009. Interview, February 21. Professor, Institute for Islamic Studies, University of the Philippines–Diliman, Quezon City.

Kandeh, Jimmy D. 2004. In Search of Legitimacy: The 1996 Elections. In *Between Democracy and Terror: The Sierra Leone Civil War,* edited by I. Abdullah. Dakar: CODESRIA.

———. 2008. Rogue Incumbents, Donor Assistance and Sierra Leone's Second Post-Conflict Elections of 2007. *Journal of Modern African Studies* 46, no. 4: 609–35.

Kanter, R. M. 1989. *When Giants Learn to Dance.* New York: Simon & Schuster.

Karame, Kari. 2009. Reintegration and the Relevance of Social Relations: The Case of Lebanon. *Conflict, Security and Development* 9, no. 4: 495–514.

Kaufmann, Chaim. 1996. Possible and Impossible Solutions to Ethnic Civil Wars. *International Security* 20, no. 1 (Spring): 136–75.

———. 1998. "When All Else Fails: Ethnic Population Transfers and Partitions in the Twentieth Century." *International Security* 23, no. 2 (Fall): 120–56.

———. 1999. "When All Else Fails: Evaluating Population Transfers and Partition as Solutions to Ethnic Conflict." In *Civil War, Insecurity, and Intervention,* edited by Jack Snyder and Barbara Walter. New York: Columbia University Press.

Kechichian, Joseph A. 1995. The Lebanese Army: Capabilities and Challenges in the 1980s. *Conflict Quarterly* 5, no. 1 (Winter).

Keen, D. 2005. *Conflict and Collusion in Sierra Leone*, Oxford: James Currey.

Khalid, Mansour. 2003. *War and Peace in Sudan: A Tale of Two Countries.* London: Keegan Paul.

Kibasomba, R. 2005. *Post-War Defence Integration in the Democratic Republic of the Congo.* Pretoria: Institute for Security Studies

King, Gary, Robert Keohane, and Sidney Verba. 1994. *Designing Social Inquiry: Scientific Inference in Qualitative Research.* Princeton, NJ: Princeton University Press.

Knight, Mark. 2009. Security Sector Reform: Post-Conflict Integration. Global Facilitation Network for Security Sector Reform, University of Birmingham, UK. www.ssrnetwork.net/publications/postconfl.php.

Knight, Mark, and Alpaslan Ozerdem. 2004. Guns, Camps and Cash: Disarmament, Demobilization and Reinsertion of Former Combatants in Transitions from War to Peace. *Journal of Peace Research* 41:499–516.

Krebs, Ronald R. 2004. A School for the Nation? How Military Service Does Not Build Nations, and How It Might. *International Security* 28, no. 4 (Spring): 85–124.

———. 2005. One Nation under Arms? Military Participation Policy and the Politics of Identity." *Security Studies* 14, no. 3 (July–September): 529–64.

———. 2006. *Fighting for Rights: Military Service and the Politics of Citizenship.* Ithaca, NY: Cornell University Press.

Kreps, D. 1990. Corporate Culture and Economic Theory. In *Perspectives on Positive Political Economy*, edited by James E. Alt and Kenneth A. Shepsle. Cambridge: Cambridge University Press.

Kriger, N. 2003. *Guerrilla Veterans in Post-War Zimbabwe: Symbolic and Violent Politics, 1980–1987.* Cambridge: Cambridge University Press.

Kuranga, David. 2011a. *Merging Militaries: Alexander the Great.* New Brunswick, NJ: Political Science Department of Rutgers University.

———. 2011b. *Merging Militaries: The British Empire.* New Brunswick, NJ: Political Science Department of Rutgers University.
Lacina, B. 2011. The Politics of Internal Security: Evidence from Language Violence in India, 1950–1989. PhD diss., Department of Political Science, Stanford University.
Lagu, Joseph. 2006. *Sudan: Odyssey through a State, from Ruin to Hope.* Omdurman, Sudan: MOB Center for Sudanese Studies at Omdurman Ahlia University.
Landasan, Arsad. 2009. Interview, April 6. Former Moro National Liberation Front combatant. General Santos City.
Lanotte, O. 2003. *République Démocratique du Congo : Guerre sans frontières.* Brussels: GRIP.
LeGrys, Barry J. 2010. British Military Involvement in Sierra Leone, 2001–2006. In *Security Sector Reform in Sierra Leone 1997–2007: Views from the Front Line*, edited by Paul Jackson and Peter Albrecht. Geneva Centre for the Democratic Control of Armed Forces. Berlin: LIT Verlag.
Lebanon. Army. 2003. Independence: Sixty Years Responsible. November 22. www.lebanesearmy.gov.lb/article.asp?ln=en&id=1465.
———. 2004. 59th Anniversary of the Lebanese Army. August 1. www.lebarmy.gov.lb/article.asp?ln=en&id=4888.
Lectuer, Martin Rupiah. 1995. Demobilisation and Integration: "Operation Merger" and the Zimbabwe National Defence Forces, 1980–1987. *African Security Review* 4, no. 3.
Lemarchand, R. 1970. *Rwanda and Burundi.* London: Pall Mall Press.
———. 1996. *Burundi: Ethnic Conflict and Genocide.* Cambridge: Cambridge University Press.
LeRoux, Len. 2005. The Post-Apartheid South African Military: Transforming with the Nation. In *Evolutions and Revolutions: A Contemporary History of Militaries in Southern Africa*, edited by Martin Rupiya. Pretoria: Institute for Security Studies.
———. 2007. The Revision of the South African Defence Review and International Trends in Force Design: Implications for the SA Army. In *South African Army Vision 2020: Security Challenges Shaping the Future South African Army*, edited by Len LeRoux. Pretoria: Institute for Security Studies.
Libération. 1995. Au Liban, les anciens miliciens se reconvertissent dans l'armée. April 14.
Licklider, Roy. 1995. The Consequences of Negotiated Settlements in Civil Wars, 1945–1993. *American Political Science Review* 89, no. 3: 681–90.
———. 2001. Obstacles to Peace Settlements. In *Turbulent Peace: The Challenges of Managing International Conflict*, edited by Chester A. Crocker, Fen Osler Hampson, and Pamela Aall. Washington, DC: US Institute of Peace Press.
———. 2008. Ethical Advice: Conflict Management vs. Human Rights in Ending Civil Wars. *Journal of Human Rights* 7:376–87.
Licklider, Roy, and Mia Bloom. 2006. *Living Together after Ethnic Killing: Exploring the Chaim Kaufmann Argument.* London: Routledge.
Lidasan, Nasser P. 2006. The Integration of MNLF and MILF Combatants into the Philippine Armed Forces: Implications to Conflict Resolution. Master's thesis, University of Bradford, Bradford, UK.

Liebenberg, Ian. 1997. The Integration of the Military in Post-Liberation South Africa: The Contribution of Revolutionary Armies. *Armed Forces and Society* 24, no. 1: 105–32.

Lingga, Abhoud Syed. 2009. Interview, March 6. Executive director, Institute of Moro Studies. Cotabato City.

Lodge, Tom. 1996. Soldiers of the Storm: A Profile of the Azanian People's Liberation Army. In *About Turn: The Transformation of South African Military and Intelligence*, edited by Jakkie Cilliers and Markus Reichardt. Pretoria: Institute for Defence Policy.

Loft, Francis. 1988. Background to the Massacres in Burundi. *Review of African Political Economy* 15, no. 43:88–93.

Lucero, Ricardo. 2008. Interview, August 7. Head of Force Integration Branch, OJ3, Armed Forces of the Philippines, AFP general headquarters, Camp Aguinaldo, Quezon City.

Luckham, Robin. 2003. Democratic Strategies for Security in Transition and Conflict. In *Governing Insecurity: Democratic Control of Military and Security Establishments in Transitional Democracies*, edited by Gavin Cawthra and Robin Luckham. London: Zed Books.

Madut-Arop, Arop. 2006. *Sudan's Painful Road to Peace: A Full Story of the Founding and Development of SPLM/SPLA*. Charleston, SC: BookSurge.

Maharaj, Mac. 2008. The ANC and South Africa's Negotiated Transition to Democracy and Peace. Berghof Transitions Series 2. Berlin: Berghof Research Center for Constructive Conflict Management.

Makinano, Merliza M., and Alfredo Lubang. 2000. Disarmament, Demobilisation and Reintegration: The Mindanao Experience. In *South Asia at Gun Point: Small Arms & Light Weapons Proliferation*, edited by Dipankar Banerjee. Colombo: Regional Centre for Strategic Studies.

Malan, Mark. 2003. Security and Military Reform. In *Sierra Leone: Building the Road to Recovery*, edited by M. Malan, S. Meek, T. Thusi, J. Ginifer, and P. Coker. ISS Monograph 80. Pretoria: Institute for Security Studies.

Malia, Joseph, 1992. *The Document of National Understanding*. Oxford: Centre for Lebanese Studies.

Mansour, Albert. 1993. *The Coup against Ta'if*. Beirut (in Arabic).

Marshall, Monte G., and Ted Robert Gurr. 2005. *Peace and Conflict, 2005: Global Survey of Armed Conflicts, Self-Determination Movements, and Democracy*. College Park, MD: Center for International Development and Conflict Management.

Mashike, Lephophotho. 2007. "Blacks Can Win Everything but the Army": The "Transformation" of the South African Military between 1994 and 2004. *Journal of Southern African Studies* 33, no. 3 (September): 601–18.

———. 2008. Age of Despair: The Unintegrated Forces of South Africa. *African Affairs* 107:433–53.

Maxwell, A. 2005. Canadian Forces Personnel Practices and Regimental System. Defence Reform Commission Secretariat, Sarajevo, memo dated May 6.

Mbeki, Moeletsi. 2000. Preface. In *South Africa and Naval Power at the Millennium*, edited by Martin Edwards and Gregg Mills. Braamfontein, South Africa: South African Institute of International Affairs.

McKenna, Thomas. 1998. *Muslim Rulers and Rebels: Everyday Politics and Armed Separatism in Southern Philippines*. Quezon City: Anvil.

McLaurin, Ronald D. 1984. Lebanon and Its Army: Past, Present and Future. In *The Emergence of a New Lebanon: Fantasy or Reality*, edited by Edward Azar. New York: Praeger.

———. 1991. From Professional to Political: The Redecline of the Lebanese Army. *Armed Forces and Society*, Summer.

Melmot, S., 2008. Candide au Congo: L'échec annoncé de la réforme du secteur de sécurité (RSS). Paris: L'Institut français des relations internationales, Laboratoire de Recherche sur la Défense.

Mgbako, Chi. 2005. *Ingando* Solidarity Camps: Reconciliation and Political Indoctrination in Post-Conflict Rwanda. *Harvard Human Rights Journal* 18:201–24.

Mills, Greg. 2008. The Boot Is Now on the Other Foot: Lessons from Both Sides of Rwanda's Insurgency. *RUSI Journal: Royal United Services Institute for Defence Studies* 153, no. 3 (June): 72–78.

Mills, Greg, and Geoffrey Wood. 1993. Ethnicity, Integration and the South African Armed Forces. *South African Defence Review* 12. www.issafrica.org/Pubs/ASR/SADR12/Mills.html.

Modelski, George. 1964. International Settlement of Internal War. In *International Aspects of Civil Strife*, edited by James N. Rosenau. Princeton, NJ: Princeton University Press.

Le Monde. 1984. Le plan de restructuration de l'armée a été accepté: Les différentes milices devraient être légalisées et prendre en charge la sécurité, chacune dans les régions qu'elles contrôlent. June 26.

Moore, David. 1992. Zimbabwean Peasants: Pissed On and Pissed Off (Review of Kriger). *Southern African Review of Books* 4:5–6.

Moss, T., G. Pettersson, and N. van de Walle. 2006. *An Aid-Institutions Paradox? A Review Essay on Aid Dependency and State Building in Sub-Saharan Africa*. Washington, DC: Center for Global Development.

Motumi, Tsepe. 1996. The Spear of the Nation: The Recent History of Umkhonto We Sizwe (MK). In *About Turn: The Transformation of South African Military and Intelligence*, edited by Jakkie Cilliers and Markus Reichardt. Pretoria: Institute for Defence Policy.

Motumi, Tsepe, and Penny Mckenzie. 1998. After the War: Demobilisation in South Africa." In *From Defence to Development: Redirecting Military Resources in South Africa*, edited by Jacklyn Cock and Penny Mckenzie. Ottawa: International Development Research Center.

Mouawad, Jad. 2011. Details, Details: Long and Complex Path to Combine Delta and Northwest. *New York Times*, May 19.

Muana, Patrick K. 1997. The Kamajoi Militia: Civil War, Internal Displacement and the Politics of Counter-Insurgency. *Africa Development* 22, nos. 3–4: 77–100.

Mutengesa, Sabiiti. 2013. Facile Acronyms and Tangled Processes: A Re-Examination of the 1990s 'DDR' in Uganda. *International Peacekeeping* 20, no. 3: 338–56.

N'Gbanda Zambo, H. 1998. *Ainsi sonne le glas : Les derniers jours du Maréchal Mobutu*. Paris: Éditions Gideppe.

Ndlovu-Gateshi, Sabelo. 2009. Making Sense of Mugabeism in Local and Global

Politics: "So Blair, Keep Your England and Let Me Keep My Zimbabwe." *Third World Quarterly* 30, no. 6.
Nelson-Williams, Alfred. 2010. Restructuring the Republic of Sierra Leone Armed Forces. In *Security Sector Reform in Sierra Leone 1997–2007: Views from the Front Line*, edited by Paul Jackson and Peter Albrecht. Geneva Centre for the Democratic Control of Armed Forces. Berlin: LIT Verlag.
Nerguizian, Aram, 2009. *The Lebanese Armed Forces: Challenges and Opportunities in Post-Syria Lebanon.* Washington, DC: Center for Strategic and International Studies.
Net Press. 2006. Burundi Army Private Kills Superior in Northwest Before Killing Self. March 16.
Ngaruko, F., and J. Nkurunziza. 2000. An Economic Interpretation of Conflict in Burundi. *Journal of African Economies* 9:370–409.
Nilsson, Anders R. 2008. Dangerous Liaisons: Why Ex-Combatants Return to Violence. Cases from the Republic of Congo and Sierra Leone. PhD diss., Department of Peace and Conflict Research, Uppsala University.
Nindorera, W. 2007. *Security Sector Reform in Burundi: Issues and Challenges for Improving Civilian Protection.* Working Paper. Bujumbura and Ottawa: Centre d'Alerte et de Prévention des Conflits and North-South Institute.
North Atlantic Treaty Organization, Defence Planning and Operations Division, Crisis Management and Operations Directorate. 2001. *BiH and PfP*. Fax to Organization for Security and Cooperation in Europe, Mission to Bosnia and Herzegovina, dated October 19. Brussels.
Nyambuya, Major General Michael. 1996. National Defence: The Experience of the Zimbabwe Defence Force, Zimbabwe Defence Force. Paper presented at Angolan National Defence Symposium, Ministry of National Defence, Luanda, March 27.
Nyathi, P. T. 2004. Reintegration of Ex-Combatants into Zimbabwean Society: A Lost Opportunity. In *Zimbabwe: Injustice & Political Reconciliation*, edited by Brian Raftopoulos and Tyrone Savage. Cape Town and Harare: Institute for Justice and Reconciliation and Weaver Press.
Obotela Rashidi, N. 2004. L'an 1 de l'accord global et inclusif en République démocratique du Congo: De la laborieuse mise en place aux incessants atermoiements. In *L'Afrique des Grands Lacs: Annuaire 2003–2004*, edited by S. Marysse and F. Reyntjens. Paris: L'Harmattan.
OHR (Office of the High Representative). 2003. Decision Establishing the Defence Reform Commission. Sarajevo, May 9. www.ohr.int/decisions/statemattersdec/default.asp?content_id=29840.
———. 2004. Decision Extending the Mandate of the Defence Reform Commission. Sarajevo, December 31. www.ohr.int/decisions/statemattersdec/default.asp?content_id=33873.
Onana, R., and H. Taylor. 2008. MONUC and SSR in the Democratic Republic of Congo. *International Peacekeeping* 15, no. 4: 501–16.
Orth, Rick. 2001. Rwanda's Hutu Extremist Genocidal Insurgency: An Eyewitness Perspective. *Small Wars and Insurgencies* 12, no. 1.
Packenham, Robert A. 1973. *Liberal America and the Third World: Political Develop-*

ment Ideas in Foreign Aid and Social Science. Princeton, NJ: Princeton University Press.
Pandian, Norman. 2009. Interview, March 5. Former Moro National Liberation Front combatant and corporal, Philippine Army. Camp Siongco, Parang, Maguindanao.
Paris, Ronald. 2004. *At War's End: Building Peace after Civil Conflict.* New York: Cambridge University Press.
Peled, Alon. 1998. *A Question of Loyalty: Military Manpower Policy In Multiethnic States.* Ithaca, NY: Cornell University Press.
Perlmutter, Amos, and Valerie Plave Bennett. 1980. *The Political Influence of the Military: A Comparative Reader.* New Haven, CT: Yale University Press.
Peters, Krijn. 2007. Reintegration Support for Young Ex-Combatants: A Right or a Privilege? *International Migration* 45, no. 5: 35–59.
Philippines, Office of the President, Office of the Adviser to the President on the Peace Process. 2007. OPAPP Annual Report: Update on the MNLF Integration Program (as of 26 September 2006).
Picard, Elizabeth. 1999. *The Demobilization of the Lebanese Militias.* Oxford: Centre for Lebanese Studies.
Pillar, Paul. 1983. *Negotiating Peace: War Termination as a Bargaining Process.* Princeton, NJ: Princeton University Press.
Porter, Jack C. 2010. *The Construction of Liberal Democracy: The Role of Civil–Military Institutions in State and Nation-Building in West Germany and South Africa.* Carlisle, PA: Strategic Studies Institute, US Army War College.
Powell, K. 2007. *Security Sector Reform and the Protection of Civilians in Burundi: Accomplishments, Dilemmas and Ideas for International Engagement.* Bujumbura and Ottawa: Centre d'Alerte et de Prévention des Conflits and North-South Institute.
Prunier, Gerard. 1998. The Rwandan Patriotic Front. In *African Guerillas*, edited by Christopher Clapham. Bloomington: Indiana University Press.
Quinlivan, James T. 1995. Force Requirements in Stability Operations. *Parameters*, Winter.
———. 2003. Burden of Victory: The Painful Arithmetic of Stability Operations. *RAND Review*, 27, no. 2 (Summer).
Raeymaekers, Timothy. 2007. The Power of Protection. Governance and Transborder Trade on the Congo-Ugandan Frontier. PhD diss., Ghent University.
Rashid, Ismail. 2000. The Lomé Peace Negotiations. In *Paying the Price: The Sierra Leone Peace Process*, edited by D. Lord. Report on Accord 9. London: Conciliation Resources.
Rasul, Amina. 2005. *Broken Peace? Assessing the 1996 GRP-MNLF Final Peace Agreement.* Manila: Philippine Council for Islam and Democracy.
Regan, Patrick, Richard W Frank, and Aysegul Aydin. 2009. Diplomatic Interventions and Civil War: A New Dataset. *Journal of Peace Research* 46, no. 1: 135–46.
Rehder, Robert Beeland, Jr.. 2008. From Guerillas to Peacekeepers: The Evolution of the Rwandan Defense Forces. Paper written for US Marine Corps Command and Staff College, Marine Corps University, April 15.
Reichardt, Markus, and Jakkie Cilliers. 1996. The History of the Homeland Armies. In *About Turn: The Transformation of South African Military and Intelligence,*

edited by Jakkie Cilliers and Markus Reichardt. Pretoria: Institute for Defence Policy.

Reiter, Dan. 2003. Exploring the Bargaining Model of War. *Perspectives on Politics* 1, no. 1: 27–43.

Republic Act 9054 (Organic Act for the Autonomous Region in Muslim Mindanao). 2000.

Reuters. 2006. Burundi's Army Denies Existence of Coup Plot. December 17.

La Revue du Liban. 1992. No. 1667, January 18–25.

Reyntjens, F. 1993. The Proof of the Pudding Is in the Eating: The June 1993 Elections in Burundi. *Journal of Modern African Studies* 31, no. 4: 563–83.

———. 2009. *The Great African War: Congo and Regional Geopolitics, 1996–2006.* Cambridge: Cambridge University Press.

Richards, Paul. 1998. *Fighting for the Rain Forest: War, Youth and Resources in Sierra Leone*, 2nd ed. London: James Currey.

Rodil, B. R. 2000. *Kalinaw Mindanao: The Story of GRP-MNLF Peace Process 1975–1996.* Davao City: Alternate Forum for Research in Mindanao.

Rondé, André. 1998. L'Armée libanaise et la restauration de l'état de droit. *Revue Droit et Defense* 2.

Rothchild, Donald. 2005. Reassuring Weaker Parties after Civil Wars: The Benefits and Costs of Executive Power-Sharing Systems in Africa. *Ethnopolitics* 4, no. 3: 247–67.

Rothchild, Donald, and Philip G. Roeder. 2005. Power Sharing as an Impediment to Peace and Democracy. In *Sustainable Peace: Power and Democracy after Civil Wars*, edited by Philip G. Roeder and Donald Rothchild. Ithaca, NY: Cornell University Press.

Rufer, Reto. 2005. Disarmament, Demobilization and Reintegration (DDR): Conceptual Approaches, Specific Settings, Practical Experiences. Working Paper, Geneva Center for Democratic Control of the Armed Forces. www.dcaf.ch/content/download/35355/525927/.../RUFER_final.pdf.

Ruhunga, Sam. 2006. Military Integration as a Factor for Post-Conflict Stability and Reconciliation: Rwanda, 1994–2005. MA thesis, Naval Postgraduate School, Monterey, CA.

Rusagara, Frank K. 2008. Rwanda: Military Integration Key to Peace-Building and Democratic Governance. *New Times* (Kigali), October 23.

Ryle, John, Justin Willis, Suliman Baldo, and Jok Madut Jok, eds. 2011. *The Sudan Handbook*. London: Rift Valley Institute and James Currey.

Sali, Rabirul. 2009. Interview, March 4. Former Moro National Liberation Front commander and captain, Philippine Army. Camp Siongco, Parang, Maguindanao.

Salim, Uttuh. 2009. Interview, April 6. Moro National Liberation Front Executive Council member. Kiamba, Sarangani Province.

Salomons, Dirk. 2005. Security: An Absolute Prerequisite. In *Postconflict Development: Meeting New Challenges*, edited by Gerd Junne and Willemijn Verkoren. Boulder, CO: Lynne Rienner.

Sambanis, Nicholas. 2004. What Is Civil War? Conceptual and Operational Complexities of an Operational Definition. *Journal of Conflict Resolution* 48, no. 6 (December): 814–58.

Samii, C. 2013. Perils or Promise of Ethnic Integration? Evidence from a Hard Case in Burundi. *American Political Science Review* 107, no. 3.

Samuels, Kristi. 2009. Postwar Constitution Building: Opportunities and Challenges. In *The Dilemmas of Statebuilding: Confronting the Contradictions of Postwar Peace Operations*, edited by Roland Paris and Timothy D. Sisk. London: Routledge.

San Juan, Epifanio, Jr. 2007. *US Imperialism and Revolution in the Philippines.* New York: Palgrave Macmillan.

Santos, Soliman, Jr. 2009. Primed and Purposeful: Armed Groups and Human Security Efforts in the Philippines. Geneva and Manila: Small Arms Survey and South-South Network on Non-State Armed Group Engagement.

Sass, Bill. 1996. The Union and South African Defence Force, 1912 to 1994. In *About Turn: The Transformation of South African Military and Intelligence*, edited by Jakkie Cilliers and Markus Reichardt. Pretoria: Institute for Defence Policy.

Sayyed, J. 1997. *Integration in the Lebanese Army: Shortlived Experiment or Longlived Policy?* (in Arabic). Unpublished study for Lebanese officers, Beirut.

Schatzberg, M. 1988. *The Dialectics of Oppression in Zaire.* Bloomington: Indiana University Press.

Scheffer, Jaap de Hoop. 2004. North Atlantic Treaty Organization, Secretary General. Letter to Presidency of Bosnia and Herzegovina, December 16.

Schoeman, Maxi. 2007. South Africa. In *Security and Democracy in Southern Africa*, edited by Gavin Cawthra, Andre du Pisani, and Abillah Omari. Johannesburg: Wits University Press.

Seegers, Annette. 1995. The Security Forces and the Transition in South Africa: 1986–1994. Africa Seminar, Centre for African Studies, University of Cape Town.

———. 1996. *The Military in the Making of Modern South Africa.* London: Tauris Academic Studies.

———. 2012. Professor of Political Studies, University of Cape Town. Interview by Roy Licklider, Cape Town, June 12.

Sema, Muslimen. 2009. Interview, March 4. Moro National Liberation Front Executive Council member. Cotabato City.

Shaw, Carolyn, Julius O. Ihonvbere, Kenneth R. Gray, and Bill Dickens. 1996. Hegemonic Participation in Peace-Keeping Operations: The Case of Nigeria and ECOMOG. *International Journal on World Peace* 13, no. 2.

Shaw, Mark. 1994. Biting the Bullet: Negotiating Democracy's Defence. *South African Review* 7:228–56.

———. 1996. Negotiating Defence for a New South Africa. In *About Turn: The Transformation of South African Military and Intelligence*, edited by Jakkie Cilliers and Markus Reichardt. Pretoria: Institute for Defence Policy.

Shleifer, A., and L. Summers. 1988. Breach of Trust on Hostile Takeovers. In *Corporate Takeovers*, edited by A. Auerbach. Chicago: University of Chicago Press.

Shwere, Masiiwa. 2010. Gukurahundi: Moving towards Objective Analysis. Manuscript. https://sites.google.com/site/masiiwashwere/gukurahundi:movingtowards objectiveanalys.

Sierra Leone. 1999. Peace Agreement between the Government of Sierra Leone and the Revolutionary United Front of Sierra Leone. www.sierra-leone.org/lomeac cord.html.

REFERENCES

Sigaud, Dominique, 1988. L'Armée libanaise: Eclatement ou destin national? *Cahiers de l'Orient*, November.

Simonsen, Sven Gunnar. 2007. Building "National" Armies—Building Nations? Determinants of Success for Post-Intervention Integration Efforts. *Armed Forces and Society* 33, no. 4: 571–90.

SIPRI (Stockholm International Peace Research Institute). 2010. *The SIPRI Military Expenditure Database*. Available at http://milexdata.sipri.org/.

Small, Melvin, and J. David Singer. 1982. *Resort to Arms: International and Civil Wars, 1816–1980*. Beverly Hills, CA: Sage.

Smith-Höhn, Judy. 2010. *Rebuilding the Security Sector in Post-Conflict Societies: Perceptions from Urban Liberia and Sierra Leone*. Geneva Centre for the Democratic Control of Armed Forces. Berlin: LIT Verlag.

Söderberg Kovacs, Mimmi. 2007. From Rebellion to Politics: The Transformation of Rebel Groups to Political Parties in Civil War Peace Processes. PhD diss., Department of Peace and Conflict Research, Uppsala University.

South Africa, Parliamentary Integration Oversight Committee. 1995. BMATT Report on Bridging Training. Appendix C to the South African National Defence Force Progress Report, June 12.

Southall, R. 2006. A Long Prelude to Peace: South African Involvement in Ending Burundi's War. In *South Africa's Role in Conflict Resolution and Peacemaking in Africa*, edited by R. Southall. Pretoria: Human Sciences Resource Council Press.

Spear, Joanna. 2006. Disarmament, Demobilization, Reinsertion and Reintegration in Africa. In *Ending Africa's Wars: Progressing to Peace*, edited by Oliver Furley and Roy May. Burlington, VT: Ashgate.

Staw, B., L. Sandelands, and J. Dutton. 1981. Threat Rigidity Effects in Organizational Behaviour: A Multilevel Analysis. *Administrative Science Quarterly* 20:345–54.

Stearns, J. 2008. Laurent Nkunda and the National Congress for the Defence of the People. In *L'Afrique des Grands Lacs: Annuaire 2003–2004*, edited by S. Marysse and F. Reyntjens. Paris : L'Harmattan.

Stedman, Stephen John. 1993. The End of the Zimbabwean Civil War. In *Stopping the Killing: How Civil Wars End*, edited by R. Licklider. New York: New York University Press.

———. 2002. Policy Implications. In *Ending Civil Wars: The Implementation of Peace Agreements*, edited by Stephen John Stedman, Donald Rothchild, and Elizabeth M. Cousens. Boulder, CO: Lynne Rienner.

Svensson, Isak. 2009. Who Brings Which Peace? Neutral versus Biased Mediation and Institutional Peace Arrangements in Civil Wars. *Journal of Conflict Resolution* 53, no. 3: 446–69.

Takiedine, Riad. 1995. *The Army Revived, 1988–1994*.

Tan, Samuel. 1993. *Internationalization of the Bangsamoro Struggle*. Quezon City: University of the Philippines Press.

Thusi, Thokozani, and Sarah Meek. 2003. Disarmament and Demobilisation. In *Sierra Leone: Building the Road to Recovery*, edited by M. Malan, S. Meek, T. Thusi, J. Ginifer, and P. Coker. ISS Monograph 80. Pretoria: Institute for Security Studies.

Tilly, Charles. 1975. Reflections on the History of European State-Making. In *The Formation of National States in Western Europe*, edited by Charles Tilly. Princeton, NJ: Princeton University Press.
Toft, Monica Duffy. 2010. *Securing the Peace: The Durable Settlement of Civil Wars*. Princeton, NJ: Princeton University Press.
Trefon, T. 2009a. Introduction: Réforme et désillusions. In *Réforme au Congo (RDC): Attentes et désillusions*, edited by T. Trefon. Paris: L'Harmattan.
———. 2009b. Public Service Provision in a Failed State: Looking beyond Predation in the Democratic Republic of Congo. *Review of African Political Economy* 36, no. 119: 9–21.
Tull, D. 2005. *The Reconfiguration of Political Order in Africa: A Case Study of North Kivu (DR Congo)*. Hamburg: Institut für Afrika-Kunde.
Tull, D., and A. Mehler. 2005. The Hidden Costs of Power-Sharing: Reproducing Insurgent Violence in Africa. *African Affairs* 104, no. 416: 375–98.
UN Group of Experts on the DRC. 2005a. *Report of the Group of Experts on the Democratic Republic of the Congo*. S/2005/30. New York: United Nations Security Council.
———. 2005b. *Report of the Group of Experts on the Democratic Republic of the Congo*. S/2005/436. New York: United Nations Security Council.
———. 2009. *Report of the Group of Experts on the Democratic Republic of the Congo*. S/2009/603. New York: United Nations Security Council.
———. 2010. *Report of the Group of Experts on the Democratic Republic of the Congo*. S/2010/596. New York: United Nations Security Council.
United Nations. 1996a. *Final Report of the Secretary-General on the United Nations Operation in Burundi*. New York: United Nations.
———. 1996b. *The United Nations and Rwanda, 1993–1996*, New York: United Nations.
———. 2004. Second Report of the Secretary-General on the United Nations Operation in Burundi. New York: United Nations.
United Nations, Office for the Coordination of Humanitarian Affairs, Integrated Regional Information Networks. 2000. Africa: IRIN Interview with UN-Secretary General Kofi Annan. February 1.
———. 2005. Burundi: Government Resumes Military Cooperation with Belgium. March 30.
———. 2008. Zimbabwe: Operation Glossary: A Guide to Zimbabwe's Internal Campaigns. Johannesburg: United Nations. www.irinnews.org/report/78003/zimbabwe-operation-glossary-a-guide-to-zimbabwe-s-internal-campaigns.
United Nations, Security Council. 1993. *Report of the Secretary-General on Rwanda*. S/26488. New York: United Nations.
US Institute of Peace. 1992a. *General Peace Agreement for Mozambique*. Peace Agreements Digital Collection: Mozambique. Washington, DC: US Institute of Peace.
———. 1992b. *Joint Declaration*. Peace Agreements Digital Collection: Mozambique. Washington, DC: United States Institute of Peace.
Vaccaro, Matthew. 1996. The Politics of Genocide: Peacekeeping and Rwanda. In *UN Peacekeeping and the Uncivil Wars of the 1990s*, edited by William J. Durch. New York, St. Martin's Press.

REFERENCES

Vankovska, Biljana, and Håkan Wiberg. 2003. *Between Past and Future: Civil–Military Relations in Post-Communist Balkan States.* London: I. B. Tauris.

Verweijen, Judith. 2013. The Ambiguity of Militarization. The Complex Interaction between the Congolese Armed Forces and Civilians in the Kivu Provinces, Eastern DR Congo. PhD diss., Utrecht University.

Vlassenroot, K. 2003. Violence et Constitution de Milices dans L'est du Congo: Le Cas des Mai-Mayi. In *L'Afrique des Grands Lacs: Annuaire 2003–2004*, edited by S. Marysse and F. Reyntjens. Paris: L'Harmattan.

———. 2004. Reading the Congolese Crisis. In *Conflict and Social Transformation in Eastern DR Congo*, edited by K. Vlassenroot and T. Raeymakers. Gent: Academia Press Scientific Publishers.

Vlassenroot, K., and T. Raeymakers. 2004. Introduction. In *Conflict and Social Transformation in Eastern DR Congo*, edited by K. Vlassenroot and T. Raeymakers. Gent: Academia Press Scientific Publishers.

Vlassenroot, K., and F. Van Acker. 2001. War as Exit from Exclusion? The Formation of Mayi-Mayi Militias in Eastern Congo. *Afrika Focus* 17, nos. 1–2: 51–78.

Vreeland, James Raymond. 2003. *The IMF and Economic Development.* Cambridge: Cambridge University Press.

Wadi, Julkipli. 2009. Interview, February 21. Professor, Institute for Islamic Studies, University of the Philippines–Diliman, Quezon City.

Walter, Barbara F. 2002. *Committing to Peace: The Successful Settlement of Civil Wars.* Princeton, NJ: Princeton University Press.

Weber, Max. 1946. The Meaning of Discipline. In *From Max Weber: Essays in Sociology*, edited by H. H. Gerth and C. Wright Mills. New York: Oxford University Press.

Werner, Suzanne. 1998. Negotiating the Terms of Settlement: War Aims and Bargaining Leverage. *Journal of Conflict Resolution* 42, no. 3: 321–43.

———. 1999. The Precarious Nature of Peace: Resolving the Issues, Enforcing the Settlement, and Renegotiating the Terms. *American Journal of Political Science* 43, no. 3: 912–34.

Werner, Suzanne, and Amy Yuen. 2005. Making and Keeping Peace. *International Organization* 59:261–92.

Willame, J.-C. 1997. *Banyrwanda et Banyamulenge : Violences ethniques* et *gestion de l'identitaire au Kivu.* Paris : L'Harmattan.

———. 1999. *L'Odyssée Kabila: Trajectoire pour un Congo nouveau?* Paris: Karthala.

———. 2007. *Les "faiseurs de paix au Congo" : Gestion d'une crise internationale dans un Etat sous tutelle.* Brussels : GRIP.

Williams, Rocky. 1992. Historical Parallel or Historical Amnesia? The Formation of the Union Defence Force. *South African Defence Review* 2.

———. 1994. *The Midnight Ride: Defence Restructuring and Lessons from the Erasmus Purges.* Occasional Paper 60. Pretoria: Africa Institute of South Africa.

———. 2002. Integration or Absorption? The Creation of the South African National Defence Force, 1993 to 1999. *African Security Review* 11, no. 2.

———. 2003. Defence in a Democracy: The South African Defence Review and the Redefinition of the Parameters of the National Defence Debate. In *Ourselves to Know: Civil–Military Relations and Defence Transformation in Southern Africa*,

edited by Rocky Williams, Gavin Cawthra, and Diane Abrahams. Pretoria: Institute for Security Studies.
———. 2005. Demobilization and Reintegration: The South African Experience. *Journal of Security Sector Management* 3, no. 2 (March).
———. 2006. *South African Guerilla Armies: The Impact of Guerilla Armies on the Creation of South Africa's Armed Forces.* Monograph 27. Pretoria: Institute of Security Studies.
Williamson, O. E. 1975. *Markets and Hierarchies: Analysis and Antitrust Implications—A Study in the Economics of Internal Organizations.* New York: Free Press.
Winkates, James. 2000. The Transformation of the South African National Defence Force: A Good Beginning. *Armed Forces and Society* 26, no. 3 (Spring): 451–72.
Wolters, S. 2004. *Continuing Instability in the Kivus: Testing the DRC Transition to the Limit.* Pretoria: Institute for Security Studies.
———. 2007. *Trouble in Eastern DRC: The Nkunda Factor.* Pretoria: Institute for Security Studies.
Wolters, S., and H. Boshoff. 2006. *The Impact of Slow Military Reform on the Transition Process in the DRC.* Pretoria: Institute for Security Studies.
Wood, J. R. T. 1995. Rhodesian Insurgency—Part II. Durban. Available at www.rhodesianforces.org.
———. 2009. *Counter-Strike from the Sky: The Rhodesian All-Arms Fireforce in the War in the Bush 1974–1980.* Johannesburg: South Publishers.
World Bank. 2007. Status of the MDRP in the Democratic Republic of the Congo—May 2007. Washington, DC: World Bank.
———. 2004. Technical Annex for a Proposed Grant of SDR 22.2 Million to the Republic of Burundi for an Emergency Demobilization, Reinsertion and Reintegration Program. Washington, DC: World Bank.
Wyrod, Christopher. 2008. Sierra Leone: A Vote for Better Governance. *Journal of Democracy* 19, no. 1: 70–83.
Young, C. 1965. *Politics in the Congo. Decolonization and Independence.* Princeton, NJ: Princeton University Press.
Young, C., and T. Turner. 1985. *The Rise and Decline of the Zairian State.* Madison: University of Wisconsin Press.
Zack-Williams, Alfred B., and Stephen P. Riley. 1993. The Military and Civil Society in Sierra Leone: The Coup and Its Consequences. *Review of African Political Economy* 56.
Zartman, I. William. 1993. The Unfinished Agenda: Negotiating Internal Conflicts. In *Stopping the Killing: How Civil Wars End*, edited by Roy Licklider. New York: New York University Press.
———. 1995a. Negotiating the South African Conflict. In *Elusive Peace: Negotiating an End to Civil Wars*, edited by I. William Zartman. Washington, DC: Brookings Institution Press.
———. 1995b. *Ripe for Resolution: Conflict and Intervention in Africa.* New York: Oxford University Press.
Zirker, Daniel, Constantine P. Danopoulos, and Alan Simpson. 2008. The Military as a Distinct Ethnic or Quasi-Ethnic Identity in Developing Countries. *Armed Forces and Society* 34:314–57.

CONTRIBUTORS

Andrea Bartoli is a senior fellow at the Center for Peacemaking Practice and the dean of the School for Conflict Analysis and Resolution at George Mason University. He works primarily on peacemaking and genocide prevention. Bartoli was the founding director of Columbia University's Center for International Conflict Resolution, and he has been involved in many conflict resolution activities as a member of the Community of Sant'Egidio (www.santegidio.org/en). The most recent books that he coedited are *Peacemaking: From Practice to Theory* (2011) and *Attracted to Conflict* (2013). His new book is *Negotiating Peace: The Role of NGOs in Peace Processes* (2013).

Stephen Burgess is professor of international security at the US Air War College. His three books are *South Africa's Weapons of Mass Destruction*, *Smallholders and Political Voice in Zimbabwe*, and *The United Nations under Boutros Boutros-Ghali, 1992–97*. He has published numerous articles and book chapters on African security and strategic issues. Burgess received a doctorate from Michigan State University and has been on the faculties of the University of Zambia and the University of Zimbabwe.

Florence Gaub is a senior analyst at the European Union's Institute for Security Studies. She works on Arab military forces, with a special focus on conflict and postconflict contexts, such as Lebanon, Libya, and Iraq. Her publications include *Military Integration after Civil Wars* (2010) and *NATO and the Arab World* (2013) as well as numerous articles. Gaub received her PhD from Humboldt University in Berlin.

Rosalie Arcala Hall is a professor of political science at the University of the Philippines Visayas in Miagao, Iloilo. She has been a recipient of research grants from the Fulbright Program, Toyota Foundation, Nippon Foundation, East Asian Development Network, and Austrian Exchange Services. Her research on postconflict civil–military relations has involved extensive fieldwork in Aceh in Indonesia, Dili in Timor Leste, and Mindanao in the Philippines. Her essays have been published in the *Philippine Political Science Journal*, *Asian Security*, and the *Korean Journal of Defense Analysis*. Hall received a PhD in public and international affairs from Northeastern University in Boston.

Caroline Hartzell is a professor of political science at Gettysburg College. Her current research focuses on the effects the terms of civil war settlements have on the quality of the postconflict peace. Her publications include *Crafting Peace: Power-Sharing Institutions and the Negotiated Settlement of Civil Wars*, coauthored with Matthew Hoddie (2007); and *Strengthening Peace in Post–Civil War States: Transforming Spoilers into Stakeholders*, coedited with Matthew Hoddie (2010).

Paul Jackson is a professor of African politics and head of the International Development Department at the University of Birmingham. He has had twenty years of international experience in research, teaching, and policy advice. He is currently a senior security and justice adviser for the UK Stabilisation Unit, a member of the Folke Bernadotte Academy working group on security-sector reform, and an advisory board member of the Geneva Centre for the Democratic Control of Armed Forces. Jackson was educated as a political economist, and his specialization has been governance, specifically security governance, justice, and decentralization, particularly in postconflict and fragile environments. He has extensive overseas experience and has taught for several years on governance, security, and postconflict reconstruction. His publications include *Reconstructing Security after Conflict: Security Sector Reform in Sierra Leone*, with Peter Albrecht (2010); *Security Sector Reform in Sierra Leone 1997–2007: Views from the Front Line*, editor (2010); and *Conflict, Security and Development*, with coauthor Danielle Beswick (2011).

Ronald R. Krebs is an associate professor in the Department of Political Science at the University of Minnesota. He is the author of *Fighting for Rights: Military Service and the Politics of Citizenship* (2006); and coeditor of *In War's Wake: International Conflict and the Fate of Liberal Democracy* (2011). He is also associate editor of *Security Studies*, and he has been named a Fulbright senior scholar (2012) and a McKnight Land-Grant Professor at the University of Minnesota (2006–8).

David D. Laitin is the James T. Watkins IV and Elise V. Watkins Professor of Political Science at Stanford University. As a student of comparative politics, he has conducted field research focusing on issues of language and religion and how these cultural phenomena link nation to state, in Somalia, Yorubaland, Nigeria; Catalonia in Spain; Estonia; and France. His books include *Politics, Language, and Thought: The Somali Experience*; *Hegemony and Culture: Politics and Religious Change among the Yoruba*; *Language Repertoires and State Construction in Africa*; *Identity in Formation: The Russian-Speaking Populations in the Near Abroad*; and *Nations, States and Violence*. In collaboration with James Fearon, he has published papers on ethnicity, ethnic cooperation, the sources of civil war, and on policies that work to settle civil wars, and he has collaborated with Alan Krueger and Eli Berman on international terrorism. Laitin has been a recipient of fellowships from the Howard Foundation, the Rockefeller Foundation, the Guggenheim Foundation, and the Russell Sage Foundation and has received several grants from the National Science Foundation. He is an elected member of the American Academy of Arts and Sciences and the National Academy of Sciences.

Matthew LeRiche is an adviser on defense education in South Sudan supporting wider security-sector reform efforts and is currently engaged by the Defence Section of the UK Embassy in Juba. Previously, at the time of writing, he was a fellow in managing humanitarianism at the International Development Department of the London School of Economics and Political Science. He received a doctorate in war studies from King's College London. His most recent publication is *South Sudan: From Revolution to Independence*, with coauthor Matt Arnold (2012). LeRiche has also produced and written several documentary films on Sudan, South Sudan, and education. As a frequent analyst of Sudan, South Sudan, and regional political and security issues, he contributes to various publications, including the *Huffington Post* and *Global Brief*, as well as those of the International Crisis Group and the Enough Project of the Centre for American Progress.

Roy Licklider is a professor of political science at Rutgers University and adjunct senior research associate at the Saltzman Institute of War and Peace Studies at Columbia University. His research focuses on how civil wars end and how people who have been killing one another with considerable skill and enthusiasm can form working political communities. His books include *Political Power and the Arab Oil Weapon*; *Stopping the Killing: How Civil Wars End*; and *Living Together after Ethnic Killing: Exploring the Chaim Kaufmann Argument* (with Mia Bloom). He has published articles in the *American Political Science Review*, *International Studies Quarterly*, *Journal of Peace Research*, *Journal of Human Rights*, *Security Studies*, and *Small Wars and Insurrections*, as well as in edited volumes.

Rohan Maxwell is a senior political-military adviser at NATO Headquarters Sarajevo. He has participated in military integration efforts in Bosnia and Herzegovina since 2003. He assumed his current position in 2005, the year that NATO brokered an agreement to replace separate armed forces, dating from the 1992–95 conflict and the subsequent peace treaty, with a single Armed Forces of Bosnia and Herzegovina. He and his staff are responsible for politico-military aspects of the implementation of that agreement. He is coauthor (with John Andreas Olsen) of *Destination NATO: Defence Reform in Bosnia and Herzegovina, 2003–13* (2013). He served with the Canadian Army from 1982 to 2003 and completed his undergraduate and postgraduate studies at the Royal Military College of Canada.

Martha Mutisi is an academic practitioner interested in issues at the critical intersection of peace, security, and development. She is currently the manager of the Interventions Department at the African Centre for the Constructive Resolution of Disputes, known as ACCORD. She is currently coediting the book *Women, Peace, and Security: Approaches to Gender Equality and Empowerment* (2013). Mutisi received a PhD from the School for Conflict Analysis and Resolution at George Mason University.

Bruce Russett is Dean Acheson Research Professor of International Politics at Yale University. He is a fellow of the American Academy of Arts and Sciences and has received honorary doctorates from Uppsala University and Williams College. He edited the *Journal of Conflict Resolution* from 1973 to 2009, and, with Paul Kennedy, served as a staff member for the Ford Foundation's 1995 report *The United Nations in Its*

Second Half-Century. He is a past president of the International Studies Association and also of the Peace Science Society (International), and in 2009 he received the society's third quadrennial Founder's Medal for "significant and distinguished lifelong scientific contributions to peace science." Of his twenty-seven books, the most recent include *Grasping the Democratic Peace* (1993); *The Once and Future Security Council* (1997); *Triangulating Peace: Democracy, Interdependence, and International Organizations*, with John Oneal (2001), which was awarded the International Studies Association's prize for best book of the decade 2000–2009; and *Hegemony and Democracy* (2011).

Cyrus Samii is assistant professor in the Wilf Family Department of Politics, New York University. He writes and teaches on quantitative social science and methodology and program evaluation as applied to the analysis of governance, development, and conflict management programs. His substantive research has looked at war-to-peace transitions, including studies of ex-combatant reintegration, transitional justice, army reform, community security, postconflict economic development, and civic education. He has designed and implemented studies in Burundi, Cote d'Ivoire, Indonesia, Liberia, and Nepal. He holds a PhD from Columbia University.

Mimmi Söderberg Kovacs is an assistant professor in the Department of Peace and Conflict Research at Uppsala University and a senior researcher at the Nordic Africa Institute in Uppsala. She has been a visiting scholar at the Belfer Center for Science and International Affairs at the John F. Kennedy School of Government at Harvard University; at the African Center for the Constructive Resolution of Disputes, known as ACCORD, in South Africa; and at the Centre for African Studies at the University of Cape Town. She has published works on rebel-to-party transformations in civil wars, spoilers in peace processes, the concept of peace beyond the absence of war, postwar democratization, and elections in war-torn societies. Geographically, her empirical focus has been on Sub-Saharan Africa, in particular the peace processes in Sierra Leone and Liberia. Kovacs is currently the project leader for a major research project on big man politics and electoral violence in West Africa.

Judith Verweijen is a senior research fellow at the Nordic Africa Institute in Uppsala and a postdoctoral research fellow at the Conflict Research Group at Ghent University. Her doctoral research, completed in 2013, was based on long-term ethnographic fieldwork on the interaction between the Congolese armed forces and civilians in the eastern Kivu provinces of the Democratic Republic of the Congo. In addition, she conducted research on various dimensions of armed group mobilization. Her present research, which is part of larger projects, focuses on the economic interactions between state security forces and civilians as well as local justice and security governance in the Democratic Republic of the Congo. Her articles have appeared in several journals, including *Review of African Political Economy*, *African Affairs*, and *Politique Africaine*.

INDEX

Information in figures and tables is indicated by f and t, respectively. Information in footnotes is indicated by n between page number and note number.

A
Absorbed Forces (Sudan), 32–33, 36, 40–41, 43–44, 45, 47, 237, 255
Abu Sayyaf, 113, 116, 117n1
accommodation capacity
   in industrial organization perspective, 241–42
   military integration and, 18–19, 20, 22
   in Philippines, 112, 113, 115
Acland, Sir John, 52
Act for Peace, 105
Addis Ababa Peace Agreement, 31–38, 41–43, 46–47, 237, 254. *See also* Sudan
ADF. *See* Allied Democratic Forces
AFBiH. *See* Armed Forces of Bosnia and Herzegovina
AFDL. *See* Alliance des forces démocratiques pour la libération du Congo-Zaïre
Afghanistan, 187, 240, 266
AFP. *See* Armed Forces of the Philippines
AFRC. *See* Armed Forces Revolutionary Council
African National Congress (ANC), 120–21, 123, 128, 129, 130, 217, 254
African National Council (ANC), 50, 53
African Union, 97, 218, 242n3
African Union Mission in Burundi, 219
African Union Mission in Somalia (AMISOM), 222
African Union-United Nations Hybrid Operation in Darfur (UNAMID), 97, 204
Afrikaner Volksfront, 129

Agreement on a Partial Ceasefire (Mozambique), 167, 177n2
AIDS, in Mozambique, 163
Ajello, Aldo, 165, 168, 171
Alexander the Great, 3
Algeria, 227n10
al-Hadi, Imam, 34
Alier, Abel, 34, 35, 36, 38
Alliance des forces démocratiques pour la libération du Congo-Zaïre (AFDL), 100, 137, 140, 159n15
Allied Democratic Forces (ADF), 158n2. *See also* Uganda
All People's Congress (APC), 196, 203, 206
al-Mahdi, Sadiq, 33, 41, 44
alternative outcomes
   in Bosnia-Herzegovina, 190–91
   in Democratic Republic of Congo (DRC), 154–55
   in South Africa, 131
   in Sudan, 43
Amal (militia), 72, 73, 74, 75, 77. *See also* Lebanon
Amin, Idi, 35
AMISOM. *See* African Union Mission in Somalia
ANC. *See* African National Congress; African National Council
Angola
   Democratic Republic of Congo and, 137, 141, 145, 147, 158n2, 159n16, 160n21
   in industrial organization perspective, 240
   Rwanda and, 97
   South Africa and, 59, 120
   Zimbabwe and, 61, 240
Annan, Kofi, 166, 171. *See also* United Nations
Anyanya, 2, 31–32, 34–38, 39t, 39–43, 45–47, 234, 237. *See also* Sudan

INDEX

Aoun, Michel, 70–71, 74, 76
APC. *See* All People's Congress
APLA. *See* Azanian People's Liberation Army
Arafat, Yassir, 72
ARBiH. *See* Armija Republike BiH
Armed Forces of Bosnia and Herzegovina (AFBiH), 180, 181–92
Armed Forces of the Philippines (AFP), 107–8, 110–11, 118n9
Armed Forces Revolutionary Council (AFRC) (Sierra Leone), 197–98, 199, 200–201, 207, 208
Armija Republike BiH (ARBiH), 179, 183, 189
ARMM. *See* Autonomous Region for Muslim Mindanao
Army of Serb Republic (VRS), 179–84, 186–88, 192, 194n10
Army of the Federation of Bosnia-Herzegovina (VF), 179, 180–84, 186–88, 192, 194n10
Arusha Peace Agreement
    Burundi and, 88, 215, 216–19, 223, 225, 263
    Democratic Republic of Congo and, 88
    Rwanda and, 88, 89–93, 101n3, 238
Austro-Hungarian military, 257n5
"autochthonous," 138
Autonomous Region for Muslim Mindanao (ARMM), 104, 111, 116, 117n2, 118n8
Azanian People's Liberation Army (APLA), 122–23, 125, 126–27, 128

B
Bagaza, Jean Baptiste, 214, 227n3, 227n6
BALIK-BARIL program, 109
Bangladesh, 243n6
bargaining model of war, 14, 17–18, 23f, 24
Bartoli, Andrea, 7, 11, 163–77, 236, 253, 293
Bashir, Omar, 33
Belgium
    Burundi and, 221, 222, 223, 239
    Democratic Republic of Congo and, 144, 145, 147, 160n21
    Lebanon and, 79
Bemba, Jean-Pierre, 143, 158n8
Bentiu, 40
Berri, Nabih, 73
best practices, 245, 248
bloating, of postmerger armies, 235

Blue Nile, 33
BMATT. *See* British Military Advisory Training Team
Boers, 120
Bol, Kerubino Kuanyan, 32
Bophutatswana, 122. *See also* Transkei, Venda, Bophutatswana, and Ciskei Defence Forces
Bor Dinka community, 36
Bor Mutiny, 32, 42, 45. *See also* Sudan
Boshoff, Henri, 228n15, 228n22
Bosnia-Herzegovina, 246t, 247t
    alternative outcomes in, 190–91
    Armed Forces of Bosnia and Herzegovina in, 180, 181–92
    Armija Republike BiH in, 179, 183, 189
    Army of Serb Republic in, 179–84, 186–88, 192, 194n10
    Army of the Federation of Bosnia-Herzegovina in, 179, 180–84, 186–88, 192, 194n10
    compromises in, 184–85
    creation of integrated military in, 185–87
    Dayton Accords and, 248
    Defence Reform Commission in, 180–81, 183, 185, 191
    efficiency and, 236–37
    emergence of military integration as issue in, 183
    employment in, 191, 262t
    ethnic groups in, 179, 183–84, 194n1
    European Union Force in, 180, 186, 188, 189
    Federation of Bosnia and Herzegovina in, 179, 190
    General Framework Agreement for Peace in, 179–80, 183, 186, 188, 190
    High Representative in, 179, 180, 183
    historic role of military in, 182–83
    Hrvatsko Vijece Obrane in, 179, 183
    Implementation Force in, 179, 190
    in industrial organization perspective, 236–37, 239, 240
    International Criminal Tribunal for the Former Yugoslavia and, 180, 194n7
    military capabilities in, 189–90
    military integration in, 183–92, 259
    military training in, 186–87
    NATO Headquarters Sarajevo and, 180–81, 186, 188

NATO in, 82, 179–90, 192, 194n2, 239
neutrals in, 183–84
Office of the High Representative in, 179–80, 184, 187, 194n2
opponents in, 183–84
Organization for Security and Cooperation in Europe and, 185, 194n2, 239
origins of military integration in, 182–85
outcome in, 188–91
Partnership for Peace and, 179, 180, 182, 183, 184–85, 187, 188–89
Peace Support Operations Training Center and, 186–87
personnel selection in, 185–86
political control in, 188–89
quota system in, 182, 184, 185–86
rank allocation in, 186–87
resumption of violence in, 191–92
Russia and, 188–89, 194n2
security in, 191–92, 262t
Serb Republic in, 179, 184, 188, 189, 191
shaping of new military in, 185
Stabilization Force in, 179, 180–81, 186–87, 194n2
supporters in, 183–84
sustainability of armed forces in, 190
symbolism and, 188, 192, 236–37
trust in, 262t
United Kingdom and, 186–87
United States and, 181, 186, 187, 189, 194n2
Yugoslav People's Army and, 182–83, 187
Botha, P. W., 121
Botswana, 51, 120
Boustani, Emile, 72
Boutros-Ghali, Boutros, 174
brassage, 145, 147, 148, 151, 160n26, 236
Brigade of the Mountain, 72
Britain. *See* United Kingdom
British Military Advisory Training Team (BMATT)
    South Africa and, 123, 125, 127
    Zimbabwe and, 49, 53–57, 64, 67n13
Buliok offensive, 113
Burgess, Stephen, 6, 7t, 10, 87–102, 238, 263, 293
Burundi, 246t, 247t
    50/50 integration in, 19, 218, 225, 228n17
    African Union Mission in Burundi, 219

Arusha Peace Agreement and, 88, 215, 216–19, 223, 225, 263
Bagaza in, 214, 227n3
Belgium and, 221, 222, 223, 239
Buyoya in, 214, 218, 227n6
China and, 221
Community of Sant'Egidio in, 215
compromises in, 217–18
Conseil national pour la défense de la démocratie in, 215, 216–17, 218–25, 228n21, 228n23, 228n25
creation of integrated military in, 218–22
Democratic Republic of Congo and, 158n2
emergence of military integration as issue in, 214–15
employment in, 262t
Forces armées burundaises in, 213, 215, 218–21, 222, 223, 226, 228n16–17, 234
Forces armées du peuple in, 214
Forces nationales de libération in, 214, 215–19, 220, 222, 223–24, 225, 228n21
Forces pour la défense de la démocratie in, 215–17, 218–23, 224, 228n25–26
Forces Technical Agreement in, 219–20, 224, 225
Front pour la démocratie au Burundi in, 214–15, 217, 218, 227n6
Front pour la libération nationale in, 214, 215, 228n21
genocide in, 213–14
historic role of military in, 213–14
Hutus in, 213–17, 218, 223, 225, 226n1, 228n17
incidence of military integrations in, 13
in industrial organization perspective, 234, 239
Institut supérieur des cadres militaires in, 214
Joint Ceasefire Commission in, 219–20, 228n16
military capabilities in, 222
military integration in, 214–25, 253, 259, 263
military training in, 221–22
National Defence Force in, 224
national unity in, 225
Netherlands and, 221
neutrals in, 215–17

Burundi (cont'd)
   opponents in, 215–17
   origins of military integration in, 213–18
   outcome in, 222–24
   Parti pour la libération du peuple hutu in, 214, 227n6
   personnel selection in, 219–20
   political control in, 222–24
   rank allocation in, 220–21
   Rwanda and, 214
   security in, 262t
   shaping of new military in, 218–19
   Somalia and, 222
   South Africa and, 216, 217, 221, 227n13
   Sudan and, 221
   supporters in, 215–17
   sustainability in, 224
   Tanzania and, 214, 216, 217, 219
   trust in, 225, 262t
   Tutsis in, 213–16, 217, 220, 222–23, 224–25, 226n1
   Twa in, 213, 226n1
   Union for National Progress in, 216, 217, 223, 227n6
   United Nations Operation in (ONUB), 221, 222–23
   United States and, 217, 221
Burundian Workers Party (UBU), 214
Burundi Leadership Training Program, 221
Bush War, 55, 59
business. See industrial organization (IO) perspective; mergers, business
Buyoya, Pierre, 214, 218, 227n6

C
Cairo Agreement, 72
Catholic Church. See Community of Sant'Egidio
Cawthra, Gavin, 119, 130
CBR. See Centre de brassage et de recyclage
CCFADM. See Joint Commission for the Formation of the Mozambican Defence Force
CDF. See Civil Defence Forces
CDR. See Coalition pour la Défense de la République
Central Intelligence Organisation (CIO) (Zimbabwe), 59, 60, 61
Centre de brassage et de recyclage (CBR), 145, 147, 148, 151, 160n26, 236

Chad, 41, 158n2
Chiluba, Fredrick, 159n13
China
   Burundi and, 221
   Democratic Republic of Congo and, 160n21, 161n36
   military integration in, 2, 95
   Mozambique and, 171
   victory in, 1
   Zimbabwe and, 51, 54, 56, 58
Chissano, Joaquim, 164, 167, 175
CIAT. See Comité international d'accompagnement de la transition
CIO. See Central Intelligence Organisation
Ciskei, 122. See also Transkei, Venda, Bophutatswana, and Ciskei Defence Forces
Civil Defence Forces (CDF) (Sierra Leone), 198, 200, 201, 207, 208, 211n6
civilian control, as dimension of success, 4, 4t. See also political control
civil-military operations (CMOs), in Philippines, 106, 113
civil wars. See also resumption of violence
   negotiated settlements and, vs. interstate wars, 14–15
   state power and, 14–15
Clinton, Bill, 217
CMF. See Commonwealth Monitoring Force
CMOs. See civil-military operations
CNDD. See Conseil national pour la défense de la démocratie
CNDP. See Congrès national pour la défense du peuple
Coalition pour la Défense de la République (CDR), 91
CODESA. See Convention for a Democratic South Africa
Cold War
   military integration and, 17
   Philippines and, 164
   United Nations Department of Political Affairs and, 165
   victory and, 1–2
Colletta, Nat J., 94
Comité international d'accompagnement de la transition (CIAT), 144, 160n21, 161n35
Commonwealth Monitoring Force (CMF) (Zimbabwe), 52

INDEX

Commonwealth Observer Group (Zimbabwe), 52
Communist Party (Lebanon), 72
Community of Sant'Egidio, 164, 174, 177n2, 215, 239
Compaoré, Blaise, 159n13
compromises
　in Bosnia-Herzegovina, 184–85
　in Burundi, 217–18
　in Democratic Republic of Congo, 141–43
　in Lebanon, 76
　in Mozambique, 168–69
　in Philippines, 107–8
　in South Africa, 123–24
　in Sudan, 35–36
Condominium (British rule of Sudan), 33
Congo. *See* Democratic Republic of Congo; Republic of the Congo
Congrès national pour la défense du peuple (CNDP) (Congo), 149–50
Conseil national pour la défense de la démocratie (CNDD) (Burundi), 215, 216–17, 218–25, 228n21, 228n23, 228n25
Conservative Party (UK), 52
control variables, 21, 23*f*
Convention for a Democratic South Africa (CODESA), 121, 124
corporate culture, 232, 235–36
Costa Rica, 26n6
Côte d'Ivoire, 203, 218, 221–22, 296
creation, of integrated military
　in Bosnia-Herzegovina, 185–87
　in Burundi, 218–22
　in Democratic Republic of Congo, 143–48
　in Lebanon, 76–79
　in Mozambique, 169–72
　in Philippines, 108–10
　in Rwanda, 92–96
　in Sierra Leone, 199–202
　in South Africa, 124–27
　in Sudan, 36–40
　in Zimbabwe, 53–57
Croatia, 13, 190. *See also* Bosnia-Herzegovina
culture, corporate, 232, 235–36
Cyprus, 13

D
Dahab, Abdallah Ateif, 38
Dai, Tobias, 167

Darfur, 33, 97, 98–99, 204, 240
Dayton Accords, 248
Defence Reform Commission (DRC) (Zimbabwe), 180–81, 183, 185, 191
de Klerk, F. W., 121, 131, 254
Democratic Republic of Congo (DRC), 246*t*, 247*t*. *See also* Forces armées de la République démocratique du Congo; Rassemblement congolais pour la démocratie
　alternative outcomes in, 154–55
　Angola and, 137, 141, 145, 147, 158n2, 159n16, 160n21
　Arusha Accords and, 88
　"autochthonous" in, 138
　Belgium and, 144, 145, 147, 160n21
　Burundi and, 158n2
　Centre de brassage et de recyclage in, 145, 147, 148, 151, 160n26, 236
　China and, 160n21, 161n36
　Comité international d'accompagnement de la transition in, 144, 160n21, 161n35
　compromises in, 141–43
　Congrès national pour la défense du peuple in, 149–50
　creation of integrated military in, 143–48
　emergence of military integration as issue in, 140
　employment in, 153, 262*t*
　EUSEC in, 144, 160n22
　Forces armées congolaises and, 140, 141, 143, 145, 146
　Forces armées zaïroises and, 139–40, 147, 155–56, 159n12, 162n41
　Forces pour la défense de la démocratie and, 158n2, 228n23
　Global and All-Inclusive Agreement and, 138, 142, 145, 158n7, 159n17
　historic role of military in, 139–40
　Hutus in, 138, 150, 158n4, 158n6, 159n16
　in industrial organization perspective, 236, 237, 239–40
　Integrated Brigades in, 145, 147, 148, 151–52, 160n26
　interahamwe militias in, 138, 158n2
　Inter-Congolese Dialogue and, 140, 141
　Kabila, Joseph and, 141–45, 158n8, 162n39
　Kabila, Laurent and, 137, 138, 140–41, 158n2

INDEX

Democratic Republic of Congo (DRC) (*cont'd*)
  Lingalaphones in, 147
  Lusaka Cease-Fire Agreement in, 140
  Mai Mai militias in, 138, 142, 145–47, 150, 154, 158n1, 158n5, 158n7
  Maison Militaire in, 143, 149
  military capabilities in, 3, 150
  military integration in, 140–56, 253, 259–60, 264–65
  Mission de l'organisation des Nations Unies en RD Congo (MONUC) and, 140, 144, 160n21, 160n25, 161n36
  Mouvement de Libération and, 137, 158n1, 158n7, 161n39, 162n39
  Mouvement pour la libération du Congo and, 137, 138, 145, 148, 158n7–8, 159n14–15
  neutrals in, 140–41
  North Korea and, 141
  opponents in, 140–41
  origins of military integration in, 139–43
  outcome in, 148–51
  personnel selection in, 145–46
  political control in, 148–50
  presidential guard in, 143
  rank allocation in, 146–47
  resumption of violence in, 151–56
  Rwanda and, 94, 95, 137
  Rwandophones in, 138, 142, 146, 148, 158n4
  security in, 262*t*
  shaping of new military in, 143–45
  South Africa and, 145, 147
  South African National Defence Force in, 129
  Structure militaire d'intégration in, 145
  Superior Defence Council in, 143, 159n17, 161n28
  supporters in, 140–41
  sustainability of armed forces in, 150–51
  Swahiliphones in, 147
  training in, 147–48
  Transitional Government in, 138, 141, 144, 158n8, 160n20, 160n22, 161n28
  trust in, 262*t*
  Tutsis in, 138, 143, 148, 149–50, 158n4, 158n6
  Uganda and, 137, 141, 158n2
  United Kingdom and, 160n21
  United States and, 160n21, 161n36
  unity in, 140, 147, 154, 159n11
  Zimbabwe and, 60–61, 137, 141
Department of Peacekeeping Operations (DPKO) (United Nations), 166, 171, 174, 175, 243n6
Department of Political Affairs (DPA) (United Nations), 165–66, 174
DeRouen, Karl, 26n10
Detective Reconnaissance Emergency Action Mission (DREAM) (Sierra Leone), 211n9
Dhlakama, Alfonso, 167, 168, 174
diamonds
  in Sierra Leone, 197
  in Zimbabwe, 60–61, 62, 63, 64, 238
Domingos, Raul, 167, 177n2
DPA. *See* Department of Political Affairs
DPKO. *See* Department of Peacekeeping Operations
DRC. *See* Defence Reform Commission; Democratic Republic of Congo
DREAM. *See* Detective Reconnaissance Emergency Action Mission
duration of conflict, in logit analysis, 23*t*
Dyck, Lionel, 66n11

E
ECOMOG. *See* Economic Community of West African States Military Observer Group
Economic Community of West African States (ECOWAS)
  Sierra Leone and, 197, 204
  South African Development Community and, 173
  Standby Force, 204
Economic Community of West African States Military Observer Group (ECOMOG), 197–98, 240, 242n3
ECOWAS. *See* Economic Community of West African States
Edelman, Murray, 251
efficacy. *See* military efficacy
efficiency, 232, 233, 236–38
el-Ghazal, Bahr, 42
emergence, of military integration as issue
  in Bosnia-Herzegovina, 183
  in Burundi, 214–15
  in Democratic Republic of Congo, 140
  in Lebanon, 73–74
  in Mozambique, 165–66

in Philippines, 107
in Rwanda, 92
in Sierra Leone, 198
in South Africa, 121
in Sudan, 34
employment
   in Bosnia-Herzegovina, 191, 262*t*
   in Burundi, 262*t*
   in Democratic Republic of Congo, 153, 262*t*
   hypothesis, 250
   in industrial organization perspective, 240–41
   in Lebanon, 76, 82, 83, 262*t*
   in Mozambique, 172, 176, 262*t*
   in Philippines, 104–5, 107, 114, 115–16, 262*t*
   resumption of violence and, 9, 44
   in Rwanda, 98–99, 262*t*
   in Sierra Leone, 206, 262*t*
   in South Africa, 262*t*
   in Sudan, 44, 262*t*
   in Zimbabwe, 63, 262*t*
ethics, 10, 12, 260, 266–67, 268
Ethiopia, 32, 34, 35, 41, 243n6
ethnic cleansing, 1. *See also* Bosnia-Herzegovina; Burundi; Rwanda
EUFOR. *See* European Union Force
European Union Force (EUFOR), in Bosnia-Herzegovina, 180, 186, 188, 189
EUSEC. *See* Mission de conseil et d'assistance de l'Union européenne en matière de réforme du secteur de la sécurité en RD Congo
Executive Outcomes (security firm), 197, 240

F
FAB. *See* Forces armées burundaises
FAC. *See* Forces armées congolaises
FADM. *See* Forcas Armadas da Defesa de Moçambique
Failed States Index, Zimbabwe in, 59
FAP. *See* Forces armées du peuple
FAR. *See* Forces armées rwandaises
FARDC. *See* Forces armées de la République démocratique du Congo
Fatah al-Islam, 79
FAZ. *See* Forces armées zaïroises
FBiH. *See* Federation of Bosnia and Herzegovina

FDD. *See* Forces pour la défense de la démocratie
FDLR. *See* Forces démocratiques pour la libération du Rwanda
FDN. *See* National Defence Force
Federation of Bosnia and Herzegovina (FBiH), 179, 190
Fifth Brigade (5B), 58–59, 62, 64, 67n12, 67n14, 238
Fire Force tactics, 66n5
FNL. *See* Forces nationales de libération
Forcas Armadas da Defesa de Moçambique (FADM), 169, 170
Forces armées burundaises (FAB), 213, 215, 218–21, 222, 223, 226, 228n16–17, 234
Forces armées congolaises (FAC), 140, 141, 143, 145, 146
Forces armées de la République démocratique du Congo (FARDC)
   coercive capacity of, 153
   employment and, 153
   failure of, 237
   in industrial organization perspective, 237
   Integrated Brigades and, 147
   makeup of, 161n33
   military capabilities of, 150
   militias and, 150
   personnel selection and, 145–46
   prewar predecessors and, 154–55
   rank allocation and, 146
   resumption of violence and, 151
   size of, 161n28
   in Superior Defense Council, 159n17
   training of, by United States and China, 161n36
Forces armées du peuple (FAP), 214
Forces armées rwandaises (FAR), 88–90, 91–100, 101n6–7, 101n9–10, 158n2. *See also* Rwanda
Forces armées zaïroises (FAZ), 139–40, 147, 155–56, 159n12, 162n41
Forces démocratiques pour la libération du Rwanda (FDLR), 150, 238
Forces Libanaises, 72, 73–77, 79, 81, 83, 240
Forces nationales de libération (FNL)
   in Burundi, 214, 215–19, 220, 222, 223–24, 225, 228n21
   Democratic Republic of Congo and, 158n2

Forces pour la défense de la démocratie (FDD)
  in Burundi, 215–17, 218–23, 224, 228n25–26
  Democratic Republic of Congo and, 158n2, 228n23
  Ndayikengurukiye and, 227n4
Forces Technical Agreement (FTA) (Burundi), 219–20, 224, 225
France
  Burundi and, 221–22
  Democratic Republic of Congo and, 160n21
  Lebanon and, 71, 74, 78, 79, 240
  military integration in, 2
  Mozambique and, 168, 239
Frelimo. See Liberation Front of Mozambique
Frente de Libertação de Moçambique (Frelimo). See Liberation Front of Mozambique
FRODEBU. See Front pour la démocratie au Burundi
FROLINA. See Front pour la libération nationale
Front of the Nationalist and Patriotic Parties (Lebanon), 72
Front pour la démocratie au Burundi (FRODEBU), 214–15, 217, 218, 227n6
Front pour la libération nationale (FROLINA), 214, 215, 228n21
FTA. See Forces Technical Agreement

G
Gabon, 160n21
Gahiro, Samuel, 228n25
Garang, John, 32, 35, 41, 42–43, 45
Gatsinzi, Marcel, 94
Gaub, Florence, 10, 69–84, 236, 252, 293
Gemayel, Amin, 70, 71
Gemayel, Bashir, 73
General Framework Agreement for Peace (GFAP) (Bosnia-Herzegovina), 179–80, 183, 186, 188, 190
genocide. See Bosnia-Herzegovina; Burundi; Rwanda
Georgia, 188–89
GFAP. See General Framework Agreement for Peace
GIA. See Global and All-Inclusive Agreement
Glassmyer, Katherine, 5, 19, 105
Global and All-Inclusive Agreement (GIA), 138, 142, 145, 158n7, 159n17

Global Peace Operations Initiative, 242n3
GNU. See Government of National Unity
Gonçalves, Jamie, 164, 177
Government of National Unity (GNU) (Zimbabwe), 60
Greater Equatoria, 31–32. See also Sudan
Green Bombers, 60, 67n15
Guardians of the Cedar, 72
Guebuza, Armando, 167, 172, 173, 175, 177n2
guerrilla forces. See militias
Guinea, 203

H
Habyarimana, Emmanuel, 94
Habyarimana, Juvénal, 89, 90, 91
Haddad, Saad, 72
Hall, Rosalie Arcala, 10, 103–18, 235, 237, 268, 293
Hartzell, Caroline, 5, 6, 7t, 10, 13–27, 15, 243n6, 260, 264, 293–94
Herzegovina. See Bosnia-Herzegovina
Hezbollah, 72, 73, 77, 80, 83, 236
High Representative (HR) (Bosnia-Herzegovina), 179, 180, 183
historic role of military
  in Bosnia-Herzegovina, 182–83
  in Burundi, 213–14
  in Democratic Republic of Congo (DRC), 139–40
  in Lebanon, 69–73
  in Philippines, 106
  in Sierra Leone, 196–98
  in South Africa, 119–21
  in Sudan, 33–34
  in Zimbabwe, 50–51
HIV/AIDS, in Mozambique, 163
Hobeika, Elie, 73
Hoddie, Matthew, 5, 6, 7t, 15
horizontal acquisitions, 231
Hoss, Salim, 70–71
HR. See High Representative
Hrawi, Elias, 70–71, 76
Hrvatsko Vijece Obrane (HVO), 179, 183
Hungary, 257n5
Hutus. See also interahamwe militias
  in Burundi, 213–17, 218, 223, 225, 226n1, 228n17
  in Democratic Republic of Congo, 138, 150, 158n4, 158n6, 159n16

in Rwanda, 89–95, 97, 98–100, 101n3, 102n12
HVO. *See* Hrvatsko Vijece Obrane

I
IBs. *See* Integrated Brigades
ICD. *See* Inter-Congolese Dialogue
identity. *See also* symbolism; unity
   in Bosnia-Herzegovina, 184
   in Democratic Republic of Congo, 138, 146
   in Lebanon, 71, 82–83
   in logit analysis, 23t
   military culture and, 257n5, 259
   in Philippines, 110, 112–14
   victory and, 1
ideology
   business mergers and, 233
   victory and, 1
IFOR. *See* Implementation Force
IMATT. *See* International Military Advisory and Training Team
Implementation Force (IFOR) (Bosnia-Herzegovina), 179, 190. *See also* Stabilization Force
India, 3, 222
Indonesia, 243n6
industrial organization (IO) perspective
   accommodation capacity in, 241–42
   Angola in, 240
   Bosnia-Herzegovina in, 236–37, 239, 240
   Burundi in, 234, 239
   competitors in, 238–40
   Democratic Republic of Congo in, 236, 237, 239–40
   efficiency and, 236–38
   employment in, 240–41
   horizontal acquisitions in, 231
   justification for mergers in, 231
   Lebanon in, 235–36
   Mozambique in, 236, 239, 240
   Nepal in, 234–35
   outsourcing in, 240–41
   overvaluation of state soldiers in, 234–35
   peacekeeping operations in, 233, 240
   Philippines in, 235, 237–38
   Rwanda in, 235, 236, 238, 240
   Sierra Leone in, 237, 239, 240
   South Africa in, 234
   Sudan in, 234, 237, 240
   symbolism and, 236–38
   transaction theory and, 232
   troop-contributing countries in, 233, 240
   Uganda in, 240
   undervaluation of militias in, 234–35
   Zimbabwe in, 235, 238, 239
ingando process, in Rwanda, 88, 93–96, 99, 101n3, 102n16. *See also* Rwanda
Inkatha Freedom Party (South Africa), 122
Institut supérieur des cadres militaires (ISCAM) (Burundi), 214
Integrated Brigades (IBs) (Democratic Republic of Congo), 145, 147, 148, 151–52, 160n26
integration. *See* military integration
interahamwe militias. *See also* Hutus
   in Democratic Republic of Congo, 138, 158n2
   in Rwanda, 90, 91, 94, 101n7
Inter-Congolese Dialogue (ICD), 140, 141
International Criminal Tribunal for the Former Yugoslavia, 180, 194n7
International Military Advisory and Training Team (IMATT), 195, 199–202, 203, 204–6, 209, 211n7
interstate wars, settlements with, *vs.* civil wars, 14–15, 26n4
IO. *See* industrial organization (IO) perspective
Iraq, 6, 187, 240, 266
ISCAM. *See* Institut supérieur des cadres militaires
Islamo-Palestino-Progressists, 72
Ismail, Aboobaker, 131
Israel
   Lebanon and, 10, 69, 71, 72, 77, 79, 80, 83, 236
   nationalism in, 3
   Sudan and, 34–35
"Italian formula," 174
Italy, Mozambique and, 164, 167, 177n2, 239

J
Jaaliyin tribe, 33
Jackson, Paul, 10, 49–67, 234, 238, 257n9, 266, 294
Jaja, Samir, 75
JCC. *See* Joint Ceasefire Commission (Burundi)
Jervis, Robert, 257n3
Jikiri, Yusuf, 118n6

INDEX

JMCC. *See* Joint Military Coordinating Committee
JNA. *See* Yugoslav People's Army
JOC. *See* Joint Operational Command
Joint Ceasefire Commission (Burundi), 219–20, 228n16
Joint Ceasefire Commission (Sudan), 37, 38
Joint Commission for the Formation of the Mozambican Defence Force (CCFADM), 168, 169, 170–71, 176, 239
Joint Communiqué, 165
Joint Military Coordinating Committee (JMCC) (South Africa), 121, 122, 125
Joint Operational Command (JOC) (Zimbabwe), 60
Jordan, 243n6
Jugoslavenska Narodna Armija (JNA), 182–83, 187
Jumblatt, Walid, 73, 74

K

Kabbah, Ahmed Tejan, 197–200, 208, 210n2, 237
Kabila, Joseph, 60–61, 141–45, 158n8, 162n39, 228n23
Kabila, Laurent, 97, 137, 138, 140–41, 158n2
Kagame, Paul, 93, 94, 97, 98, 101n3, 101n9, 236
Kenya, 41, 173, 176, 224
Khalid, Mansour, 31, 35
Khatib, Sami, 70
Kiir, Salva, 40
Kordofan, 33
Korea. *See* North Korea
Koroma, Johnny Paul, 197–98, 200, 203
Kosovo, 82, 119
Krebs, Ronald, 3, 7, 11, 245–57, 259, 261, 263, 294
Kwa Zulu Self Protection Force (KZSPF), 122, 125, 126–27, 128
KZSPF. *See* Kwa Zulu Self Protection Force

L

Lagu, Joseph, 31, 32, 34–36, 37–40, 41, 43, 45
Lahoud, Emile, 70, 71, 79, 82, 235, 236
Laitin, David, 11, 231–43, 265, 268, 294
Lancaster House Constitutional Conference, 52
Laos, 13
Lebanese Youth, 72
Lebanon, 246t, 247t
 Amal militia in, 72, 73, 74, 75, 77
 approval ratings of state and institutions in, 81t
 Belgium and, 79
 compromises in, 76
 creation of integrated military in, 76–79
 emergence of military integration as issue in, 73–74
 employment in, 76, 82, 83, 262t
 Forces Libanaises in, 72, 73–77, 79, 81, 83, 240
 Hezbollah and, 72, 73, 77, 80, 83, 236
 historic role of military in, 69–73
 in industrial organization perspective, 235–36
 Israel and, 10, 69, 71, 72, 77, 79, 80, 83, 236
 military budget in, 82
 military capabilities in, 80
 military integration in, 73–84, 259
 military power in, 80
 military training in, 79
 militias in, 72–73
 neutrals in, 74–75
 opponents in, 74–75
 origins of military integration in, 69–76
 outcome in, 79–83
 personnel selection in, 77–78
 political control in, 79–80
 rank allocation in, 77–78
 resumption of violence in, 80, 83–84
 security in, 262t
 shaping of new military in, 76–77
 supporters in, 74–75
 sustainability of armed forces in, 81–82
 symbolism and, 83, 236
 Syria and, 70–73, 75–77, 79–80, 83
 trust in, 262t
 United States and, 79
 unity in, 76, 79, 83
LeRiche, Matthew, 10, 31–48, 237, 255, 294–95
Lesotho, 120, 129
Liberation Front of Mozambique (Frelimo)
 Mozambique and, 163–73, 175, 176, 236, 254
 Zimbabwe and, 53, 59
Liberia, 6, 196, 202–3, 210n4
Libya, 41

Licklider, Roy, 1–12, 7t, 11–12, 119–32, 254, 257n8, 259–68, 295
Linas-Marcoussis Agreement, 218
Lingalaphones, 147
Lomé Peace Accord, 195, 196, 198–200, 205, 207–8, 209, 210n2
Long, Austin, 131
Lord Resistance Army (LRA), 158n2
LRA. *See* Lord Resistance Army
Lusaka Cease-Fire Agreement, 140, 159n14

M

MACP. *See* Military Aid to Civil Power
Mai Mai militias, 138, 142, 145–47, 150, 154, 158n1, 158n5, 158n7
Maison Militaire, 143, 149
Malaguiok, Ronnie, 117n1
Malawi, 50
Mandela, Nelson, 119, 131, 216, 254
Marxism, 58, 59
Masuku, Lookout, 56
Matabeleland, 58–59
Maxwell, Rohan, 11, 179–94, 248, 295
Mazula, Aguiar, 172
Mbeki, Thabo, 159n13, 227n13
MDC. *See* Movement for Democratic Change
mediation
    in Burundi, 215, 217, 219, 227n13, 228n15, 234
    in Democratic Republic of Congo, 140, 154, 167
    in Lebanon, 73
    in logit analysis, 23t
    military integration and, hypothesis of, 20–21
    in Mozambique, 164, 166, 169, 177n2
    negotiated settlements and, 20
    in South Africa, 121, 123–24
    in Sudan, 32, 34, 35
    in Zimbabwe, 52, 61
Meiring, George, 122
mergers, business. *See also* industrial organization (IO) perspective
    corporate culture and, 232
    effects on targets in, 232
    efficiency and, 232, 233
    horizontal acquisitions in, 231
    ideology and, 233
    justification for, 231
    military integration *versus*, 233

psychological factors in, 232
    symbolism and, 233
    transaction theory and, 232
MILF. *See* Moro Islamic Liberation Front
Military Aid to Civil Power (MACP) (Sierra Leone), 203
military capabilities. *See* military efficacy
military efficacy
    in Bosnia-Herzegovina, 189–90
    in Burundi, 222
    in Democratic Republic of Congo, 3, 150
    in integration success, 3, 4t
    in Lebanon, 80
    in Mozambique, 172–73
    in Philippines, 113–14
    in Rwanda, 97–98
    in Sierra Leone, 202–4
    in South Africa, 129
    in Sudan, 41–43
    in Zimbabwe, 61–62
military historic role
    in Bosnia-Herzegovina, 182–83
    in Burundi, 213–14
    in Democratic Republic of Congo (DRC), 139–40
    in Lebanon, 69–73
    in Philippines, 106
    in Sierra Leone, 196–98
    in South Africa, 119–21
    in Sudan, 33–34
    in Zimbabwe, 50–51
military integration
    after victories, 259, 260
    bargaining model of war and, 17–18
    best practices, 245, 248
    bloating of postmerger armies in, 235
    in Bosnia-Herzegovina, 183–92, 259
    in Burundi, 214–25, 253, 259, 263
    business mergers *versus*, 233
    in China, 2, 95
    corporate culture and, 235–36
    defined, 3
    in Democratic Republic of Congo, 140–56, 253, 259–60, 264–65
    economic motivation and, 19–20, 44
    explaining, 17–22
    incidence of, from 1946 to 2006, 13, 16t
    as indicative of commitment to peace, 249–50
    in Lebanon, 73–84, 259

military integration (*cont'd*)
  mediation and, 20–21
  in Mozambique, 165–76, 253–54, 259
  nation building and, 255–56
  negotiated settlements and, 2, 14–17, 16*t*
  in Philippines, 106–15, 259, 260, 263
  political primacy and, 252–55
  power-sharing, 89–92
  rebel strength and, 22, 24
  resumption of violence and, 7–9, 8*t*, 261–65, 262*t*
  in Rwanda, 89–99, 259
  security and, 2, 44
  in Sierra Leone, 198–209, 259, 260
  signaling hypothesis and, 249–50
  in South Africa, 121–32, 254, 259, 263
  state capacity and, 18–19, 20, 22, 25
  success of, defining, 3–4
  in Sudan, 31–47, 254–55, 259, 263
  symbolism and, 250–51
  in United Kingdom, 2
  in United States, 2, 265
  unity and, 250–51
  in Zimbabwe, 53–64, 259
military power. *See also* military efficacy
  balance of, 17–18, 25
  bargaining model of war and, 17–18
  state power and, 15
Military Reintegration Programme (MRP) (Sierra Leone), 199, 200–202, 208, 209
military training
  in Bosnia-Herzegovina, 186–87
  in Burundi, 221–22
  in Democratic Republic of Congo, 147–48
  in Lebanon, 79
  in Mozambique, 171–72
  in Philippines, 109–10
  in Rwanda, 94–96
  in Sierra Leone, 201–2
  in South Africa, 127
  in Zimbabwe, 55–57
militias
  in Democratic Republic of Congo, 138, 142, 145–47, 150, 154, 158n1, 158n2, 158n5, 158n7
  interahamwe, 90, 91, 94, 101n7, 138, 158n2
  in Lebanon, 72–73
  in Rwanda, 90, 91, 94, 101n7

Sudanese armed forces as, 33
undervaluation of, 234–35
Mills, Greg, 101n7
Mission de conseil et d'assistance de l'Union européenne en matière de réforme du secteur de la sécurité en RD Congo (EUSEC), 144, 160n22
Mission de l'organisation des Nations Unies en RD Congo (MONUC), 140, 144, 154, 160n21, 160n25, 161n36
Misuari, Nur, 104, 107, 109, 111, 114, 117n2
MK. *See* Umkhonto We Sizwe (Spear of the Nation)
MLC. *See* Mouvement pour la libération du Congo
MNLF. *See* Moro National Liberation Front
Mobutisme, 159n11
Mobutu Sese Seko, 94, 97, 100, 137, 159n11. *See also* Forces armées zaïroises
Modise, Joe, 129
MONUC. *See* Mission de l'organisation des Nations Unies en RD Congo
Morais, Hermínio, 167
Moro Islamic Liberation Front (MILF), 103–5, 107, 113–16, 117n1, 261, 268
Moro National Liberation Front (MNLF), 19, 103–16, 117n1–2, 118n6–8, 235, 260, 261, 263. *See also* Philippines
Mouvement de Libération, 137, 158n1, 158n7, 161n39, 162n39
Mouvement populaire de la révolucion (DRC political party), 159n11
Mouvement pour la libération du Congo (MLC), 137, 138, 145, 148, 158n7–8, 159n14–15
Movement for Democratic Change (MDC) (Zimbabwe), 49, 59–60, 63
Mozambique, 246*t*, 247*t*
  50/50 integration in, 19
  Agreement on a Partial Ceasefire in, 167, 177n2
  China and, 171
  Chissano in, 164, 167, 175
  Community of Sant'Egidio in, 164, 174, 177n2
  compromises in, 168–69
  creation of integrated military in, 169–72
  Dhlakama in, 167, 168, 174
  Domingos in, 167, 177n2

emergence of military integration as issue in, 165–66
employment in, 172, 176, 262*t*
Forcas Armadas da Defesa de Moçambique and, 169, 170
Frelimo in, 163–73, 175, 176, 236, 254
Guebuza in, 167, 172, 173, 175, 177n2
HIV/AIDS in, 163
in industrial organization perspective, 236, 239, 240
Italy and, 164, 177n2, 239
Joint Commission for the Formation of the Mozambican Defence Force and, 168, 169, 170–71
Joint Communiqué in, 165
military capabilities in, 172–73
military expenditures in, 172
military integration in, 165–76, 253–54, 259
military training in, 171–72
neutrals in, 166–68
opponents in, 166–68
origins of military integration in, 163–69
outcome in, 172–75
personnel selection in, 169–70
political control in, 172
Portugal and, 120, 163, 171
rank allocation in, 170
Renamo and, 59, 163–75, 177n2, 236, 239, 254
resumption of violence in, 175–76
Rhodesia and, 163
security in, 262*t*
shaping of new military in, 169
South Africa and, 163
South African Defence Force and, 120
South African Development Community and, 173
Soviet Union and, 163, 166
supporters in, 166–68
sustainability of armed forces in, 173
symbolism and, 236
trust in, 262*t*
United Kingdom and, 168, 170, 171, 239
United Nations Department of Peacekeeping Operations and, 166, 171, 174, 175
United Nations Department of Political Affairs and, 165–66, 174
United Nations Organization for Mozambique in, 166, 170, 171, 173, 174

United States and, 165, 168, 239
unity in, 165, 171, 236
Zimbabwe African National Army and, 50–51
Zimbabwe and, 167
Zimbabwe Defense Force and, 61
MRP. *See* Military Reintegration Programme
Mugabarbona, Alain, 223
Mugabe, Robert. *See also* Zimbabwe; Zimbabwe African People's Union
  African National Council and, 53
  assassination plot on, 66n7
  in industrial organization perspective, 235, 238
  Marxism of, 58
  Muzorewa and, 51
  Nkomo and, 49
  Operation Quartz and, 66n7
  overview of, 49
  Rhodesian Security Force and, 53, 55
  South Africa and, 53
  Zimbabwe African People's Union and, 58
  Zimbabwe People's Revolutionary Army and, 51
Muggah, Robert, 94
Mukasi, Charles, 216, 227n6. *See also* Union for National Progress
Mulele rebellion, 158n5
Munyakazi, Laurent, 94
Murr, Michel, 73
Museveni, Yoweri, 101n4
Mutisi, Martha, 11, 163–77, 236, 295
Muzorewa, Abel, 50, 51, 53, 61

N
Namibia, 51, 120, 158n2, 159n13
National Commission for Disarmament, Demobilization, and Reintegration (NCDDR) (Sierra Leone), 199, 201
National Defence Force (FDN) (Burundi), 224
National Liberal party (Lebanon), 72
National Movement (Lebanon), 72
National Peace Accords (South Africa), 121
National Provisional Ruling Council (NPRC) (Sierra Leone), 196–97
National Resistance Army (NRA) (Uganda), 101n4–5
NATO. *See* North Atlantic Treaty Organization
NATO Headquarters Sarajevo (NHQSa), 180–81, 186, 188

INDEX

NCDDR. *See* National Commission for Disarmament, Demobilization, and Reintegration
Ndadaye, Melchior, 214, 215
Ndayikengurukiye, Jean-Bosco, 214, 227n4
Ndayizeye, Domitien, 218, 223
Ndengeyinka, Balthazar, 94
Ndombasi, Abdoulaye Yerodia, 158n8
negotiated settlements
 amnesty and, 260
 and bargaining model of war, 14
 commitment and, 250
 in Democratic Republic of Congo, 264
 disarmament and, 2
 gamble of, 264–65
 mediation and, 20
 military integration and, 2, 14–17, 16t
 nation building after, 2
 resumption of violence and, 9, 260
 in Rwanda, 89, 90, 97
 in South Africa, 119, 120, 254
 stalemate and, 1–2
 state power and, 18
 victory and, 18
Nepal, 6, 218, 221–22, 234–35, 243n6
Netherlands, 221, 223, 239
neutrals
 in Bosnia-Herzegovina, 183–84
 in Burundi, 215–17
 in Democratic Republic of Congo, 140–41
 in Lebanon, 74–75
 in Mozambique, 166–68
 in Philippines, 107
 in Rwanda, 92
 in Sierra Leone, 198–99
 in South Africa, 122–23, 124
 in Sudan, 34–35
N'Goma, Arthur Zahidi, 158n8
Ngulinzira, Boniface, 91, 102n11
Nhongo, Rex, 56
NHQSa. *See* NATO Headquarters Sarajevo
Nigeria, 65n2, 197, 198, 242n3
Nimeiri, Jaafar, 31–32, 34–37, 40–41, 43–45, 47, 237, 254–55. *See also* Sudan
Niyombare, Godfroid, 228n25
Niyongabo, Prime, 227n14
Niyonguruza, Juvenal, 222
Niyoyankana, Germain, 223
Nkomo, Joshua, 49, 50, 58
Nkunda, Laurent, 142, 149–50

nonstatutory forces (NSF) (South Africa), 122, 123–24, 127, 132. *See also* Pan-Africanist Congress; Umkhonto We Sizwe (Spear of the Nation)
North Atlantic Treaty Organization (NATO), in Bosnia-Herzegovina, 82, 179–90, 192, 194n2, 239. *See also* NATO Headquarters Sarajevo; Partnership for Peace
Northern Rhodesia, 50
North Korea. *See also* Fifth Brigade
 Democratic Republic of Congo and, 141
 Zimbabwe and, 58, 238
NPRC. *See* National Provisional Ruling Council
NRA. *See* National Resistance Army
NSF. *See* nonstatutory forces
Ntigurirwa, Silas, 227n14
Nyadzonya, 55
Nyamwisi, Antipas Mbusa, 162n39
Nyanda, Siphiwe, 122
Nyangoma, Léonard, 217
Nyasaland, 50
Nyerere, Julius, 215, 216
Nzobonimpa, Manasse, 227n14

O
Oduho, Joseph, 38
Office of the High Representative (OHR) (Bosnia-Herzegovina), 179–80, 184, 187, 194n2
OHR. *See* Office of the High Representative
oil, in Sudan, 40, 41
Oleya, Renaldo, 38
ONUMOZ. *See* United Nations Organization for Mozambique
Operation Drive Out Rubbish, 60
Operation Gukarahundi, 59
Operation Hectic, 66n7
Operation Makovotera Papi?, 60
Operation Murmbatsvina, 60
Operation Quartz, 53, 66n7
Operation Rabbit, 60
Operation SEED (Soldiers Employed in Economic Development) (Zimbabwe), 56, 67n13, 235
Operation Tsuro, 60
Operation Who Did You Vote For?, 60
opponents
 in Bosnia-Herzegovina, 183–84

INDEX

in Burundi, 215–17
in Democratic Republic of Congo, 140–41
in Lebanon, 74–75
in Mozambique, 166–68
in Philippines, 107
in Rwanda, 92
in Sierra Leone, 198–99
in South Africa, 122–23
in Sudan, 34–35
Organization for Security and Cooperation in Europe (OSCE), 185, 194n2, 239
Organization of the Islamic Conference, 194n2, 239
origins, of integrated military
in Bosnia-Herzegovina, 182–85
in Burundi, 213–18
in Democratic Republic of Congo, 139–43
in Lebanon, 69–76
in Mozambique, 163–69
in Philippines, 106–8
in Sierra Leone, 196–99
in South Africa, 119–24
in Sudan, 33–36
in Zimbabwe, 50–53
Orth, Rick, 95
OSCE. *See* Organization for Security and Cooperation in Europe
outcome
in Bosnia-Herzegovina, 188–91
in Burundi, 222–24
in Democratic Republic of Congo, 148–51
in Lebanon, 79–83
in Mozambique, 172–75
in Philippines, 110–12
in Rwanda, 97–98
in Sierra Leone, 202–6
in South Africa, 128–31
in Sudan, 40–43
in Zimbabwe, 57–62
outsourcing, 240–41, 243n5

P

PAC. *See* Pan-Africanist Congress
Pakistan, 57, 243n6
Palestine, 83
Palestine Liberation Army, 72
PALIPEHUTU. *See* Parti pour la libération du peuple hutu
Pan-Africanist Congress (PAC), 122, 126. *See also* nonstatutory forces

parity, military integration and, hypothesis of, 18, 22
Parti pour la libération du peuple hutu (PALIPEHUTU), 214, 227n6
Partisan (Yugoslav guerrilla force), 182
Partnership for Peace (PfP), 179, 180, 182, 183, 184–85, 187, 188–89
Path t Lao, 13
peace. *See* resumption of violence
Peacekeeping and Stability Operations Institute, 6
peacekeeping operations (PKOs), 233, 240, 242n3. *See also* United Nations Department of Peacekeeping Operations
Peace Support Operations Training Center (Bosnia-Herzegovina), 186–87
People's Liberation Party (PLP) (Sierra Leone), 204
personnel selection
in Bosnia-Herzegovina, 185–86
in Burundi, 219–20
in Democratic Republic of Congo, 145–46
in Lebanon, 77–78
in Mozambique, 169–70
in Philippines, 108–9
in Rwanda, 93
in Sierra Leone, 200–201
in South Africa, 125–26
in Sudan, 38–40, 39*t*
in Zimbabwe, 53–54
PfP. *See* Partnership for Peace
Phalangists, 72
Philippines, 246*t*, 247*t*
accommodation capacity in, 112, 113, 115
as archetypal integration, 19
Armed Forces of the Philippines and, 107–8, 110–11, 118n9
Autonomous Region for Muslim Mindanao and, 104, 111, 116, 117n2, 118n8
civil-military operations in, 106, 113
Cold War and, 164
compromises in, 107–8, 108–10
emergence of military integration as issue in, 107
employment in, 104–5, 107, 114, 115–16, 262*t*
historic role of military in, 106
identity inside armed forces in, 112–14
in industrial organization perspective, 235, 237–38

INDEX

Philippines(*cont'd*)
　military capabilities in, 113–14
　military integration in, 106–15, 259, 260, 263
　military training in, 109–10
　Moro Islamic Liberation Front and, 103–5, 107, 113–16, 117n1, 261, 268
　Moro National Liberation Front and, 19, 103–16, 117n1–2, 118n6–8, 235, 260, 261, 263
　neutrals in, 107
　opponents in, 107
　origins of military integration in, 106–8
　outcome in, 110–12
　personnel selection in, 108–9
　resumption of violence in, 114–15
　revolutionary committees in, 108–9
　security in, 262*t*
　shaping of new military in, 108
　Special Regional Security Force and, 107, 111, 115
　supporters in, 107
　sustainability in, 104
　symbolism and, 104, 106, 114, 115, 237–38
　Tripoli Agreement and, 107
　trust in, 262*t*
PKOs. *See* peacekeeping operations
political control
　in Bosnia-Herzegovina, 188–89
　in Burundi, 222–24
　in Democratic Republic of Congo, 148–50
　in Lebanon, 79–80
　in Mozambique, 172
　in Sierra Leone, 204–5
　in South Africa, 128–29
　in Sudan, 40–41
　in Zimbabwe, 57–61
Portugal, 50–51, 120, 163, 168, 171, 239
Portuguese East Africa, 50
poverty
　in Lebanon, 82
　military integration and, 20, 24
　in Philippines, 116
power. *See* military power; state power
power sharing
　in Burundi, 218–19
　in Democratic Republic of Congo, 138, 140–41, 152, 154, 159n13
　forms of, 15
　mediation and, 20
　motivation and, 15
　in Mozambique, 172, 176
　in Philippines, 104
　in Rwanda, 88, 89–92, 97
　in Sierra Leone, 198, 207
　in Zimbabwe, 64
private security, 197, 240
Progressive Socialist Party (Lebanon), 72

Q
quota(s)
　in Bosnia-Herzegovina, 182, 184, 185–86
　in Burundi National Defence Force, 19, 218, 224, 225, 228n17
　in Democratic Republic of Congo, 141, 143, 145
　in Lebanese Army, 76, 78
　in Philippines, 108–9, 114
　success of, 260–61

R
Radetzky March (Roth), 233
Raffaelli, Mario, 164, 177n2
Ramos, Fidel, 107
rank allocation
　in Bosnia-Herzegovina, 186–87
　in Burundi, 220–21
　in Democratic Republic of Congo, 146–47
　in Lebanon, 77–78
　in Mozambique, 170
　in Rwanda, 94
　in South Africa, 126–27
　in Sudan, 38–40, 39*t*
　in Zimbabwe, 54
Rassemblement congolais pour la démocratie (RCD), 137, 138. *See also* Democratic Republic of Congo
　Goma (RCD-G), 138, 142, 144–45, 149–50, 158n1, 158n6–8, 160n24
　Kisangani/Mouvement de Libération (RCD-K/ML), 137, 158n1, 158n7, 161n39, 162n39
　National (RCD-N), 137, 148, 158n1, 158n7
RCs. *See* revolutionary committees
RDF. *See* Rwandan Defence Force
RDRC. *See* Rwandan Reintegration, Development and Resettlement Commission

Renamo. *See* Resistência Nacional Moçambicana (Renamo)
Republic of Sierra Leone Armed Forces (RSLAF), 195, 196, 200–209, 211n9
Republic of South Sudan, 33. *See also* Sudan
Republic of the Congo, 13. *See also* Democratic Republic of Congo
Republika Srpska (RS), 179, 184, 188, 189, 191
research design, 3–6, 4*t*, 7*t*
Resistência Nacional Moçambicana (Renamo), 59, 163–75, 177n2, 236, 239, 254
resumption of violence
    in Bosnia-Herzegovina, 191–92
    in Burundi, 224–25
    in Democratic Republic of Congo, 151–56
    employment and, 9, 44
    in Lebanon, 80, 83–84
    military integration and, 7–9, 8*t*, 261–65, 262*t*
    in Mozambique, 175–76
    negotiation process and, 9, 260
    in Philippines, 114–15
    in Rwanda, 98–99
    security and, 9
    in Sierra Leone, 207–9
    in South Africa, 131–32
    success and, 4–6, 4*t*
    in Sudan, 31–33, 42–46
    symbolism and, 9
    in Zimbabwe, 62–64
revolutionary committees (RCs), in Philippines, 108–9
Revolutionary United Front of Sierra Leone (RUF), 196–208, 210n4, 211n6
Rhodesia, 50–51, 66n2, 163, 239. *See also* Zimbabwe
Rhodesia Defence Regiment, 53
Rhodesian African Rifles, 54
Rhodesian Front Party, 50
Rhodesian Light Infantry, 53
Rhodesian Security Forces (RSF), 50, 51, 52–57, 62–64, 67n14
Rhodesian War, 50–52
Riccardi, Andrea, 164, 177n2
Roth, Joseph, 233
RPA. *See* Rwandan Patriotic Front and Army
RPF. *See* Rwandan Patriotic Front and Army
RS. *See* Serb Republic
RSF. *See* Rhodesian Security Forces

RSLAF. *See* Republic of Sierra Leone Armed Forces
Ruberwa, Azarias, 158n8
Rusagara, Frank, 96
Russett, Bruce, xi–xii, 295–96
Russia. *See also* Soviet Union
    Bosnia-Herzegovina and, 188–89, 194n2
    Democratic Republic of Congo and, 160n21
    nationalism in, 3
    Zimbabwe and, 51
Rwanda, 246*t*, 247*t*. *See also* Rassemblement congolais pour la démocratie
    Arusha Peace Agreement and, 88, 89–93, 101n3, 238
    Burundi and, 214
    Coalition pour la Défense de la République in, 91
    creation of integrated military in, 92–96
    Darfur and, 98–99
    Democratic Republic of Congo and, 94, 95, 137
    emergence of military integration as issue in, 92
    employment in, 98–99, 262*t*
    Forces armées rwandaises and, 88–90, 91–100, 101n6–7, 101n9–10, 158n2
    Hutus in, 89–95, 97, 98–100, 101n3, 102n12
    in industrial organization perspective, 235, 238, 240
    ingando process in, 88, 93–96, 99, 101n3, 102n16
    interahamwe militias in, 90, 91, 94, 101n7
    military capabilities in, 97–98
    military integration in, 89–99, 259
    military training in, 94–96
    neutrals in, 92
    opponents in, 92
    outcome in, 97–98
    personnel selection in, 93
    power-sharing integration in, 89–92
    rank allocation in, 94
    resumption of violence in, 98–99
    security in, 262*t*
    shaping of new military in, 92–93
    supporters in, 92
    sustainability in, 94
    trust in, 262*t*
    Tutsis in, 89–90, 95, 98–99, 101n3–5

INDEX

Rwanda (cont'd)
   Uganda and, 90, 101n4
   United Nations in, 89, 91
   unity in, 98, 99
   Zimbabwe and, 97
Rwandan Defence Force (RDF), 88, 95–96, 98, 99–100, 102n16
Rwandan Patriotic Front and Army (RPF/RPA), 88–99, 101n3–4, 102n12, 102n15, 238
Rwandan Reintegration, Development and Resettlement Commission (RDRC), 102n16
Rwandophones, 138, 142, 146, 148, 158n4
Rwarakabije, Paul, 102n16

S
SADC. *See* South African Development Community
SADF. *See* South African Defence Force
SAF. *See* Sudanese Armed Forces
Salamat, Hashim, 117n1
Sambanis, Nicholas, 5, 19, 105
Samii, Cyrus, 11, 213–28, 234, 239, 252, 257n7, 296
SANDF. *See* South African National Defence Force
Sankoh, Foday, 198, 208
SAS. *See* Special Air Service
Second Congo War, 137, 158n3. *See also* Democratic Republic of Congo
security
   in Bosnia-Herzegovina, 191–92, 262*t*
   in Burundi, 262*t*
   in Democratic Republic of Congo, 153, 262*t*
   economic development and, 44
   in Lebanon, 262*t*
   military integration and, 2, 44–45
   in Mozambique, 262*t*
   in Philippines, 262*t*
   private firms in, 197, 240
   resumption of violence and, 9
   in Rwanda, 262*t*
   in Sierra Leone, 262*t*
   in South Africa, 262*t*
   in Sudan, 44–45, 262*t*
   in Zimbabwe, 63, 262*t*
Security Force Auxiliaries (SFA) (Zimbabwe), 54
Selassie, Haile, 32, 34, 35

selection bias, 26n2
Selous Scouts, 53, 66n5, 66n11
Sema, Muslimen, 104
Serbia, 182, 189, 190, 237
Serb Republic (RS), 179, 184, 188, 189, 191, 193
Serufuli, Eugène, 160n24
settlements. *See* negotiated settlements
SFA. *See* Security Force Auxiliaries
SFOR. *See* Stabilization Force
shaping, of integrated military
   in Bosnia-Herzegovina, 185
   in Burundi, 218–19
   in Democratic Republic of Congo, 143–45
   in Lebanon, 76–77
   in Mozambique, 169
   in Philippines, 108
   in Rwanda, 92–93
   in Sierra Leone, 199–200
   in South Africa, 124–25
   in Sudan, 36–38
   in Zimbabwe, 52–53
Shihab, Fouad, 71, 78, 235
Shihabisme, 71, 82, 235–36
Sierra Leone, 246*t*, 247*t*
   All People's Congress in, 196, 203, 206
   Armed Forces Revolutionary Council in, 197–98, 199, 200–201, 207, 208
   Civil Defence Forces in, 198, 200, 201, 207, 208, 211n6
   Commission for the Management of Strategic Resources, National Reconstruction, and Development in, 198
   creation of integrated military in, 199–202
   Darfur and, 204
   Detective Reconnaissance Emergency Action Mission Team and, 211n9
   diamonds in, 197
   Economic Community of West African States Military Observer Group in, 197–98
   emergence of military integration as issue in, 198
   employment in, 206, 262*t*
   historic role of military in, 196–98
   in industrial organization perspective, 237, 239, 240

INDEX

International Military Advisory and Training Team and, 195, 199–202, 203, 204–6, 209, 211n7
Kabbah, Ahmed and, 197–200, 208, 210n2
Koroma, Johnny Paul in, 197–98, 200, 203
Liberia and, 196, 202–3
Lomé Peace Accord and, 195, 196, 198–200, 205, 207–8, 209, 210n2
Mano River region in, 202–3, 206
Military Aid to Civil Power policy in, 203
military capabilities in, 202–4
military integration in, 198–209, 259, 260
Military Reintegration Programme in, 199, 200–202, 208, 209
military training in, 201–2
National Commission for Disarmament, Demobilization, and Reintegration in, 199, 201
National Provisional Ruling Council in, 196–97
neutrals in, 198–99
opponents in, 198–99
origins of military integration in, 196–99
outcome in, 202–6
People's Liberation Party in, 204
personnel selection in, 200–201
political control in, 204–5
Republic of Sierra Leone Armed Forces in, 195, 196, 200–209, 211n9
resumption of violence in, 207–9
Revolutionary United Front of Sierra Leone and, 196–208, 210n4, 211n6
Sankoh and, 198, 208
security in, 262t
shaping of new military in, 199–200
South African private security and, 197, 240
supporters in, 198–99
sustainability in, 205–6
Territorial Defence Force in, 210n5
trust in, 262t
United Kingdom and, 11, 201, 209, 210n3, 239, 260
and United Nations Mission in Sierra Leone, 200, 201
unity in, 208
Zambia and, 200
Sierra Leonean Army (SLA), 196, 198, 200–202, 207, 208, 210n1
Sierra Leone People's Party (SLPP), 197, 198–99, 203

signaling hypothesis, 249–50
Sithole, Ndabaningi, 66n3
SLA. *See* Sierra Leonean Army
Slimane, Michel, 82
SLPP. *See* Sierra Leone People's Party
SMI. *See* Structure militaire d'intégration
Smuts, Jan, 120
Soames, Sir Charles, 52
Söderberg Kovacs, Mimmi, 11, 195–211, 237, 296
Soldiers Employed in Economic Development (SEED) (Zimbabwe), 56, 67n13, 235
Soldiers of God, 72
Somalia, 222, 240
South Africa, 3, 6, 246t, 247t
African National Congress and, 120–21, 123, 128, 129, 130, 254
Afrikaner Volksfront and, 128
alternative outcomes in, 131
Angola and, 59, 120
Azanian People's Liberation Army and, 122–23, 125, 126–27, 128
Boers and, 120
British Military Advisory Training Team and, 123, 125, 127
Burundi and, 216, 217, 219, 221, 227n13
compromises in, 123–24, 124–27
Convention for a Democratic South Africa and, 121, 124
Democratic Republic of Congo and, 145, 147
emergence of military integration as issue in, 121
employment in, 262t
historic role of military in, 119–21
in industrial organization perspective, 234
Joint Military Coordinating Committee in, 121, 122, 125
Kwa Zulu Self Protection Force and, 122, 125, 126–27, 128
Mandela and, 119, 131, 216, 254
military capabilities in, 129
military integration in, 121–32, 254, 259, 263
military training in, 127
Mozambique and, 163
Mugabe and, 53
National Peace Accords in, 121
neutrals in, 122–23, 124

INDEX

South Africa (cont'd)
  nonstatutory forces in, 122, 123–24, 127, 132
  opponents in, 122–23
  origins of military integration in, 119–24
  outcome in, 128–31
  Pan-Africanist Congress and, 122, 126
  personnel selection in, 125–26
  political control in, 128–29
  rank allocation in, 126–27
  resumption of violence in, 131–32
  Rhodesian Security Forces and, 50, 54
  security in, 262t
  shaping of new military in, 124–25
  Sierra Leone and, 197
  supporters in, 122–23
  sustainability of armed forces in, 130–31
  symbolism and, 128
  Transkei, Venda, Bophutatswana, and Ciskei Defence Forces and, 126, 127, 128
  trust in, 262t
  Union of South Africa in history of, 120
  United Kingdom and, 26n8, 124, 125, 132
  United States and, 240
South African Defence Force (SADF), 54, 119–32, 234, 235
South African Development Community (SADC), 173
South African National Defence Force (SANDF), 122–25, 127–32, 254
South African War, 120
Southern Philippines Council for Peace and Development (SPCPD), 117n2
Southern Rhodesia, 50
Southern Sudanese Army, 32. *See also* Sudan
South Lebanese Army, 72
Soviet Union. *See also* Russia
  Mozambique and, 163, 166
  nationalism in, 3
  nation building after revolution in, 2
  Zimbabwe and, 56
SPCPD. *See* Southern Philippines Council for Peace and Development
Spear of the Nation. *See* Umkhonto We Sizwe (Spear of the Nation)
Special Air Service (SAS) (Zimbabwe), 53
Special Regional Security Force (SRSF) (Philippines), 107, 111, 115, 118n8
SPLA. *See* Sudan People's Liberation Army

SPLM/A. *See* Sudan People's Liberation Movement/Army
Sri Lanka, 243n6
SRSF. *See* Special Regional Security Force
Stabilization Force (SFOR) (Bosnia-Herzegovina), 179, 180–81, 186–87, 194n2. *See also* Implementation Force
state capacity
  in logit analysis, 23f
  measurement of, 22
  military integration and, 18–19, 25
state power
  civil war and, 14–15
  military power and, 15
  negotiated settlements and, 18
Stedman, Stephen, 167, 173–74
Structure militaire d'intégration (SMI) (Democratic Republic of Congo), 145
success. *See also* resumption of violence
  civilian control and, 4, 4t
  dimensions of, 3–4, 4t
  efficacy and, 3, 4t
  and likelihood of resumption of violence, 4–6, 4t
Sudan, 246t, 247t. *See also* Lagu, Joseph
  Absorbed Forces in, 32–33, 36, 40–41, 43–44, 45, 47, 237, 255
  Addis Ababa Peace Agreement and, 31–38, 41–43, 46–47, 237, 254
  alternative outcomes in, 43
  Anyanya in, 31–32, 34–38, 39t, 40–43, 45–47, 234, 237
  compromises in, 35–36
  creation of integrated military in, 36–40
  emergence of military integration as issue in, 34
  employment in, 44, 262t
  historic role of military in, 33–34
  in industrial organization perspective, 234, 237, 240
  military capabilities in, 41–43
  military integration in, 31–47, 254–55, 259, 263
  neutrals in, 34–35
  Nimeiri and, 31–32, 34–37, 40–41, 43–45, 47, 237, 254–55
  oil in, 40, 41
  opponents in, 34–35
  origins of military integration in, 33–36
  outcome in, 40–43

personnel selection in, 38–40, 39*t*
political control in, 40–41
rank allocation in, 38–40, 39*t*
resumption of violence in, 31–33, 42–46
security in, 262*t*
shaping of new military in, 36–38
supporters in, 34–35
symbolism and, 35, 237
trust in, 262*t*
United Kingdom and, 33–34, 39
unity in, 45
Sudan Defense Forces, 33, 34
Sudanese Armed Forces (SAF), 31–32, 34, 35, 36–37, 38, 42–43, 46–47
Sudan People's Liberation Army (SPLA), 42
Sudan People's Liberation Movement/Army (SPLM/A), 31, 41, 45
Suliman, Migrani, 38
Superior Defence Council (Democratic Republic of Congo), 143, 159n17, 161n28
supporters
    in Bosnia-Herzegovina, 183–84
    in Burundi, 215–17
    in Democratic Republic of Congo, 140–41
    in Lebanon, 74–75
    in Mozambique, 166–68
    in Philippines, 107
    in Rwanda, 92
    in Sierra Leone, 198–99
    in South Africa, 122–23
    in Sudan, 34–35
sustainability
    of Armed Forces of Bosnia and Herzegovina, 190
    in Burundi, 224
    of Forces armées de la République démocratique du Congo, 150–51
    of Lebanese Army, 81–82
    of Mozambican military, 173
    in Philippines, 104
    of Republic of Sierra Leone Armed Forces, 205–, 205–6
    in Rwanda, 94
    of South African National Defence Force, 130–31
    of Sudanese Armed Forces, 44
    of Zimbabwe National Army, 64
Swahiliphones, 147
symbolic politics, 251
symbolism. *See also* unity

Bosnia-Herzegovina armed forces and, 188, 192, 236–37
Burundi National Defence Force and, 225
business mergers and, 233
hypothesis, 250–51
Lebanese Army and, 69–70, 83, 236
Mozambique and, 236
Philippines and, 104, 106, 114, 115, 237–38
resumption of violence and, 9
South Africa and, 128
Sudan and, 35, 237
unity and, 9
value of, 236–38
Syria, 70–73, 75–77, 79–80, 83
Syrian National Socialist Party, 72

T
Ta'if Treaty (Lebanon), 70, 73, 75, 76, 77
Tanzania, 53, 65n2, 214, 216, 217, 219
Taylor, Charles, 210n4. *See also* Liberia
TCCs. *See* troop-contributing countries (TCCs)
territorial conflict, 21, 23*t*
Territorial Defence Force (Sierra Leone), 210n5
Tilly, Charles, 251
Tito, Josip Broz, 182, 192
Togo, 198. *See also* Lomé Peace Accord
Torit Mutiny, 38
training. *See* military training
transaction theory, 232
Transkei, 122
Transkei, Venda, Bophutatswana, and Ciskei Defence Forces (TVBC), 126, 127, 128
Tripartite Accords (Lebanon), 73
Tripoli Agreement, 107
troop-contributing countries (TCCs), 233, 240, 242n3
trust
    in Bosnia-Herzegovina, 192, 262*t*
    in Burundi, 225, 262*t*
    in Democratic Republic of Congo, 152–53, 154, 262*t*
    in Lebanon, 262*t*
    in Mozambique, 262*t*
    in Philippines, 262*t*
    resumption of violence and, 9
    in Rwanda, 262*t*
    in Sierra Leone, 262*t*
    in South Africa, 262*t*
    in Sudan, 46, 262*t*
    in Zimbabwe, 64, 262*t*

INDEX

Turkey, 194n2
Tut, Gai, 39–40
Tutsis
   in Burundi, 213–16, 217, 220, 222–23, 224–25, 226n1
   in Democratic Republic of Congo, 138, 143, 148, 149–50, 158n4, 158n6
   in Rwanda, 89–90, 95, 98–99, 101n3–5
   in Uganda, 101n4
Twa, 213, 226n1
Twagiramungu, Faustin, 91

U

UBU. *See* Burundian Workers Party
UDI. *See* Unilateral Declaration of Independence
Uganda. *See also* Rassemblement congolais pour la démocratie
   Democratic Republic of Congo and, 137, 141, 158n2
   disarmament and, 2
   in industrial organization perspective, 240
   Mouvement pour la libération du Congo and, 137, 138, 145, 148, 158n7–8, 159n14–15
   Rwanda and, 90, 101n4
   Sudan and, 35, 41
   Tutsis in, 101n4
Ugandan National Resistance Front, 101n4
Umkhonto We Sizwe (Spear of the Nation) (MK), 119, 120–28, 131–32, 217, 234. *See also* nonstatutory forces
Umma party (Sudan), 44
UNAMID. *See* African Union-United Nations Hybrid Operation in Darfur
UNAMIR. *See* United Nations Assistance Mission for Rwanda
UNAMSIL. *See* United Nations Mission in Sierra Leone
União nacional para a independência total de Angola (UNITA), 158n2, 159n16
Unilateral Declaration of Independence (UDI) (Rhodesian Front Party), 50
Union for National Progress (UPRONA) (Burundi), 216, 217, 223, 227n6
Union of South Africa, 120
UNITA. *See* União nacional para a independência total de Angola

United Kingdom. *See also* British Military Advisory Training Team
   Bosnia-Herzegovina and, 186–87
   Conservative Party in, 52
   Democratic Republic of Congo and, 160n21
   India and, 3
   military integration in, 2
   Mozambique and, 168, 170, 171, 239
   Rhodesia and, 50
   Sierra Leone and, 11, 199–201, 209, 210n3, 239, 260
   South Africa and, 26n8, 124, 125, 132
   Sudan and, 33–34, 39
   Zimbabwe and, 50, 52, 55, 239
United Nations Assistance Mission for Rwanda (UNAMIR), 89, 91, 93, 99
United Nations Department of Peacekeeping Operations (DPKO), 166, 171, 174, 175, 243n6
United Nations Department of Political Affairs (DPA), 165–66
United Nations Mission in Sierra Leone (UNAMSIL), 200, 201
United Nations Mission to the Democratic Republic of Congo (MONUC), 140, 144, 160n21, 160n25, 161n36
United Nations Operation in Burundi (ONUB), 221, 222–23
United Nations Organization for Mozambique (ONUMOZ), 166, 170, 171, 173, 174–75
United States. *See also* US Agency for International Development
   Bosnia-Herzegovina and, 181, 186, 187, 189, 194n2
   Burundi and, 217
   in Central America, 266
   Democratic Republic of Congo and, 160n21, 161n36
   Lebanon and, 79
   military integration in, 2, 265
   Mozambique and, 165, 168, 239
   South Africa and, 240
   troop-contributing countries and, 242n3
unity. *See also* symbolism
   in Burundi, 225
   in Democratic Republic of Congo, 140, 147, 154, 159n11

hypothesis, 250–51
in Lebanon, 76, 79, 83
military efficacy and, 3
in Mozambique, 165, 171, 236
in Rwanda, 98, 99
in Sierra Leone, 208
in Sudan, 45, 237
symbolism and, 9
in Zimbabwe, 63
Unity Accord, 59
UPRONA. *See* Union for National Progress
US Agency for International Development (USAID), 105, 221
US Army War College, 6

V
variables, control, 21
Venda, 122. *See also* Transkei, Venda, Bophutatswana, and Ciskei Defence Forces
Verweijen, Judith, 11, 137–62, 237, 239, 253, 264, 296
VF. *See* Army of the Federation of Bosnia-Herzegovina
victory
  decline in, 17
  as increasingly difficult to achieve, 1–2
  military integration after, 259, 260
  negotiated settlements and, 18
violence, ongoing. *See* resumption of violence
Vojska Federacije BiH (VF), 179, 180–84, 186–88, 192, 194n10
Vojska Republike Srpske (VRS), 179–84, 186–88, 192, 194n10
VRS. *See* Army of Serb Republic

W
Walls, Peter, 51, 53, 55
Walter, Barbara, 2, 5
war. *See* civil wars; interstate wars
Warsaw Pact, 183
War Veterans Welfare Organisation (Zimbabwe), 56
Weber, Max, 251
West Nile Bank Front (WNBF), 158n2
West Side Boys, 200
Williams, Rocky, 120
WNBF. *See* West Nile Bank Front
World Council of Churches, 35

Y
Yemen, 82
Youth Brigades (Zimbabwe), 61
Youth of Ali, 72
Yugoslavia, 179, 182–83, 194n7. *See also* Bosnia-Herzegovina; Croatia; Serbia
Yugoslav People's Army (JNA), 182–83, 187

Z
Zaire, 100, 137, 139, 159n11. *See also* Democratic Republic of Congo
Zambia, 50, 51, 53, 65n2, 159n13, 160n21, 200
Zambo, N'Gbanda, 159n11
ZANLA. *See* Zimbabwe African National Army
ZANU-PF. *See* Zimbabwe African National Union-Patriotic Front
ZAPU. *See* Zimbabwe African People's Union
Zghorta Liberation Army, 72
Zimbabwe, 246*t*, 247*t*. *See also* Mugabe, Robert; Rhodesia
  African National Council and, 50, 53
  Angola and, 61, 240
  British Military Advisory Training Team and, 49, 53–57, 64, 67n13
  Central Intelligence Organisation in, 59, 61
  China and, 51, 54, 56, 58
  creation of integrated military in, 53–57
  Democratic Republic of Congo and, 60–61, 137, 141
  diamonds in, 60–61, 62, 63, 64, 238
  economic development in, 56, 62
  employment in, 63, 262*t*
  in Failed States Index, 59
  Fifth Brigade in, 58–59, 62, 64, 67n12, 67n14, 238
  Government of National Unity in, 60
  historic role of military in, 50–51
  in industrial organization perspective, 235, 238, 239
  Joint Operational Command in, 60
  military capabilities in, 61–62
  military integration in, 53–64, 259
  military training in, 55–57
  Movement for Democratic Change and, 49, 59–60, 63
  Mozambique and, 50–51, 61, 167
  Muzorewa and, 50, 51, 53, 61
  Nkomo and, 49
  North Korea and, 58, 238

Zimbabwe (cont'd)
　origins of military integration in, 50–53
　outcome in, 57–62
　personnel selection in, 53–54
　political control in, 57–61
　rank allocation in, 54
　resumption of violence in, 62–64
　Russia and, 51
　Rwanda and, 97
　security in, 63, 262t
　shaping of new military in, 52–53
　Soldiers Employed in Economic Development (SEED) program in, 56, 67n13, 235
　South African Defence Force and, 120
　sustainability in, 64
　trust in, 64, 262t
　United Kingdom and, 50, 52, 55, 239
　Unity Accord in, 59
Zimbabwe African National Army (ZANLA), 50–59, 62, 64, 67n12, 238
Zimbabwe African National Union-Patriotic Front (ZANU-PF), 49, 50–53, 57–61, 62–63, 64, 66n3, 66n7
Zimbabwe African People's Union (ZAPU), 50–53, 57–59, 61, 64, 66n3
Zimbabwe Defense Force (ZDF), 56, 59–62, 238
Zimbabwe National Army (ZNA), 57, 58–59, 64
Zimbabwe National Liberation War Veterans Association, 61
Zimbabwe People's Militia (ZPM), 58–59
Zimbabwe People's Revolutionary Army (ZIPRA), 50–59, 62–64, 66n7, 67n12, 238
Zimbabwe Prison Service, 61
Zimbabwe Reconstruction and Development Conference (ZIMCORD), 56
Zimbabwe Republic Police, 61
ZIMCORD. *See* Zimbabwe Reconstruction and Development Conference
ZIPRA. *See* Zimbabwe People's Revolutionary Army
ZNA. *See* Zimbabwe National Army
ZPM. *See* Zimbabwe People's Militia
Zuma, Jacob, 227n13
Zuppi, Don Matteo Maria, 164, 177n2